Chemistry of Air Pollutants and Their Health Impacts

CHEMISTRY OF AIR POLLUTANTS AND THEIR HEALTH IMPACTS

Dr. Gábor Kovács

ARCLER
PRESS

www.arclerpress.com

Chemistry of Air Pollutants and Their Health Impacts

Dr. Gábor Kovács

Arcler Press
4164 Lakeshore Road
Burlington, ON L7L 1A4
Canada
www.arclerpress.com
Email: orders@arclereducation.com

ISBN: 978-1-77956-413-9 (Hardcover)

ABOUT THE AUTHOR

Dr. Gábor Kovács is a lecturer at the Sapientia Hungarian University of Transylvania, Faculty of Technical and Human Sciences, Department of Horticulture in Târgu Mureș, Romania. He has been in this role since 2021. Kovács obtained his B.Sc. and M.Sc. in Chemistry/Applied Electrochemistry from Babeș-Bolyai University in Cluj-Napoca, Romania in 2007 and 2008, respectively. He later earned his PhD in Chemistry from Babeș-Bolyai University in 2012. Prior to his current lecturing position, Kovács had several research roles. He was an (assistant) Researcher at the Babeș-Bolyai University Faculty of Physics / Institute of Interdisciplinary Bio-Nano Sciences in Cluj-Napoca. Kovács also held postdoctoral research positions, including at the Department of Applied and Environmental Chemistry, University of Szeged in Hungary, and the Institute of Environmental Science and Technology, University of Szeged in Hungary. Kovács has taught a variety of chemistry-related courses at B.Sc. and M.Sc. level, including General Chemistry, Electrochemistry, Corrosion and Anticorrosive Protection, Organic Chemistry, Biochemistry and Phytosanitary Chemistry and Toxicology. Kovács has led several research projects funded by the Hungarian Academy of Sciences and the National Research, Development and Innovation Office of Hungary. These projects have focused on areas such as synthesis, characterization and applicability of photocatalytic nanostructures, carbon nanotube-based composites, and the development of noble metal-TiO2 based photoactive catalysts.

TABLE OF CONTENTS

LIST OF FIGURES

LIST OF TABLES

PREFACE

Air is one of the most fundamental necessities for sustaining human life and other living organisms. However, poor air quality in large urban areas poses significant threats to both human health and the environment. Numerous studies have demonstrated that both short-term and long-term exposure to air pollutants can lead to a variety of acute and chronic health issues, reduce public well-being, decrease gross domestic product, and increase mortality rates. The Environmental Protection Agency (EPA) has identified pollutants such as particulate matter (PM), sulfur dioxide, carbon monoxide, lead, and nitrogen dioxide as key contributors to air pollution.

Among these, PM10 plays a major role in air pollution. Exposure to PM10 is associated with serious health complications, including lung cancer, cardiovascular and respiratory diseases, chronic obstructive pulmonary disease, lung inflammation, and asthma. Additionally, the rise in deaths due to respiratory illnesses has been linked to a decrease in life expectancy. For instance, research indicates that a 310 μg/m3 increase in PM10 concentration can lead to a nearly 1% increase in mortality rates.

Nitrogen compounds, such as NO2 and NOx, are another category of atmospheric pollutants primarily emitted by motor vehicles. Exposure to these pollutants can cause symptoms like coughing, wheezing, viral infections, lung and bronchial inflammation, asthma, and even premature death. Volatile organic compounds (VOCs) are also significant as they are precursors to the formation of ground-level ozone. Studies by Medina-Ramon et al. and Burnett et al. have shown that exposure to ozone can trigger asthma and lead to higher hospitalization rates.

In response to these concerns, various programs such as Aphekom, BenMAP, AirQ, and AirQ+ have been developed to assess the health impacts of air pollution. Among these, AirQ+ is particularly effective in evaluating the adverse health effects of air pollutants over specific periods and locations. It can estimate pollution-related mortality by utilizing data on total population, population at risk, baseline incidence rates, and relative risk.

The book *"Chemistry of Air Pollutants and Their Health Impacts"* is a comprehensive guide that covers the intricate relationship between air pollutants and their effects on human health and the environment. It is structured into 10 chapters, each focusing on a critical aspect of air pollution, its chemistry, and its consequences.

Chapter 1 provides a broad overview of air pollution, its history, and its significance. It discusses the evolution of air quality concerns, from the industrial revolution to modern-day challenges. The chapter introduces the key pollutants and their sources, offering a foundation for understanding the complexities of air pollution and its impact on human health.

The second chapter explores the composition and structure of the Earth's atmosphere, which is crucial for understanding how pollutants interact with various atmospheric layers. It covers the troposphere, stratosphere, mesosphere, and thermosphere, explaining how each layer plays a role in the dispersion and transformation of pollutants. The chapter also addresses the natural processes that maintain atmospheric balance and how human activities disrupt this equilibrium.

In the third chapter, the focus shifts to the cycling of primary gaseous pollutants, such as carbon monoxide, sulfur dioxide, nitrogen oxides, and volatile organic compounds (VOCs). It examines the sources, transformation, and eventual fate of these pollutants in the atmosphere. The chapter also discusses the concept of atmospheric residence time and how it affects the global distribution of pollutants.

Chapter 4 highlights the natural trace components of the atmosphere, including water vapor, carbon dioxide, methane, and ozone. It explains their roles in atmospheric chemistry and how they interact with anthropogenic pollutants. The chapter also covers the significance of greenhouse gases and their contribution to global warming and climate change.

The chemistry of the atmosphere is the central theme of the fifth chapter. It explores the chemical reactions that occur in the atmosphere, including photochemical reactions, oxidation processes, and the formation of secondary pollutants like ozone and particulate matter. The chapter also addresses the role of sunlight, temperature, and humidity in influencing these chemical processes.

Ozone plays a dual role in the atmosphere, acting as both a protective shield in the stratosphere and a harmful pollutant in the troposphere. Chapter 7 focuses on the chemistry of ozone, detailing the processes that lead to its formation and destruction. It also discusses the implications of ozone depletion and the health risks associated with ground-level ozone exposure.

Chapter 8 categorizes the various sources of air pollution, including natural sources like wildfires and volcanic eruptions, as well as anthropogenic sources such as industrial emissions, vehicle exhaust, and agricultural activities. It highlights the differences between point sources, non-point sources, and area sources, providing a comprehensive overview of where pollutants originate.

The effects of air pollution on human health, ecosystems, and the environment are examined in Chapter 9. It covers both short-term and long-term health impacts, including respiratory and cardiovascular diseases, cancer, and premature death. The chapter also discusses the broader environmental consequences, such as acid rain, eutrophication, and loss of biodiversity.

Chapter 10 explores the relationship between air pollution and rainwater quality. It discusses the processes that lead to acid rain, its impact on water bodies, soil, and vegetation, and the methods used to measure and mitigate its effects. The chapter also touches on the implications of acid rain for human health and infrastructure.

The final chapter focuses on the methods used to analyze and control air pollution. It covers monitoring techniques, such as air quality indices and sampling methods, as well as control strategies like emission reduction, pollution prevention, and regulatory policies. The chapter emphasizes the importance of a multidisciplinary approach to tackling air pollution and protecting public health.

Thus, *"Chemistry of Air Pollutants and Their Health Impacts"* provides an in-depth exploration of air pollution from a chemical perspective, while also addressing its significant health and environmental impacts. Through its 10 chapters, the book offers a comprehensive understanding of the sources, chemistry, and consequences of air pollutants, as well as the strategies for mitigating their harmful effects.

—Author

CHAPTER 1

INTRODUCTION TO AIR POLLUTANTS

Contents

INTRODUCTION

With far-reaching effects, air pollution poses serious risks to human health and the environment. Pollutants that damage ecosystems include particulate matter (PM), sulfur dioxide (SO_2), nitrogen oxides (NOx), and volatile organic compounds (VOCs). These pollutants also exacerbate the decline in air quality. These contaminants have the potential to produce acid rain, which damages aquatic life and vegetation by acidifying soil and water bodies. Because it interferes with photosynthesis, ground-level ozone, a primary cause of smog, can stunt plant growth and lower agricultural yields. Pollutants can also alter the chemistry of the atmosphere, which can result in worldwide problems like ozone layer thinning and climate change. Air pollution poses a serious threat to public health from a health standpoint.

Cardiovascular and respiratory conditions have been related to ground-level ozone and fine particulate matter (PM 2.5). These pollutants can worsen asthma, cause chronic bronchitis, raise the risk of heart attacks and strokes, and more. Prolonged exposure is linked to early death and reduced lung function. There is an increased risk for vulnerable groups, such as children, the elderly, and those with underlying medical conditions. Furthermore, new research suggests that air pollution may have an effect on mental health as well, possibly exacerbating neurological disorders and cognitive decline. Changing to cleaner energy sources, increasing public awareness, and enforcing stricter emission regulations are just a few of the comprehensive strategies needed to address air pollution. Pollution control initiatives not only enhance the environment but also have a positive impact on public health, which may lower medical expenses and improve quality of life.

1.1. OVERVIEW OF ENVIRONMENTAL AND HEALTH IMPACTS OF AIR POLLUTION

Because air pollution damages ecosystems and degrades natural resources, it has a substantial negative influence on the environment. Pollutants that harm plant life and soil quality, impacting agriculture and wildlife, include sulfur dioxide and nitrogen oxides. These pollutants cause acid rain, which corrodes water bodies and upsets aquatic ecosystems. Sea levels rise, ice caps melt, and other effects of global warming are caused by greenhouse gases like carbon dioxide and methane. Furthermore, air pollution increases exposure to UV radiation by decreasing the ozone layer and reducing visibility. The cumulative impact of these effects compromises biodiversity and environmental health, underscoring the pressing need for efficient pollution control strategies. Air pollution is one of the biggest problems of our time because of its effects on both public and individual health due to rising morbidity and mortality rates, as well as its influence on climate change. Numerous pollutants play a significant role in human disease. Among these are PM, which are particles with varying but extremely small diameters that enter the respiratory system through inhalation and can lead to cancer, reproductive, cardiovascular, and central nervous system disorders, as well as respiratory and other diseases. Ozone protects against UV radiation in the stratosphere, but when it is concentrated too much below ground, it can be hazardous and negatively impact the cardiovascular and respiratory systems. Moreover, dioxins, sulfur dioxide, nitrogen oxide, VOCs, and polycyclic aromatic hydrocarbons (PAHs) are all regarded as hazardous air pollutants for people. High levels of carbon monoxide can even cause direct poisoning through inhalation. Depending on the extent of exposure, heavy metals like lead can cause either acute poisoning or long-term intoxication when absorbed into the body. The diseases brought on by the substances mentioned above are mostly related to the respiratory system and include asthma, bronchiolitis, lung cancer, cardiovascular events, central nervous system dysfunctions, and cutaneous diseases. Last but not least, both natural disasters and climate change brought on by pollution in the environment have an impact on the geographic distribution of numerous infectious diseases. Public awareness and a multidisciplinary approach from scientific experts are the only ways to address this issue; national and international organizations must address the emergence of this threat and offer long-term solutions.

Air pollution is harming not only human health but also the environment in which we live. The most important environmental effects are as follows:

1. **Acid rain**: It is wet (rain, fog, snow) or dry (particulates and gas) precipitation containing toxic amounts of nitric and sulfuric acids. They are able to acidify the water and soil environments, damage trees and plantations, and even damage buildings and outdoor sculptures, constructions, and statues.
2. **Haze** is produced when fine particles are dispersed in the air and reduce the transparency of the atmosphere. It is caused by gas emissions in the air coming from industrial facilities, power plants, automobiles, and trucks.
3. **Ozone**, as discussed previously, occurs both at ground level and in the upper level (stratosphere) of the Earth's atmosphere. Stratospheric ozone is protecting us from the Sun's harmful ultraviolet (UV) rays. In contrast, ground-level ozone is harmful to human health and is a pollutant. Unfortunately, stratospheric ozone is gradually damaged by ozone-depleting substances (i.e., chemicals, pesticides, and aerosols). If this protecting stratospheric ozone layer is thinned, then UV radiation can reach our Earth, with harmful effects for human life (skin cancer) and crops. In plants, ozone penetrates through the

stomata, inducing them to close, which blocks CO_2 transfer and induces a reduction in photosynthesis.

4. **Global Climate Change** is an important issue that concerns mankind. As is known, the "greenhouse effect" keeps the Earth's temperature stable. Unhappily, anthropogenic activities have destroyed this protecting temperature effect by producing large amounts of greenhouse gases, and global warming is mounting, with harmful effects on human health, animals, forests, wildlife, agriculture, and the water environment. A report states that global warming is adding to the health risks of poor people.

   People living in poorly constructed buildings in warm-climate countries are at high risk for heat-related health problems as temperatures mount.

5. **Wildlife** is burdened by toxic pollutants coming from the air, soil, or the water ecosystem and, in this way, animals can develop health problems when exposed to high levels of pollutants. Reproductive failure and birth effects have been reported.

6. **Eutrophication** occurs when elevated concentrations of nutrients (especially nitrogen) stimulate the blooming of aquatic algae, which can cause a disequilibration in the diversity of fish and their deaths.

   The ability of an ecosystem to neutralize acidity is unquestionably linked to a critical concentration of pollution that it can withstand without collapsing. This load was determined to be 20 kg/ha/yr by the Canada Acid Rain Program. Hence, soil, and water are negatively impacted by air pollution.

   There have been reports on the effects of PM pollution on crop yield and food productivity. Its effects on aquatic environments are linked to fish productivity potential and the survival of living things. Plants exposed to ozone exhibit abnormalities in photosynthetic rhythm and metabolism.

   Acid rain is caused by sulfur and nitrogen oxides, which are toxic to marine life and plants alike. Not to mention, as was already mentioned, the primary danger to our ecosystems (soil, water, and air) and to living things is the toxicity linked to lead and other metals (Moores, 2009; Villányi et al., 2010).

1.1.1. Approach to the Problem of Pollution

Since many different human activities have an impact on the environment, the interactions between humans and their physical surroundings have been thoroughly studied. The biotic (living things and microorganisms) and abiotic (atmosphere, lithosphere, and hydrosphere) are coupled to form the environment. The introduction of substances into the environment that are hazardous to humans and other living things is known as pollution. Pollutants are dangerous substances that are produced in excess of normal concentrations and lower the quality of our environment.

They can be solids, liquids, or gases. Because of the pollution that human activity causes to the soil, water, and air that we breathe, the environment is negatively impacted. The Industrial Revolution brought about the production of massive amounts of airborne pollutants that are harmful to human health, even though it was also a great success in terms of technology, society, and the provision of numerous services. Without a doubt, environmental pollution on a global scale is regarded as a multifaceted international public health concern. This significant issue is linked to social, economic, and legal issues as well as lifestyle choices. Undoubtedly, global urbanization and industrialization are escalating to previously unheard-of and unsettling levels in our day and age. Considering that anthropogenic air pollution causes approximately 9 million deaths annually, it ranks among the world's greatest threats to public health.

All of the above are unquestionably closely related to climate change, and if they become dangerous, humanity may suffer dire consequences. Numerous ecosystems are negatively impacted by climate change and the effects of global warming, leading to issues with food safety, ice, and iceberg melting, animal extinction, and damage to plants.

Air pollution affects people's health in many ways. Even days with low levels of air pollution can have an impact on the health of vulnerable and sensitive people. The following conditions are strongly linked to short-term exposure to air pollution: asthma, respiratory diseases, wheezing, cough, shortness of breath, and high rates of hospitalization (a measure of morbidity).

Chronic asthma, pulmonary insufficiency, cardiovascular diseases, and cardiovascular mortality are the long-term effects linked to air pollution. A Swedish cohort study suggests that prolonged exposure to air pollution may cause diabetes. Furthermore, it appears that early-life exposure to air pollution causes a range of harmful health outcomes, including respiratory, cardiovascular, mental, and perinatal illnesses that can result in chronic disease in adulthood or infant death.

There has been mention of the elevated risk of morbidity and mortality in national reports. These studies, which were carried out in various locations across the globe, demonstrate a relationship between daily mortality and PM concentration ranges. Changes in the climate and global warming may make matters worse. In addition, higher rates of hospitalization (a measure of morbidity) have been noted for particular causes among the elderly and vulnerable people. Because fine and ultrafine PM can more readily enter the bloodstream and penetrate the deepest areas of the airways, it appears to be linked to more serious illnesses.

Large metropolitan areas are the primary victims of air pollution, with vehicle emissions being the main cause of the deterioration of air quality. Additionally, there is the risk of industrial mishaps, in which the local population could perish due to the spread of a toxic fog. Numerous factors influence the dispersion of pollutants, but wind and atmospheric stability are the most important ones.

The issue is particularly severe in developing nations because of overcrowding, unchecked urbanization, and the advancement of industrialization. This results in poor air quality, particularly in nations with high levels of social inequality and low knowledge of environmentally sustainable management practices. People who use fuels like wood or solid fuel for household needs, because they have low incomes, are exposed to contaminated, poor-quality air at home. It is noteworthy that 3 billion people use the aforementioned energy sources for cooking and heating on a daily basis. Due to their prolonged exposure to indoor air pollution, women in developing nations appear to be at the highest risk of developing diseases. China is one of the Asian nations with the worst air pollution issues because of its rapid industrial development and overpopulation. Fine particles have been linked to China's observed lung cancer mortality rate. As was previously mentioned, prolonged exposure is linked to negative consequences on the cardiovascular system. It's interesting to note, though, that developed and high-income nations have seen a greater prevalence of cardiovascular diseases than developing, low-income nations that are heavily exposed to air pollution.

There are reports of extreme air pollution in India, where dangerously low air quality is observed. One of the most polluted cities in India is New Delhi. As a result of air pollution, flights into and out of New Delhi International Airport are frequently canceled. India is experiencing pollution in both its urban and rural areas as a result of the country's rapid industrialization, urbanization, and growth in motorcycle usage. However, one of the main causes of household air pollution in Nepal and India is the burning of biomass related to cooking and heating needs and practices. India exhibits spatial heterogeneity, whereby regions with distinct climatic conditions, populations, and educational attainment levels produce varying indoor air quality. North Indian states have been found to have higher PM2.5 levels (557–601 $\mu g/m^3$) in comparison to the Southern States (183–214 $\mu g/m^3$). Compared to the tropical climate of Southern India, the colder climate in the North Indian regions may be the primary cause of this, requiring longer stays at home and more heating. In India, indoor air pollution from households is linked to serious health consequences, particularly for women and young children who spend more time indoors. While acute lower respiratory diseases are more common in young children under the age of five, chronic obstructive respiratory disease (CORD) and lung cancer are primarily observed in women.

Air pollution, primarily sulfur dioxide and smoke, accumulated to 1,500 mg/m^3, which increased the number of deaths in December 1952 in London and in New York City in 1963 (400 deaths). Based on monitoring of outdoor pollution in six U.S. metropolitan cities, a correlation between pollution and mortality was reported. Mortality appears to be more closely correlated in all cases with the levels of fine, inhalable, and sulfate particles than with the levels of aerosol acidity, nitrogen dioxide, sulfur dioxide, or total particulate pollution. Various types of interventions should undoubtedly be considered depending on the extent of the impact on public health.

There have been reports of success and effectiveness in reducing air pollution, particularly locally. Sufficient technological measures are implemented, taking into account the origin and type of the emission, along with its effects on the environment and human health. The significance of controlling air pollution from both point and non-point sources. Without a doubt, every source in a certain area needs to be listed in a thorough emission inventory. In addition to the sources mentioned above and their characteristics, topography, and meteorology should be taken into account, as previously mentioned. Evaluation of the control strategies and tactics is frequently extended from the local to the regional and ultimately to the international levels. It is possible for air pollution to spread and travel from one remote location to another. The term "air pollution management" refers to bringing air pollutants that have an adverse effect on human health or the environment's ecosystem down to acceptable levels or perhaps completely eliminating them. Actions are taken by governmental and private organizations and authorities to guarantee the quality of the air. The WHO and EPA established air quality standards and guidelines for various pollutants as a means of managing air quality. To identify the trouble spots, these standards need to be compared to the emissions inventory standards using dispersion modeling and causal analysis. Typically, a combination of direct measurements and emissions modeling is used to create inventories.

Here, for instance, we outline the source-level control strategies involving automobile catalytic converters. These are devices that use catalysis and redox reactions to convert the toxic gases and pollutants released by combustion engines into less toxic pollutants. To lessen traffic during rush hour, private vehicle usage in Greece was limited by registering license plate numbers. Collectors and closed systems can reduce industrial emissions to the legally mandated minimum levels of air pollution. Evaluating the financial worth of the advantages derived from suggested initiatives is necessary for the current air quality improvement strategies. These suggested programs by government agencies come with guidelines that must be followed. Planning claims in Europe are initiated by air quality limit values, or AQLVs (Air Quality Limit Values). The national air quality limit values in the USA are set by the NAAQS (National Ambient Air Quality Standards). Significant progress has been made in lowering overall emissions and the consequences they have on human health and the environment, even though standards and directives are based on different mechanisms. While the USA establishes global geographical air quality criteria based on the severity of their air quality problem and records all sources of pollutants and their precursors, the European Directive identifies geographical areas of risk exposure as monitoring/assessment zones to record the emission sources and levels of air pollution.

In a similar spirit, funds have been used to directly or indirectly support air quality projects as well as the technological infrastructure needed to maintain acceptable air quality. These plans center on a database inventory from environmental planning awareness campaigns related to air quality. Moreover, industries, machines, and automobiles in metropolitan areas may be subject to air pollution controls. Innovation in technology can only succeed if it can satisfy societal demands.

In order to help decision-makers make the best choices possible, technology must both mirror the methods and practices of those engaged in risk assessment and evaluation and serve as a facilitator by offering data and assessments. In conclusion, many factors need to be taken into account when designing an effective air quality control strategy. These include engineering factors and the characteristics of air pollutants, environmental factors and ambient air quality conditions, and finally, economic operating costs for technological advancement as well as administrative and legal expenses. Environmental issues can be resolved by using neoliberal concepts of competitiveness and economic considerations. The deployment of a dialogue has been sparked by the advancement of environmental governance and technological advancement. Political parties, scientists, the media, and governmental and non-governmental organizations have all expressed opposition to and objections to environmental politics. There are now movements and acts of radical environmental activism. It is frequently investigated whether and how the emergence of new information and communication technologies (ICTs) has affected social movements like activism and communication channels. The term "digital activism" has become more widely used across a wide range of disciplines since the 1990s.

These days, a variety of digital technologies can be employed to generate environmental digital activism. To be more precise, online-capable gadgets like computers and smartphones are being used to push for social and political change (Moores, 2009; Villányi et al., 2010).

1.1.2. Sources of Exposure

It is well known that the majority of pollutants released into the environment come from large-scale human activities like driving cars, using combustion engines, power plants, and industrial machinery. Due to their massive scale, these activities are by far the main causes of air pollution; cars are thought to be responsible for about 80% of the pollution that exists today. In addition to several natural sources like soil and volcanic eruptions and forest fires, some other human activities like field cultivation methods, gas stations, fuel tanks, heaters, and cleaning practices also have a minor impact on our environment.

The classification of air pollutants is based mainly on the sources producing pollution. Therefore, it is worth mentioning the four main sources, following the classification system: Major sources, Area sources, Mobile sources, and Natural sources.

1. **Major Sources** include the emission of pollutants from power stations, refineries, and petrochemicals, the chemical and fertilizer industries, metallurgical, and other industrial plants, and, finally, municipal incineration.
2. **Indoor Area Sources** include domestic cleaning activities, dry cleaners, printing shops, and petrol stations.
3. **Mobile Sources** include automobiles, cars, railways, airways, and other types of vehicles.
4. **Natural Sources** include, as stated previously, physical disasters such as forest fires, volcanic erosion, dust storms, and agricultural burning

However, many classification systems have been proposed. Another type of classification is a grouping according to the recipient of the pollution, as follows:

1. **Air Pollution** is determined as the presence of pollutants in the air in large quantities for long periods. Air pollutants are dispersed particles, hydrocarbons, CO, CO_2, NO, NO_2, SO_3, so forth,.
2. **Water Pollution** is organic and inorganic charge and biological charge at high levels that affect the water quality.
3. **Soil Pollution** occurs through the release of chemicals or the disposal of wastes, such as heavy metals, hydrocarbons, and pesticides.

By contaminating precipitation and settling into soil and water environments, air pollution can have an adverse effect on the quality of soil and water bodies. Notably, acid precipitation can modify the chemistry of the soil by impacting water quality, plant life, and cultures. Furthermore, soil acidity promotes the movement of heavy metals, which leads to the metals' entry into wet environments. It is well known that fish and other wildlife are poisoned by heavy metals like aluminum. Given that acid rain increases the risk to soils with low calcium carbonate levels, soil quality appears to be important. PM, snow, and rainfall into bodies of water.

Lastly, pollution is classified following type of origin:

1. **Radioactive and Nuclear Pollution**, releases radioactive and nuclear pollutants into water, air, and soil during nuclear explosions and accidents, from nuclear weapons, and through handling or disposal of radioactive sewage.

   Radioactive materials can contaminate surface water bodies and, are noxious to the environment, plants, animals, and humans. It is known that several radioactive substances such as radium and uranium concentrate in the bones and can cause cancers.

2. **Noise pollution** is produced by machines, vehicles, traffic noises, and musical installations that are harmful to our hearing.

   The term "DALYs" was first used by the World Health Organization. A disease or health condition's DALYs are calculated by adding the Years of Life Lost (YLL) resulting from population-wide premature mortality and the Years Lost due to Disability (YLD) for those who are affected by the illness or its aftereffects. Disability-adjusted life years lost (DALYs) are primarily caused by air pollution in Europe, with noise pollution coming in second. Research has been done on the possible links between air pollution and noise pollution and health. The effects of ambient noise on cardiovascular disease were found to be independent of air pollution, meaning that DALYs associated with noise were more significant than those associated with air pollution. One should consider ambient noise to be a separate public health risk.

3. **Environmental pollution** occurs when changes in the physical, chemical, or biological constituents of the environment (air masses, temperature, climate, etc.) are produced.

   Either by introducing hazardous substances or by raising levels above normal, pollutants damage our ecosystem. The aforementioned sources directly produce primary pollutants, which are then released as secondary pollutants as a result of the primary ones. As previously mentioned, pollutants can be either naturally occurring or man-made, and they can also be biodegradable or non-biodegradable. Furthermore, their source may be distributed or a single point (point-source).

The variations in the ability of pollutants to cause harmful effects can be attributed to differences in their physical and chemical characteristics. For instance, we can say that because aerosol compounds are smaller in the atmosphere—either solid or liquid—and can penetrate deeper than gaseous compounds, they are more toxic. Our respiratory system has an easier time getting rid of gaseous substances. Millions of people die young each year as a result of these particles' ability to harm the lungs and even enter the bloodstream. Additionally, it appears that the aerosol acidity ([H+]) significantly increases the generation of secondary organic aerosols (SOA); however, other scientific groups do not concur with this last finding (Moores, 2009; Villányi et al., 2010).

1.1.3. Climate and Pollution

Climate change and air pollution are closely associated. The other side of the same coin that degrades Earth's quality is climate. The amount of sunlight that enters the atmosphere is influenced by pollutants like aerosols, methane, black carbon, and tropospheric ozone. As a result, the Earth's temperature is rising, which causes ice to melt and create icebergs and glaciers. Similarly, the incidence and prevalence of imported and residual infections in Europe will be impacted by climate change.

Climate and weather have a significant impact on the length, timing, and severity of outbreaks as well as altering the global distribution of infectious diseases. The transmission of parasitic or viral diseases by mosquitoes is highly susceptible to climate change, as it can shorten the incubation period of the pathogen and change the vector's geographic distribution. Similar to this, a high frequency of waterborne infections results from water warming brought on by climatic changes

Recently, it appears that eradicated diseases like malaria, poliomyelitis, cholera, and tick-borne encephalitis are resurfacing in Europe as a result of population migration. Storms and other natural climatic disasters are linked to the spread of epidemics, and they appear to be happening more frequently these days.

The new infections impacting public health are also linked to malnourishment and immune system dysregulation. Chikungunya outbreaks were reported in Italy, and autochthonous cases were reported in France, indicating that the virus "took the airplane" from the Indian Ocean to Europe. Following flooding, there appears to have been a rise in cryptosporidiosis in the Czech Republic and the United Kingdom.

As stated previously, aerosols are tiny in size and considerably affect the climate. They are able to dissipate sunlight (the albedo phenomenon) by dispersing a quarter of the sun's rays back to space and have cooled the global temperature over the last 30 years (Moores, 2009; Villányi et al., 2010).

1.2. AIR POLLUTANTS

Any physical, chemical, or biological alteration to the air is referred to as air pollution. The air pollution caused by dangerous gases, dust, and smoke has a significant impact on humans, animals, and plants. The atmosphere contains a specific percentage of gases. It is detrimental to survival for these gases' composition to change. Global warming is the term for the increase in the earth's temperature caused by this imbalance in the gaseous composition.Lead, sulfur oxides, nitrogen oxides, carbon monoxide, particle pollution, and ground-level ozone are the six main air pollutants that the World Health Organization (WHO) monitors.

Air pollution can have a catastrophic impact on soil, groundwater, and other environmental elements. It also presents a significant risk to living things. Thus, since these pollutants are linked to more serious and pervasive issues with human health and environmental impact, our primary focus is on them. There are significant ecological effects of acid rain, global warming, the greenhouse effect, and climate change on air pollution.

1.2.1. Types of Air Pollutants

There are various categories of air pollutants, each with unique sources and effects on the environment and human health. Tiny airborne particles known as PM, which includes PM2 and PM10, can enter the respiratory system deeply and lead to health issues like asthma and cardiovascular disorders. NOx are harmful to ecosystems and human health because they cause smog and acid rain. They are mainly produced by combustion processes in vehicles and industrial activities. Burning fossil fuels releases SO_2, which can cause respiratory problems and acid rain, which harms aquatic life and plants.

The colorless, odorless gas known as CO, which is produced when incomplete combustion occurs, can harm blood oxygen delivery and is especially hazardous in small areas. VOCs, or volatile organic compounds, are released into the atmosphere by cars, factories, and paints, among other things.

These emissions cause ground-level ozone to form, which is harmful to plants and human health. Last but not least, crop yields can be lowered and respiratory problems can result from ground-level ozone (O_3), which is created when NOx and VOCs react in sunlight. Since the sources and effects of each type of pollutant are unique, tailored mitigation strategies are required.

There are two types of air pollutants:

1. **Primary Pollutants:** The pollutants that directly cause air pollution are known as primary pollutants. Sulfur dioxide emitted from factories is a primary pollutant.

2. **Secondary Pollutants:** The pollutants formed by the intermingling and reaction of primary pollutants are known as secondary pollutants. Smog, formed by the intermingling of smoke and fog, is a secondary pollutant.

1.2.2. Causes of Air Pollution

Numerous human activities as well as natural processes contribute to air pollution, with industrial, agricultural, and residential sources accounting for a large portion of the problem. The burning of fossil fuels like coal and oil by industrial activities results in the emission of pollutants, making them significant contributors. Numerous pollutants, such as (SO_2, (NOx, and PM, are released by factories and power plants.

These pollutants can deteriorate air quality and cause smog and acid rain. Transportation is also very important because cars with gasoline and diesel engines release a lot of CO, NOx, and VOCs. In addition to having an impact on air quality, these emissions help create ground-level ozone, a dangerous component of smog. Through the release of ammonia (NH_3) from fertilizers and animal waste, as well as methane (CH_4) from rice paddies and livestock, agricultural practices contribute to air pollution.

A powerful greenhouse gas is methane, and ammonia can lead to the formation of PM. Burning biomass or fossil fuels for domestic purposes like cooking and heating releases pollutants like PM and carbon monoxide. Furthermore, using some household products—like paints and cleaners—can release VOCs that worsen air pollution. Air pollution is also caused by natural sources, such as volcanic eruptions that release sulfur dioxide and other pollutants into the atmosphere, and wildfires that release copious amounts of gases and PM into the atmosphere. In general, the interaction of these man-made and natural sources creates intricate situations of air pollution that have an effect on ecosystems, climate, and human health. Targeting both natural and anthropogenic sources, comprehensive strategies are needed to address air pollution.

The following are the important causes of air pollution:

1. **Burning of Fossil Fuels:** The combustion of fossil fuels emits a large amount of sulfur dioxide. Carbon monoxide released by incomplete combustion of fossil fuels also results in air pollution.
2. **Automobiles:** The gases emitted from vehicles such as jeeps, trucks, cars, buses, so forth,. pollute the environment. These are the major sources of greenhouse gases and also result in diseases among individuals.
3. **Agricultural Activities:** NH_3 is one of the most hazardous gases

emitted during agricultural activities. Insecticides, pesticides, and fertilizers emit harmful chemicals in the atmosphere and contaminate it.

4. **Factories and Industries:** Factories and industries are the main sources of carbon monoxide, organic compounds, hydrocarbons, and chemicals. These are released into the air, degrading its quality.
5. **Mining Activities:** In the mining process, the minerals below the earth are extracted using large pieces of equipment. The dust and chemicals released during the process not only pollute the air but also deteriorate the health of the workers and people living in the nearby areas.
6. **Domestic Sources:** Household cleaning products and paints contain toxic chemicals that are released in the air. The smell from the newly painted walls is the smell of the chemicals present in the paints. It not only pollutes the air but also affects breathing (Moores, 2009; Villányi et al., 2010).

1.3. EFFECT OF AIR POLLUTION ON HEALTH

The most common air pollutants are ground-level ozone and PM. Air pollution is distinguished into two main types:

1. **Outdoor pollution** is the ambient air pollution.
2. **Indoor pollution** is the pollution generated by household combustion of fuels.

Individuals who are subjected to elevated levels of air pollutants may exhibit symptoms of various diseases, varying in severity. These effects are divided into categories that affect health over short and long terms. Ages 65 and older, people with diabetes, people who are predisposed to heart or lung diseases, particularly asthma, and children are among the vulnerable groups that need to be aware of health protection measures.

As previously mentioned in great detail, a recent epidemiological study from the Harvard School of Public Health indicates that exposure errors and disparate epidemiological methodologies have left the relative magnitudes of the short- and long-term effects unclear.

To more effectively evaluate data on both short- and long-term human exposure, new models are suggested. The short-term effects are transient and can include more serious conditions like asthma, pneumonia, bronchitis, lung, and heart issues, or more minor ones like irritation of the eyes, nose, skin, and throat, wheezing, coughing, and chest tightness. In addition to headaches and nausea, short-term exposure to air pollution can also result in dizziness.

Long-term exposure to the pollutants can exacerbate these issues, as can damage the neurological, reproductive, and respiratory systems and increase the risk of cancer and, in rare cases, even death. Chronic long-term effects can even result in death and last for years or a lifetime. Moreover, a number of cancers may eventually develop as a result of the toxicity of various air pollutants.

As mentioned previously, breathing in air pollutants is directly linked to respiratory illnesses. These toxins will enter the body through the respiratory system and gather at the cell level. The type, source, and dose of the pollutant component that is causing the damage to the target cells should all be considered. Time, place, season, and nation all have a significant impact on health. Long-term health effects from prolonged exposure to the pollutant should also be associated with the previously mentioned factors. The respiratory system is seriously harmed by dust, benzene, O_3, PMs, and others.

Additionally, there is an increased risk if you already have a respiratory condition like asthma. Individuals with a predisposing disease state are more likely to experience long-term consequences. Acute exposure to pollutants in the trachea can cause noticeable changes in voice quality. Air pollution can cause chronic obstructive pulmonary disease (COPD), which raises morbidity and mortality rates. Transportation-related long-term impacts, industrial air pollution, and fuel combustion are the main causes of COPD risk.

Exposure to air pollutants has been linked to multiple cardiovascular effects. Long-term exposure-induced blood cell changes may have an impact on cardiac function. Long-term exposure to traffic emissions has been linked to the development of coronary arteriosclerosis, whereas short-term exposure has been linked to the development of hypertension, stroke, myocardial infarctions, and heart failure. Humans are known to experience ventricle hypertrophy following prolonged exposure to NO_2.

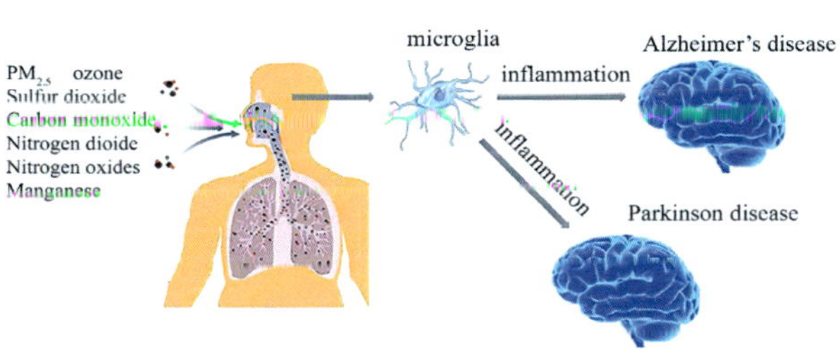

Figure 1.1. *Impact of air pollutants on the brain.*

Source: Fu C, Kuang D, Zhang H, Ren J and Chen J (2022) Different components of air pollutants and neurological disorders. Front. Public Health 10:959921. doi: 10.3389/fpubh.2022.959921.

Adults and children who have been exposed to air pollutants for an extended period of time have shown neurological effects. Long-term air pollution appears to be linked to retinopathy, low birth weight, fetal growth, psychological issues, and autism. Although the exact cause of the neurodegenerative diseases (Parkinson's and Alzheimer's) is unknown, prolonged exposure to air pollution is thought to play a role. In particular, food, and metals and pesticides are mentioned as etiological factors. The pathogenesis of neurodegenerative diseases involves multiple mechanisms, such as oxidative stress, protein aggregation, inflammation, and impairment of the neuronal mitochondria (Figure 1.1).

Dogs that spent a significant amount of time in a highly polluted area of Mexico showed signs of brain inflammation. Adult humans were observed to exhibit an immediate response to PNC on the IL-6 level, which resulted in an increase in markers of systemic inflammation (IL-6 and fibrinogen), potentially resulting in the production of acute-phase proteins. The development of oxidative stress and atherosclerosis appear to be the mechanisms underlying the neurological abnormalities brought on by chronic air pollution. After oxidative stress, inflammation appears to be a contributing factor in the hindrance of developmental maturation, impacting various organs. In a similar vein, it appears that additional factors influence the developmental maturation that determines an individual's susceptibility to long-term air pollution. These include education level, birth weight, maternal smoking, socioeconomic background, and genetic background.

But nutrition—which begins with breastfeeding—is another important consideration. Antioxidants are primarily found in food and are essential for shielding our bodies from air pollution. As scavengers of free radicals, antioxidants restrict the interaction of free radicals within the brain. Similar to this, a person's genetic background may cause them to be more or less susceptible to the oxidative stress pathway. For instance, in asthmatic children homozygous for the GSTM1 null allele, antioxidant supplementation with vitamins C and E appears to modulate the effect of ozone. In the periphery, inflammatory cytokines are released (e.g., respiratory epithelia) to stimulate the Toll-like receptor 2 of the innate immune system. In lung lavage, mice exposed to ambient PM in Los Angeles, California, USA, have recently shown signs of such activation and the events that follow, ultimately leading to neurodegeneration.

Lead exposure has been linked to neurodevelopmental morbidities in children. These kids became hyperactive, exhibited diminished intelligence, displayed aggressive and delinquent behavior, and had trouble learning. The scientific community has requested that the Centers for Disease Control and Prevention (CDC) lower the current screening guideline of 10 μg/dl because there doesn't seem to be a level of lead exposure that is "safe." It is crucial to note that poor air quality has an effect on the immune system, leading to dysfunction and neuroinflammation. Nonetheless, elevated serum levels of complement component C3 and immunoglobulins (IgA, IgM) are noted. Air pollution also affects antigen presentation because it causes macrophages to overexpress costimulatory molecules like CD80 and CD86.

Since skin is the outermost layer of the body, it is well known that skin acts as a shield against ultraviolet radiation (UVR) and other pollutants. Pollutants from traffic, including oxides, PM, VOCs, and PAHs, can cause pigmented spots on our skin. As mentioned previously, some pollutants are carcinogenic and mutagenic, specifically affecting the liver and lungs. As a result, damage to the organs is seen when pollutants enter the body through the skin or are inhaled. However, in contaminated urban areas, air pollutants (as well as those in the troposphere) lessen the harmful effects of UVR.

Air pollution that is absorbed by the skin can cause atopic dermatitis, psoriasis, acne, urticaria, eczema, and skin aging. Photochemical smoke and oxide exposure are the main causes of these conditions. Smoking and PM exposure are two factors that age the skin, resulting in wrinkles, dyschromia, and spots.

Finally, there is evidence linking pollutants to skin cancer. When fetuses and children are exposed to the aforementioned risks, higher morbidity rates are reported. There have been reports of autism, low birth weight, and fetal growth impairment. The eye is another external organ that could be impacted. The most common source of contamination is suspended pollutants, which can cause irritation, retinopathy, dry eye syndrome, or asymptomatic eye outcomes.

1. **Effects of Air Pollution:** The hazardous effects of air pollution on the environment include:

i. **Diseases:** Air pollution has resulted in several respiratory disorders and heart diseases among humans. The cases of lung cancer have increased in the last few decades. Children living near polluted areas are more prone to pneumonia and asthma. Many people die every year due to the direct or indirect effects of air pollution.

ii. **Global Warming:** The air's gaseous composition is unbalanced as a result of greenhouse gas emissions. The earth's temperature has increased as a result of this. Global warming is the term used to describe this rise in the earth's temperature. As a consequence, sea levels have risen and glaciers have melted. There are lots of underwater areas.

iii. **Acid Rain:** The burning of fossil fuels releases harmful gases such as nitrogen oxides and sulfur oxides into the air. The water droplets combine with these pollutants, become acidic and fall as acid rain which damages human, animal, and plant life.

iv. **Ozone Layer Depletion:** The primary cause of the ozone layer's thinning is the atmospheric release of hydrochlorofluorocarbons, halons, and chlorofluorocarbons. People who live in areas where the ozone layer is thinning may experience skin and vision issues as a result of the sun's harmful UV rays.

v. **Effect on Animals:** The air pollutants suspend in the water bodies and affect aquatic life. Pollution also compels the animals to leave their habitat and shift to a new place. This renders them stray and has also led to the extinction of a large number of animal species (Moores, 2009; Villányi et al., 2010).

1.3.1. Particulate Matter (PM) and Health

Research has demonstrated a connection between PM and harmful health outcomes, with an emphasis on acute (short-term) or chronic (long-term) PM exposure. Typically, chemical reactions between the various pollutants result in the formation of PM in the atmosphere. The size of the particles has a direct impact on how well they penetrate. The U.S Environmental Protection Agency defined particles as part of the category known as PM. PM pollution comprises two types of particles: extremely fine particles, typically with diameters of 2 to 5 micrometers (μm) and smaller, and particles with diameters of 10 micrometers (μm) or less, known as PM_{10}. PM contains tiny liquid or solid droplets that can be inhaled and cause serious health effects. Particles <10 μm in diameter (PM_{10}) after inhalation can invade the lungs and even reach the bloodstream. Fine particles, $PM_{2.5}$, pose a greater risk to health (Table 1.1).

Table 1.1. *Penetrability According to Particle Size*

Particle Size	Penetration Degree in Human Respiratory System
>11 μm	Passage into nostrils and upper respiratory tract
7–11 μm	Passage into nasal cavity
4.7–7 μm	Passage into larynx
3.3–4.7 μm	Passage into trachea-bronchial area
2.1–3.3 μm	Secondary bronchial area passage
1.1–2.1 μm	Terminal bronchial area passage
0.65–1.1 μm	Bronchioles penetrability
0.43–0.65 μm	Alveolar penetrability

Multiple epidemiological studies have been performed on the health effects of PM. A positive relation was shown between both short-term and long-term exposures of $PM_{2.5}$ and acute nasopharyngitis. In addition, long-term exposure to PM for years was found to be related to cardiovascular diseases and infant mortality.

Due to the lack of spatially resolved daily $PM_{2.5}$ concentration data, those studies are limited in terms of study area and city area and thus do not represent the entire population. Instead, they rely on $PM_{2.5}$ monitors. The Department of Environmental Health at Harvard School of Public Health recently conducted an epidemiological study that revealed an exposure error (Berkson error) as $PM_{2.5}$ concentrations vary spatially. The study also revealed that the relative magnitudes of the short- and long-term effects are still not fully understood. To evaluate both short- and long-term human exposures, the team created an exposure model for $PM_{2.5}$ based on data from remote sensing.

This model allows for both the assessment of long-term effects across the entire population and spatial resolution in short-term effects. In addition, long-term chronic effects include immune system affection and respiratory disorders. It is important to remember that those who have diabetes, asthma, pneumonia, or other respiratory or cardiovascular conditions are particularly vulnerable to the negative effects of PM. Because $PM_{2.5}$ and PM_{10} are small enough to penetrate interior spaces, they are strongly linked to a variety of respiratory system diseases. The chemical and physical characteristics of the particles cause harmful effects. Both organic (polycyclic aromatic hydrocarbons, dioxins, benzene, and 1–3 butadiene) and inorganic (carbon, chlorides, nitrates, sulfates, and metals) substances can be found in PM_{10} and $PM_{2.5}$. PM is divided into four main categories according to type and size (Table 1.2).

Table 1.2. *Types and Sizes of Particulate Matter (PM)*

Type		**PM Diameter [Mm]**
Particulate contaminants	Smog	0.01–1
	Soot	0.01–0.8
	Tobacco smoke	0.01–1
	Fly ash	1–100
	Cement Dust	8–100
Biological Contaminants	Bacteria and bacterial spores	0.7–10
	Viruses	0.01–1

	Fungi and molds	2–12
	Allergens (dogs, cats, pollen, household dust)	0.1–100
Types of Dust	Atmospheric dust	0.01–1
	Heavy dust	100–1000
	Settling dust	1–100
Gases	Different gaseous contaminants	0.0001–0.01

i. **Gas contaminants** include PM in aerial masses.

ii. **Particulate contaminants** include contaminants such as smog, soot, tobacco smoke, oil smoke, fly ash, and cement dust.

iii. **Biological Contaminants** are microorganisms (bacteria, viruses, fungi, mold, and bacterial spores), cat allergens, house dust and allergens, and pollen.

iv. **Types of Dust** include suspended atmospheric dust, settling dust, and heavy dust.

Last but not least, the fact that PM_{10} and $PM_{2.5}$ particles are so small that their half-lives in the atmosphere are prolonged also means that they can remain suspended in the atmosphere for a long time and even travel to far-off places where the environment and people may be exposed to the same level of pollution. They have the power to harm crops and forests, alter the nutrient balance in wet ecosystems, and acidify water sources.

As stated, $PM_{2.5}$, due to their tiny size, are causing more serious health effects. These aforementioned fine particles are the main cause of the "haze" formation in different metropolitan areas.

1.3.2. Ozone Impact in the Atmosphere

A gas called ozone (O_3) is created when high-voltage electric discharge occurs from oxygen. It is an extremely potent oxidant, 52% more potent than chlorine. It originates in the stratosphere but may also develop as a result of photochemical smog in the troposphere chain reactions.

Ozone moves with air masses and can reach far-off places from its original source. Surprisingly, ozone levels in cities are low compared to elevated levels that occur in urban areas. Because ozone reduces carbon assimilation, it may become harmful to cultures, forests, and vegetation. Ozone's antimicrobial properties impact plant microflora and decrease growth and yield. In this sense, ozone affects other natural ecosystems, causing changes in the species composition of animals and microflora. Ozone impairs cellular function by increasing DNA damage in epidermal keratinocytes. Ground-level ozone (GLO) is produced chemically when nitrogen oxides and VOCs released by man-made or natural sources react. Inhalation is the usual method of absorbing ozone. Ozone has an impact on the tear ducts and the epidermis. In addition to depleting vitamins C and E, a study involving mice exposed to high levels of ozone for a brief period of time revealed the formation of malondialdehyde in the epidermis (upper skin).

It is unlikely that ozone levels are compromising the integrity and function of the skin barrier to increase the risk of skin disease. Ozone can enter the lungs deeply through inhalation because it is not soluble in water. Ozone-induced toxic effects have been documented in urban areas worldwide, leading to immunological, morphologic, functional, and biochemical disorders.

The immediate effects of ambient ozone concentrations on mortality are the main focus of the European project (APHEA2). Over a three-year period, daily reports from various European cities were made regarding the ratio of ozone concentrations to the number of deaths per day. An increase in ozone concentration during the warm season of the year was linked to a rise in the number of deaths per day (0.33%), respiratory deaths (1.13%), and cardiovascular deaths (0.45%). There was no discernible impact in the winter.

1.3.2.1. Carbon Monoxide (CO)

When fossil fuels burn too slowly, CO is released. The symptoms of CO poisoning from inhalation include headache, weakness, nausea, vomiting, and, in the end, unconsciousness. Compared to oxygen, CO has a far stronger affinity for hemoglobin. In a similar line, prolonged exposure to high CO levels can result in serious poisoning.

The loss of oxygen brought on by CO's competitive binding results in hypoxia, ischemia, and cardiovascular disease. The greenhouse gases that are closely linked to climate change and global warming are impacted by CO

Temperatures in the soil and water should rise as a result, and severe weather or storms might happen. Nonetheless, it has been observed to result in higher plant growth in lab and field studies.

1.3.2.2. Nitrogen Oxide (NO_2)

Since nitrogen oxide is released from car engines, it is a pollutant associated with traffic. It irritates the respiratory system because it goes deep into the lungs, causing bronchospasm, coughing, wheezing, dyspnea, and even pulmonary edema at high concentrations. Humans appear to be adversely affected by concentrations greater than 0 ppm, whereas T-lymphocytes—specifically, the

CD8+ cells and NK cells that generate our immune response—are affected by concentrations greater than 2 ppm. Chronic lung disease has been linked to high levels of nitrogen dioxide exposure over an extended time. Long-term NO_2 exposure can damage one's ability to smell.

Nevertheless, since symptoms like irritation of the nose, throat, and eyes have been reported, systems other than the respiratory one may be affected. Excessive nitrogen dioxide concentrations have been shown to lower crop yield and plant growth efficiency, which is harmful to vegetation and crops. Furthermore, NO_2 can taint textiles and lessen visibility.

1.3.2.3. Sulfur Dioxide (SO_2)

The use of fossil fuels and industrial processes are the main sources of sulfur dioxide emissions, which is a hazardous gas. The SO_2 annual standard is 0.03 parts per million. It has an impact on plant, animal, and human life. Those who are susceptible include those who have lung diseases, the elderly, and small children, as they are more likely to suffer harm. The main health issues linked to sulfur dioxide emissions in industrialized regions are bronchitis, bronchospasm, mucus production, and respiratory irritation. Sulfur dioxide is a sensory irritant that enters the lung deeply, converts into bisulfite, and interacts with sensory receptors to cause bronchoconstriction.

Additionally, redness of the skin, harm to mucous membranes and the eyes (lacrimation and corneal opacity), and a worsening of pre-existing cardiovascular disease have all been reported. Acid rain and soil acidification are two negative environmental effects that appear to be connected to sulfur dioxide emissions.

1.3.2.4. Lead

Lead is a heavy metal that is utilized in various industrial facilities and released from waste incinerators, batteries, radiators, gasoline-powered cars, and waste waters. Additionally, metals, ore, and piston-engine aircraft are the main causes of airborne lead pollution. Because lead poisoning harms people, animals, and the environment—particularly in developing nations—it poses a risk to public health. Lead exposure can happen through eating, breathing, or absorbing it through the skin.

Lead is reported to be transplacentally transported as it crosses the placenta unimpeded. The harmful effects of toxins increase with the age of the fetus. The fetal nervous system is impacted by lead poisoning; edema, or swelling of the brain, is seen. When inhaled, lead builds up in the blood, soft tissues, liver, lungs, bones, and nervous, reproductive, and cardiovascular systems. Adults also reported experiencing pain in their muscles and joints, as well as memory loss and concentration problems.

Due to the neurotoxic properties of lead, which can result in learning disabilities, memory loss, hyperactivity, and even mental retardation, children, and newborns are particularly vulnerable to even low doses of the metal. High environmental lead concentrations are detrimental to crop growth in plants. Elevated lead levels have been linked to neurological effects in animals and vertebrates.

1.3.2.5. Polycyclic Aromatic Hydrocarbons (PAHs)

Due to the atmosphere's primary role in PAH dispersal, the compounds are widely distributed throughout the environment. They can be discovered in tar and coal sediments. Additionally, they are produced when organic matter is not completely burned, as occurs in engine, incineration, and forest fire situations. Benzopyrene, acenaphthylene, anthracene, and fluoranthene are examples of PAH compounds that are known to be harmful, mutagenic, and carcinogenic. They pose a significant risk of developing lung cancer.

1.3.2.6. Volatile Organic Compounds (VOCs)

Human cancer has been linked to VOCs, including xylene, toluene, benzene, and ethylbenzene. In fact, the use of novel materials and products has led to higher VOC concentrations. VOCs can be harmful to human health and contaminate indoor air. Adverse effects on human health, both short- and long-term, are documented. VOCs) cause odors in indoor air. It has been discovered that short-term exposure can irritate the throat, nose, eyes, and mucosal membranes, while long-term exposure can result in toxic reactions. Since these pollutants can have antagonistic, indifferent, or synergistic effects, it is challenging to estimate the toxic effects of complex VOC mixtures predictably.

1.3.2.7. Dioxins

They are produced by both natural and industrial processes, including volcanic eruptions and forest fires. They gather in foods like fish, shellfish, meat, and dairy products, and particularly in the fatty tissue of animals. Short-term exposure to elevated dioxin concentrations can cause skin lesions and dark spots. Prolonged exposure to dioxins can lead to cancer, immune system, endocrine, and nervous system dysfunction, infertility, and developmental abnormalities.

There is no question that the use of fossil fuels contributes significantly to air pollution. In addition to contamination from natural sources, this contamination may be caused by human activity, such as transportation or agricultural and industrial processes. It's interesting to note that, in contrast to the WHO guidelines, which are stricter, the air quality standards established by the European Air Quality Directive are a little looser (Moores, 2009; Villányi et al., 2010).

1.3.3. Air Pollution Control

Air pollution control refers to the methods used to lessen or stop the release of pollutants into the atmosphere that may be harmful to human health or the environment. Among the main focuses of pollution control are solid waste management, hazardous waste management, wastewater treatment, and air pollution control. When specific substances are present in air for long enough periods of time and in high enough concentrations to be harmful or have undesirable effects, the air is deemed polluted. These include detrimental impacts on property, air visibility, and human health. Both human activity and natural sources of pollution can affect the atmosphere. Certain natural occurrences, like forest fires and volcanic eruptions, may have long-lasting global repercussions in addition to local and regional ones.

However, the only pollution that can be mitigated and controlled is that which results from human activity, like transportation and industry. The majority of air pollutants are produced by combustion processes. In London and other major European cities during the Middle Ages, the burning of coal for fuel frequently resulted in air pollution issues. Following the Industrial Revolution, the 19th century saw an increase in the use of fossil fuels, which worsened the frequency and severity of air pollution episodes. Mobile sources of air pollution have emerged, that is gasoline-powered highway cars—had a significant influence on urban air quality issues. However, significant, and long-lasting efforts to control air quality at both the regional and local levels, as well as to regulate or limit emissions of air pollutants from both stationary and mobile sources, were not made until the middle of the 20th century.

In industrialized nations, ambient, or outside, air quality protection has been the main goal of air pollution regulation. This entails limiting the amount of particular "criteria" pollutants that are known to be a factor in chronic public health issues and urban smog. Lead, sulfur dioxide, nitrogen dioxide, ozone, fine particulates, and CO are among the criteria pollutants. Since the turn of the 20th century, many other air pollutants known as "air toxics" have also been known to have dangerous side effects in trace amounts. The majority of air toxics are organic compounds, which are made up of molecules with atoms of carbon, hydrogen, and other elements. To combat those pollutants, specific emission regulations have been put in place. Furthermore, coordinated international efforts have been made to control those pollutants in light of the long-term and extensive effects of the "greenhouse gases" on atmospheric chemistry and climate. Methane, nitrous oxide, ozone, carbon dioxide, and chlorofluorocarbons (CFCs) are examples of greenhouse gases. 2009 saw the U.S. According to an EPA ruling, greenhouse gases could be regulated as air pollutants because they are harmful to human health.

The best strategy to preserve air quality is to cut back on pollution emissions by switching to cleaner fuels and manufacturing techniques. If pollutants cannot be eliminated in this manner, they must be captured or contained as they are produced and before they have a chance to enter the atmosphere by suitable air-cleaning equipment. Here is a description of these devices. This article focuses on air pollution control technology, which is used to eliminate gaseous and particulate pollutants from stationary sources such as industrial facilities and power plants.

A variety of techniques and technologies are used in air pollution control in an effort to lessen or completely stop the release of pollutants into the atmosphere. Maintaining the environment, safeguarding human health, and improving air quality all depend on efficient air pollution control.

Following are the measures one should adopt, to control air pollution:

1.3.3.1. Avoid Using Vehicles

People should avoid using vehicles for shorter distances. Rather, they should prefer public modes of transport to travel from one place to another. This not only prevents pollution but also conserves energy.

1.3.3.2. Energy Conservation

A large number of fossil fuels are burnt to generate electricity. Therefore, do not forget to switch off the electrical appliances when not in use. Thus, you can save the environment at the individual level. The use of energy-efficient devices such as CFLs also controls pollution to a greater level.

1.3.3.3. Use of Clean Energy Resources

The use of solar, wind, and geothermal energies reduces air pollution at a larger level. Various countries, including India, have implemented the use of these resources as a step towards a cleaner environment.

Other air pollution control measures include:

1 By minimizing and reducing the use of fire and fire products.

2 Since industrial emissions are one of the major causes of air pollution, the pollutants can be controlled or treated at the source itself to reduce their effects. For example, if the reactions of a certain raw material yield a pollutant, then the raw materials can be substituted with other less polluting materials.

3 Fuel substitution is another way of controlling air pollution. In many parts of India, petrol, and diesel are being replaced by CNG –

Compressed Natural Gas fueled vehicles. These are mostly adopted by vehicles that aren't fully operating with ideal emission engines.

4 Although there are many practices in India, which focus on repairing the quality of air, most of them are either forgotten or not being enforced properly. There are still a lot of vehicles on roads that haven't been tested for vehicle emissions.

5 Another way of controlling air pollution caused by industries is to modify and maintain existing pieces of equipment so that the emission of pollutants is minimized.

6 Sometimes controlling pollutants at the source is not possible. In that case, we can have process control equipment to control the pollution.

7 A very effective way of controlling air pollution is by diluting the air pollutants.

8 The last and the best way of reducing the ill effects of air pollution is tree plantation. Plants and trees reduce a large number of pollutants in the air. Ideally, planting trees in areas of high pollution levels will be extremely effective (Moores, 2009; Villányi et al., 2010).

REFERENCES

1. Abdel-Shafy, H. I., & Mansour, M. S. M. (2016). A review on polycyclic aromatic hydrocarbons: Source, environmental impact, effect on human health, and remediation. *Egyptian Journal of Petroleum, 25*(1), 107–123. https://doi.org/10.1016/j.ejpe.2015.03.011.
2. Alexopoulos, A., Plessas, S., Ceciu, S., Lazar, V., Mantzourani, I., Voidarou, C., et al. (2013). Evaluation of ozone efficacy on the reduction of microbial population of fresh cut lettuce (*Lactuca sativa*) and green bell pepper (*Capsicum annuum*). *Food Control, 30*(2), 491–496. https://doi.org/10.1016/j.foodcont.2012.09.018.
3. Alexopoulos, A., Plessas, S., Kourkoutas, Y., Stefanis, C., Vavias, S., Voidarou, C., et al. (2017). Experimental effect of ozone upon the microbial flora of commercially produced dairy fermented products. *International Journal of Food Microbiology, 246*, 5–11. https://doi.org/10.1016/j.ijfoodmicro.2017.01.018.
4. Ashfaq, A., & Sharma, P. (2012). Environmental effects of air pollution and application of engineered methods to combat the problem. *Journal of Industrial Pollution Control, 29*.
5. Assi, M. A., Hezmee, M. N. M., Haron, A. W., Sabri, M. Y. M., & Rajion, M. A. (2016). The detrimental effects of lead on human and animal health. *Veterinary World, 9*(6), 660–671. https://doi.org/10.14202/vetworld.2016.660-671.
6. Balbo, P., Silvestri, M., Rossi, G. A., Crimi, E., & Burastero, S. E. (2001). Differential role of CD80 and CD86 on alveolar macrophages in the presentation of allergen to T lymphocytes in asthma. *Clinical and Experimental Allergy, 31*(4), 625–636. https://doi.org/10.1046/j.1365-2222.2001.01068.x.
7. Bellinger, D. C. (2008). Very low lead exposures and children's neurodevelopment. *Current Opinion in Pediatrics, 20*(2), 172–177. https://doi.org/10.1097/MOP.0b013e3282f4f97b.
8. Bezirtzoglou, C., Dekas, K., & Charvalos, E. (2011). Climate changes, environment and infection: Facts, scenarios, and growing awareness from the public health community within Europe. *Anaerobe, 17*(6), 337–340. https://doi.org/10.1016/j.anaerobe.2011.05.016.
9. Bezirtzoglou, E., & Alexopoulos, A (2009). Ozone history and ecosystems: A goliath from impacts to advance industrial benefits and interests, to environmental and therapeutic strategies. In *Ozone Depletion, Chemistry and Impacts* (pp. 135–145).

10. Bezirtzoglou, E., Dimitriou, D., & Panagiou, A. (1996). Occurrence of *Clostridium perfringens* in river water by using a new procedure. *Anaerobe, 2*(3), 169–173. https://doi.org/10.1006/anae.1996.0022.
11. Bonavigo, L., Zucchetti, M., & Mankolli, H. (2009). Water radioactive pollution and related environmental aspects. *Journal of International Environmental Application & Science, 4*(4), 357–363.
12. Boschi, N. (2012). Defining an educational framework for indoor air sciences education. In *Education and Training in Indoor Air Sciences* (p. 245). Springer Science & Business Media.
13. Bourdrel, T., Bind, M. A., Béjot, Y., Morel, O., & Argacha, J. F. (2017). Cardiovascular effects of air pollution. *Archives of Cardiovascular Diseases, 110*(11), 634–642. https://doi.org/10.1016/j.acvd.2017.05.003.
14. Burroughs Peña, M. S., & Rollins, A. (2017). Environmental exposures and cardiovascular disease: A challenge for health and development in low- and middle-income countries. *Cardiology Clinics, 35*(1), 71–86. https://doi.org/10.1016/j.ccl.2016.09.001.
15. Calba, C., Guerbois-Galla, M., Franke, F., Jeannin, C., Auzet-Caillaud, M., Grard, G., et al. (2017). Preliminary report of an autochthonous chikungunya outbreak in France, July to September 2017. *Eurosurveillance, 22*(39), 17–00647. https://doi.org/10.2807/1560-7917.ES.2017.22.39.17-00647.
16. Calderon-Garciduenas, L., Azzarelli, B., Acuna, H., et al. (2002). Air pollution and brain damage. *Toxicologic Pathology, 30*(3), 373–389. https://doi.org/10.1080/01926230252929954.
17. Castelli, F., & Sulis, G. (2017). Migration and infectious diseases. *Clinical Microbiology and Infection, 23*(5), 283–289. https://doi.org/10.1016/j.cmi.2017.03.012.
18. Chen, T. M., Gokhale, J., Shofer, S., & Kuschner, W. G. (2007). Outdoor air pollution: Nitrogen dioxide, sulfur dioxide, and carbon monoxide health effects. *The American Journal of the Medical Sciences, 333*(4), 249–256. https://doi.org/10.1097/MAJ.0b013e31803b900f.
19. Cheung, K., Daher, N., Kam, W., Shafer, M. M., Ning, Z., Schauer, J. J., et al. (2011). Spatial and temporal variation of chemical composition and mass closure of ambient coarse particulate matter (PM10–2.5) in the Los Angeles area. *Atmospheric Environment, 45*(16), 2651–2662. https://doi.org/10.1016/j.atmosenv.2011.02.066.
20. Colbeck, I., & Lazaridis, M. (2009). Aerosols and environmental pollution. *The Science of Nature, 97*(2), 117–131. https://doi.org/10.1007/s00114-009-0594-x.
21. D'Amato, G., Pawankar, R., Vitale, C., & Maurizia, L. (2016). Climate change and air pollution: Effects on respiratory allergy. *Allergy, Asthma*

& Immunology Research, 8(5), 391–395. https://doi.org/10.4168/aair.2016.8.5.391.

22. Dherani, M., Pope, D., Mascarenhas, M., Smith, K. R., & Weber, M. B. N. (2008). Indoor air pollution from unprocessed solid fuel use and pneumonia risk in children aged under five years: A systematic review and meta-analysis. *Bulletin of the World Health Organization, 86*(5), 390–394. https://doi.org/10.2471/BLT.07.044529.
23. Dockery, D. W., Pope, C. A., Xu, X., Spengler, J. D., Ware, J. H., Fay, M. E., et al. (1993). An association between air pollution and mortality in six U.S. cities. *New England Journal of Medicine, 329*(24), 1753–1759. https://doi.org/10.1056/NEJM199312093292401.
24. Drakaki, E., Dessinioti, C., & Antoniou, C. (2014). Air pollution and the skin. *Frontiers in Environmental Science, 15*(1), 2–8. https://doi.org/10.3389/fenvs.2014.00011.
25. Ebersviller, S., Lichtveld, K., Sexton, K. G., Zavala, J., Lin, Y. H., Jaspers, I., et al. (2012). Gaseous VOCs rapidly modify particulate matter and its biological effects–Part 1: Simple VOCs and model PM. *Atmospheric Chemistry and Physics Discussions, 12*(11), 5065–5105. https://doi.org/10.5194/acpd-12-5065-2012.
26. Emberson, L. D., Pleijel, H., Ainsworth, E. A., van den Berg, M., Ren, W., Osborne, S., et al. (2018). Ozone effects on crops and consideration in crop models. *European Journal of Agronomy, 100*, 19–34. https://doi.org/10.1016/j.eja.2018.06.002.
27. Eze, I. C., Schaffner, E., Fischer, E., Schikowski, T., Adam, M., Imboden, M., et al. (2014). Long-term air pollution exposure and diabetes in a population-based Swiss cohort. *Environment International, 70*, 95–105. https://doi.org/10.1016/j.envint.2014.05.014.
28. Fares, S., Vargas, R., Detto, M., Goldstein, A. H., Karlik, J., & Paoletti, E. (2013). Tropospheric ozone reduces carbon assimilation in trees: Estimates from analysis of continuous flux measurements. *Global Change Biology, 19*(8), 2427–2443. https://doi.org/10.1111/gcb.12222.
29. Farhat, A., Mohammadzadeh, A., Balali-Mood, M., Aghajanpoor-Pasha, M., & Ravanshad, Y. (2013). Correlation of blood lead level in mothers and exclusively breastfed infants: A study on infants aged less than six months. *Asia Pacific Journal of Medical Toxicology, 2*(4), 150–152.
30. Genc, S., Zadeoglulari, Z., Fuss, S. H., & Genc, K. (2012). The adverse effects of air pollution on the nervous system. *Journal of Toxicology, 2012*, 782462. https://doi.org/10.1155/2012/782462.

31. Gibson, R., & Ward, S. (2009). Parties in the digital age: A review. *Journal of Representation and Democracy, 45*(1), 87–100. https://doi.org/10.1080/00344890802710888.

32. Gryparis, A., Forsberg, B., Katsouyanni, K., Analitis, A., Touloumi, G., Schwartz, J., et al. (2004). Acute effects of ozone on mortality from the "air pollution and health: A European approach" project. *American Journal of Respiratory and Critical Care Medicine, 170*(10), 1080–1087. https://doi.org/10.1164/rccm.200403-333OC.

33. Guarnieri, M., & Balmes, J. R. (2014). Outdoor air pollution and asthma. *The Lancet, 383*(9928), 1581–1592. https://doi.org/10.1016/S0140-6736(14)60617-6.

34. Guo, Y., Zeng, H., Zheng, R., Li, S., Pereira, G., Liu, Q., et al. (2017). The burden of lung cancer mortality attributable to fine particles in China. *Science of the Total Environment, 579*, 1460–1466. https://doi.org/10.1016/j.scitotenv.2016.11.147.

35. Harmens, H., Mills, G., Hayes, F., Jones, L., Norris, D., & Fuhrer, J. (2012). *Air Pollution and Vegetation*. ICP Vegetation Annual Report 2006/2007.

36. Hashim, D., & Boffetta, P. (2014). Occupational and environmental exposures and cancers in developing countries. *Annals of Global Health, 80*(5), 393–411. https://doi.org/10.1016/j.aogh.2014.10.002.

37. Hatch, G. E., Slade, R., Harris, L. P., McDonnell, W. F., Devlin, R. B., Koren, H. S., et al. (1994). Ozone dose and effect in humans and rats: A comparison using oxygen-18 labeling and bronchoalveolar lavage. *American Journal of Respiratory and Critical Care Medicine, 150*(3), 676–683. https://doi.org/10.1164/ajrccm.150.3.8087337.

38. Heal, M. R., Kumar, P., & Harrison, R. M. (2012). Particles, air quality, policy and health. *Chemical Society Reviews, 41*(19), 6606–6630. https://doi.org/10.1039/c2cs35076a.

39. Hesterberg, T. W., Bunn, W. B., McClellan, R. O., Hamade, A. K., Long, C. M., & Valberg, P. A. (2009). Critical review of the human data on short-term nitrogen dioxide (NO_2) exposures: Evidence for NO_2 no-effect levels. *Critical Reviews in Toxicology, 39*(9), 743–781. https://doi.org/10.3109/10408440903294945.

40. Hoffmann, B., Moebus, S., Möhlenkamp, S., Stang, A., Lehmann, N., Dragano, N., et al. (2007). Residential exposure to traffic is associated with coronary atherosclerosis. *Circulation, 116*(5), 489–496. https://doi.org/10.1161/CIRCULATIONAHA.107.693622.

41. Hou, Q., An, X. Q., Wang, Y., & Guo, J. P. (2010). An evaluation of resident exposure to respirable particulate matter and health economic loss in

Beijing during the 2008 Olympic Games. *Science of the Total Environment, 408*(19), 4026–4032. https://doi.org/10.1016/j.scitotenv.2009.12.030.

42. Incecik, S., Gertler, A., & Kassomenos, P. (2014). Aerosols and air quality. *Science of the Total Environment, 355*, 488–489. https://doi.org/10.1016/j.scitotenv.2014.04.012.
43. Jacobson, M. Z., & Jacobson, P. M. Z. (2002). *Atmospheric Pollution: History, Science, and Regulation* (p. 206). Cambridge University Press. https://doi.org/10.1256/wea.243.02.
44. Jiang, X. Q., Mei, X. D., & Feng, D. (2016). Air pollution and chronic airway diseases: What should people know and do? *Journal of Thoracic Disease, 8*(1), E31–E40.
45. Kan, H., Chen, R., & Tong, S. (2012). Ambient air pollution, climate change, and population health in China. *Environment International, 42*, 10–19. https://doi.org/10.1016/j.envint.2011.03.003.
46. Kankaria, A., Nongkynrih, B., & Gupta, S. (2014). Indoor air pollution in India: Implications on health and its control. *Indian Journal of Community Medicine, 39*(4), 203–207. https://doi.org/10.4103/0970-0218.143019.
47. Kappos, A. D., Bruckmann, P., Eikmann, T., Englert, N., Heinrich, U., Höppe, P., et al. (2004). Health effects of particles in ambient air. *International Journal of Hygiene and Environmental Health, 207*(4), 399–407. https://doi.org/10.1078/1438-4639-00306.
48. Kassomenos, P., Kelessis, A., Petrakakis, M., Zoumakis, N., Christides, T., & Paschalidou, A. K. (2012). Air quality assessment in a heavily polluted urban Mediterranean environment through air quality indices. *Ecological Indicators, 18*, 259–268. https://doi.org/10.1016/j.ecolind.2011.11.021.
49. Katholi, R. E., & Couri, D. M. (2011). Left ventricular hypertrophy: Major risk factor in patients with hypertension: Update and practical clinical applications. *International Journal of Hypertension, 2011*, 495349. https://doi.org/10.4061/2011/495349.
50. Kaun, A., & Uldam, J. (2017). Digital activism: After the hype. *New Media & Society, 20*(6), 2099–2106. https://doi.org/10.1177/14614448177319.
51. Kelishadi, R., & Poursafa, P. (2010). Air pollution and non-respiratory health hazards for children. *Archives of Medical Science, 6*(4), 483–495. https://doi.org/10.5114/aoms.2010.14458.
52. Kelly, F. J. (2004). Dietary antioxidants and environmental stress. *Proceedings of the Nutrition Society, 63*(4), 579–585. https://doi.org/10.1079/PNS2004388.
53. Kjellstrom, T., Lodh, M., McMichael, T., Ranmuthugala, G., Shrestha, R., & Kingsland, S. (n.d.). Air and water pollution: Burden and strategies for

control. In *Disease Control Priorities in Developing Countries* (2nd ed., Chap. 43, pp. 817–832).

54. Kloog, I., Ridgway, B., Koutrakis, P., Coull, B. A., & Schwartz, J. D. (2013). Long- and short-term exposure to PM2.5 and mortality using novel exposure models. *Epidemiology, 24*(4), 555–561. https://doi.org/10.1097/EDE.0b013e318294beaa.

55. Klopfer, J. (1989). Effects of environmental air pollution on the eye. *Journal of the American Optometric Association, 60*(10), 773–778.

56. Kumar, A., Singh, B. P., Punia, M., Singh, D., Kumar, K., & Jain, V. K. (2014). Assessment of indoor air concentrations of VOCs and their associated health risks in the library of Jawaharlal Nehru University, New Delhi. *Environmental Science and Pollution Research, 21*(3), 2240–2248. https://doi.org/10.1007/s11356-013-2150-7.

57. Kurt, O. K., Zhang, J., & Pinkerton, K. E. (2016). Pulmonary health effects of air pollution. *Current Opinion in Pulmonary Medicine, 22*(2), 138–143. https://doi.org/10.1097/MCP.0000000000000248.

58. Leary, P. J., Kaufman, J. D., Barr, R. G., Bluemke, D. A., Curl, C. L., Hough, C. L., et al. (2014). Traffic-related air pollution and the right ventricle: The multi-ethnic study of atherosclerosis. *American Journal of Respiratory and Critical Care Medicine, 189*(9), 1093–1100. https://doi.org/10.1164/rccm.201312-2298OC.

59. Lindh, E., Argentini, C., Remoli, M. E., Fortuna, C., Faggioni, G., Benedetti, E., et al. (2019). The Italian 2017 outbreak of Chikungunya virus belongs to an emerging *Aedes albopictus*–adapted virus cluster introduced from the Indian subcontinent. *Open Forum Infectious Diseases, 6*(1), ofy321. https://doi.org/10.1093/ofid/ofy321.

60. Lippmann, M. (1989). Health effects of ozone: A critical review. *Journal of the Air Pollution Control Association, 39*(5), 672–695. https://doi.org/10.1080/08940630.1989.10466554.

61. Lorenzini, G., & Saitanis, C. (2003). Ozone: A novel plant "pathogen." In L. Sanitá di Toppi & B. Pawlik-Skowrońska (Eds.), *Abiotic Stresses in Plants* (pp. 205–229). Springer. https://doi.org/10.1007/978-94-017-0255-3_8.

62. Maggio, A., & Fagnano, M. (2008). Ozone damages to Mediterranean crops: Physiological responses. *Italian Journal of Agronomy, 3*(1), 13–20. https://doi.org/10.4081/ija.2008.13.

63. Maipa, V., Alamanos, Y., & Bezirtzoglou, E. (2001). Seasonal fluctuation of bacterial indicators in coastal waters. *Microbial Ecology in Health and Disease, 13*(3), 143–146. https://doi.org/10.1080/089106001750462687.

64. Mannucci, P. M., & Franchini, M. (2017). Health effects of ambient air pollution in developing countries. *International Journal of Environmental Research and Public Health, 14*(9), 1048. https://doi.org/10.3390/ijerph14091048.
65. Marlon, J. R., Bloodhart, B., Ballew, M. T., Rolfe-Redding, J., Roser-Renouf, C., Leiserowitz, A., et al. (2019). How hope and doubt affect climate change mobilization. *Frontiers in Communication, 4*, 20. https://doi.org/10.3389/fcomm.2019.00020.
66. McCarthy, J. T., Pelle, E., Dong, K., Brahmbhatt, K., Yarosh, D., & Pernodet, N. (2013). Effects of ozone in normal human epidermal keratinocytes. *Experimental Dermatology, 22*(5), 360–361. https://doi.org/10.1111/exd.12125.
67. Ministers of Energy and Environment Federal/Provincial/Territorial Ministers of Energy and Environment (Canada). (1999). *The Canada-Wide Acid Rain Strategy for Post-2000*. Halifax: The Ministers.
68. Mo, Z., Fu, Q., Lyu, D., Zhang, L., Qin, Z., Tang, Q., et al. (2019). Impacts of air pollution on dry eye disease among residents in Hangzhou, China: A case-crossover study. *Environmental Pollution, 246*, 183–189. https://doi.org/10.1016/j.envpol.2018.11.109.
69. Molhave, L., Clausen, G., Berglund, B., Ceaurriz, J., Kettrup, A., Lindvall, T., et al. (1997). Total volatile organic compounds (TVOC) in indoor air quality investigations. *Indoor Air, 7*(4), 225–240. https://doi.org/10.1111/j.1600-0668.1997.00002.x.
70. Möller, L., Schuetzle, D., & Autrup, H. (1994). Future research needs associated with the assessment of potential human health risks from exposure to toxic ambient air pollutants. *Environmental Health Perspectives, 102*(Suppl. 4), 193–210. https://doi.org/10.1289/ehp.94102s4193.
71. Moores, F. C. (2009). Climate change and air pollution: Exploring the synergies and potential for mitigation in industrializing countries. *Sustainability, 1*(1), 43–54. https://doi.org/10.3390/su1010043
72. Nakano, T., & Otsuki, T. (2013). Environmental air pollutants and the risk of cancer. *Gan To Kagaku Ryoho, 40*(11), 1441–1445.
73. Newlands, M. (2015). Environmental activism, environmental politics, and representation: The framing of the British environmental activist movement (Doctoral dissertation). University of East London, United Kingdom.
74. Parajuli, I., Lee, H., & Shrestha, K. R. (2016). Indoor air quality and ventilation assessment of rural mountainous households of Nepal. *International Journal of Sustainable Built Environment, 5*(2), 301–311. https://doi.org/10.1016/j.ijsbe.2016.08.003.

75. Pathak, R. K., Wang, T., Ho, K. F., & Lee, S. C. (2011). Characteristics of summertime PM2.5 organic and elemental carbon in four major Chinese cities: Implications of high acidity for water-soluble organic carbon (WSOC). *Atmospheric Environment, 45*(2), 318–325. https://doi.org/10.1016/j.atmosenv.2010.10.021.
76. Peters, A., Veronesi, B., Calderón-Garcidueñas, L., Gehr, P., Chen, L. C., Geiser, M., et al. (2006). Translocation and potential neurological effects of fine and ultrafine particles: A critical update. *Particle and Fiber Toxicology, 3*(1), 13. https://doi.org/10.1186/1743-8977-3-13.
77. Richmont-Bryant, J., Owen, R. C., Graham, S., Snyder, M., McDow, S., Oakes, M., et al. (2017). Estimation of on-road NO2 concentrations, NO2/NOX ratios, and related roadway gradients from near-road monitoring data. *Air Quality, Atmosphere & Health, 10*(5), 611–625. https://doi.org/10.1007/s11869-016-0455-7.
78. Rückerl, R., Greven, S., Ljungman, P., Aalto, P., Antoniades, C., Bellander, T., et al. (2007). Air pollution and inflammation (interleukin-6, C-reactive protein, fibrinogen) in myocardial infarction survivors. *Environmental Health Perspectives, 115*(7), 1072–1080. https://doi.org/10.1289/ehp.10021.
79. Saud, T., Gautam, R., Mandal, T. K., Gadi, R., Singh, D. P., & Sharma, S. K. (2012). Emission estimates of organic and elemental carbon from household biomass fuel used over the Indo-Gangetic Plain (IGP), India. *Atmospheric Environment, 61*, 212–220. https://doi.org/10.1016/j.atmosenv.2012.07.030.
80. Schneider, S. H. (1989). The greenhouse effect: Science and policy. *Science, 243*(4892), 771–781. https://doi.org/10.1126/science.243.4892.771.
81. Schwela, D. H., & Köth-Jahr, I. (1994). Leitfaden für die Aufstellung von Luftreinhalteplänen [Guidelines for the implementation of clean air implementation plans]. State Environmental Service of the State of North Rhine-Westphalia.
82. Singh, D. P., Gadi, R., Mandal, T. K., Saud, T., Saxena, M., & Sharma, S. K. (2013). Emissions estimates of PAH from biomass fuels used in the rural sector of Indo-Gangetic Plains of India. *Atmospheric Environment, 68*, 120–126. https://doi.org/10.1016/j.atmosenv.2012.11.042.
83. Singh, E., Tiwari, S., & Agrawal, M. (2009). Effects of elevated ozone on photosynthesis and stomatal conductance of two soybean varieties: A case study to assess impacts of one component of predicted global climate change. *Plant Biology, 11*(Suppl. 1), 101–108. https://doi.org/10.1111/j.1438-8677.2009.00263.x.

84. Sivitanides, M., & Shah, V. (2011). The era of digital activism. In *2011 Conference for Information Systems Applied Research (CONISAR) Proceedings*. Wilmington, North Carolina, USA.

85. Soon, W., Baliunas, S. L., Robinson, A. B., & Robinson, Z. W. (1999). Environmental effects of increased atmospheric carbon dioxide. *Climate Research, 13*(3), 149–164. https://doi.org/10.1260/0958305991499694.

86. Stansfeld, S. A. (2015). Noise effects on health in the context of air pollution exposure. *International Journal of Environmental Research and Public Health, 12*(10), 12735–12760. https://doi.org/10.3390/ijerph121012735.

87. Stover, R. H. (1979). Flooding of soil for disease control. In D. Mulder (Ed.), *Developments in Agricultural and Managed Forest Ecology* (Vol. 3, pp. 19–28). Elsevier.

88. Teramura, A. (2006). Effects of UV-B radiation on the growth and yield of crop plants. *Physiologic Plantarum, 58*(3), 415–427. https://doi.org/10.1111/j.1399-3054.1983.tb04203.x.

89. Thiele, J. J., Traber, M. G., Tsang, K., Cross, C. E., & Packer, L. (1997). *In vivo* exposure to ozone depletes vitamins C and E and induces lipid peroxidation in epidermal layers of murine skin. *Free Radical Biology and Medicine, 23*(3), 365–391. https://doi.org/10.1016/S0891-5849(96)00617-X.

90. Villányi, V., Turk, B., Franc, B., & Csintalan, Z. (2010). Ozone pollution and its bioindication. In V. Villányi (Ed.), *Air Pollution* (pp. 10–20). Intech Open. https://doi.org/10.5772/10047.

91. Watson, J. T., Gayer, M., & Connolly, M. A. (2007). Epidemics after natural disasters. *Emerging Infectious Diseases, 13*(1), 1–5. https://doi.org/10.3201/eid1301.060779.

92. Weisskopf, M. G., Kioumourtzoglou, M. A., & Roberts, A. L. (2015). Air pollution and autism spectrum disorders: Causal or confounded? *Current Environmental Health Reports, 2*(4), 430–439. https://doi.org/10.1007/s40572-015-0073-9.

93. Wilson, W. E., & Suh, H. H. (1997). Fine particles and coarse particles: Concentration relationships relevant to epidemiologic studies. *Journal of the Air & Waste Management Association, 47*(12), 1238–1249. https://doi.org/10.1080/10473289.1997.10464074.

94. Zhang, L., Yang, Y., Li, Y., Qian, Z. M., Xiao, W., Wang, X., et al. (2019). Short-term and long-term effects of PM2.5 on acute nasopharyngitis in 10 communities of Guangdong, China. *Science of the Total Environment, 688*, 136–142. https://doi.org/10.1016/j.scitotenv.2019.05.470.

95. Zuhara, S., & Isaifan, R. (2018). The impact of criteria air pollutants on soil and water: A review. *Journal of Environmental Science and Pollution Research*, *4*, 278–284. https://doi.org/10.30799/jespr.133.18040205.

CHAPTER 2

THE ATMOSPHERE

Contents

INTRODUCTION

Substances in the air that have the potential to negatively impact human health, the environment, and the climate are known as atmospheric pollutants. These pollutants are produced by both natural and man-made processes. Volcanic eruptions, forest fires, and dust storms are examples of natural sources that discharge gases and particulate matter into the atmosphere. However, the main causes of air pollution are human activities like transportation emissions, industrial processes, and agricultural methods. The combustion of fossil fuels to generate energy and power vehicles is a major contributor to the emission of pollutants such as particulate matter (PM), carbon monoxide (CO), sulfur dioxide (SO_2), and nitrogen oxides (NO_x). In addition to lowering air quality, these pollutants also cause secondary pollutants like acid rain and ground-level ozone to form, which worsen the situation for ecosystems and public health.

Air pollution has serious and wide-ranging effects on human health. Short-term exposure to high pollution levels can lead to cardiovascular problems, eye irritation, and respiratory problems. Lung cancer, heart disease, and chronic respiratory disorders are linked to long-term exposure. Children, the elderly, and people with underlying medical conditions are among the vulnerable groups that are most vulnerable. Moreover, negative pregnancy outcomes, such as preterm birth and low birth weight, are associated with air pollution. Air pollution is one of the biggest environmental health hazards in the world, according to estimates from the World Health Organization (WHO), which links it to millions of preventable deaths annually. Air pollution has serious negative effects on the environment in addition to health. Acid rain is a result of pollution such as sulfur dioxide and nitrogen oxides, which harms aquatic ecosystems, forests, and soils. Particulate matter can cling to the surfaces of plants, preventing photosynthesis, and reducing visibility, which can impact travel and tourism. Even though they are safe for human health in small amounts, greenhouse gases like carbon dioxide (CO_2) and methane (CH_3) contribute to climate change and global warming, which affect biodiversity and cause more extreme weather events and rising sea levels.

Initiatives to lower air pollution include public awareness campaigns, technological advancements, and regulatory changes. To reduce the amount of dangerous pollutants released into the atmosphere, governments all over the world have enacted emissions limits and air quality standards. Technological developments that have led to the creation of pollution control equipment and greener energy sources have also been instrumental in lowering emissions. To mitigate the effects of air pollution, public awareness and behavioral changes are crucial. Examples of these include cutting back on car use and supporting clean energy initiatives. Air pollution, however, continues to be a serious problem despite these initiatives, especially in areas that are developing quickly due to increased industrialization and urbanization. To address this issue and safeguard both the environment and human health, international cooperation and a dedication to sustainable practices are needed.

2.1. OVERVIEW OF ATMOSPHERE

The gas and aerosol envelope that envelops a planet's surface from the ocean to the land and ice-covered areas is called the atmosphere. It extends into space. Because the planet's gravitational pull, which draws gases and aerosols—microscopic suspended particles of dust, soot, smoke, or chemicals—inward, is strongest near the surface, the density of the atmosphere decreases outward. Certain planetary bodies, like Mercury, have nearly nonexistent atmospheres because the planet's primordial atmosphere was expelled into space due to the planet's relatively weak gravitational pull. There are still atmospheres on some planets in the solar system, including Mars, Venus, Earth, and the large outer planets. Furthermore, water has been able to exist in Earth's atmosphere in all three of its phases—solid, liquid, and gas—which has been necessary for the emergence of life on the planet. There is still much to learn about the current Earth's atmosphere's evolution. Unlike the primordial atmosphere, which formed by outgassing (venturing) during the planet's initial formation, it is believed that the current atmosphere resulted from a gradual release of gases from both the planet's interior and from the metabolic activities of life forms. Water vapor (H_2O), CO_2, SO_2, hydrogen sulfide (H_2S), CO, chlorine (Cl), fluorine (F), diatomic nitrogen (N_2; two atoms in a single molecule), and traces of other substances are among the gases currently emitted by volcanoes. The majority of volcanic emissions—about 85%—come in the form of water vapor. On the other hand, carbon dioxide makes up roughly 10% of the wastewater.

2.1.1. Composition of Atmosphere

All of the planet's living things depend heavily on the atmosphere, which is a mechanical mixture of various gases and other materials. Together, the four main gases—nitrogen, oxygen, argon, and carbon dioxide—make up 99.99% of the dry air's total volume. Nitrogen has a maximum concentration of over 78%, while oxygen has a concentration of slightly less than 21% (Table 2.1).

Table 2.1. *Atmospheric Gases*

Major Constituents of Air, by Mole Fraction			
Dry air			
Gas		Mole fraction(A)	
Name	Formula	in ppm(B)	in %
Nitrogen	N_2	780,840	78.084
Oxygen	O_2	209,460	20.946
Argon	Ar	9340	0.9340
Carbon dioxide (April, 2022)(C)[14]	CO_2	417	0.0417
Neon	Ne	18.18	0.001818
Helium	He	5.24	0.000524
Methane (2022)(C)[15]	CH_4	1.91	0.000191
Krypton	Kr	1.14	0.000114
If air is not dry:			
Water vapor(D)	H_2O	0–30,000(D)	0–3%(E)

Source: https://en.wikipedia.org/wiki/Atmosphere_of_Earth.

Notes:

(A) Mole fraction is sometimes referred to as volume fraction; these are identical for an ideal gas only.

(B) Ppm: parts per million by molecular count.

The total ppm above adds up to more than 1 million (currently 83.43 above it) due to experimental error.

(C) The concentration of CO_2 has been increasing in recent decades, as has that of CH_4.

(D) Water vapor is about 0.25% *by mass* over full atmosphere.

(E) Water vapor varies significantly locally.

2.1.1.1. Nitrogen

The most prevalent gas in the atmosphere, nitrogen makes up 78.084% of the total volume of dry gases (Figure 2.1). This is virtually chemically inert and unrelated to any chemical activity occurring in the atmosphere. Because it is not easily combined with other elements, it is referred to as a neutral substance.

This gas can be found above 100 km in altitude, but its concentration is lower than 50 km above sea level. This gas is important for plant and animal growth and reproduction. A very tiny amount of atmospheric nitrogen can be converted by certain soil bacteria into nitrates, which are then fixed to the soil and water bodies where it can be ingested by plants and animals. We refer to this process as "nitrogen fixation." Denitrification is the chemical process by which bacteria transform the nitrogen fixed in the earth's surface and return it to the atmosphere.

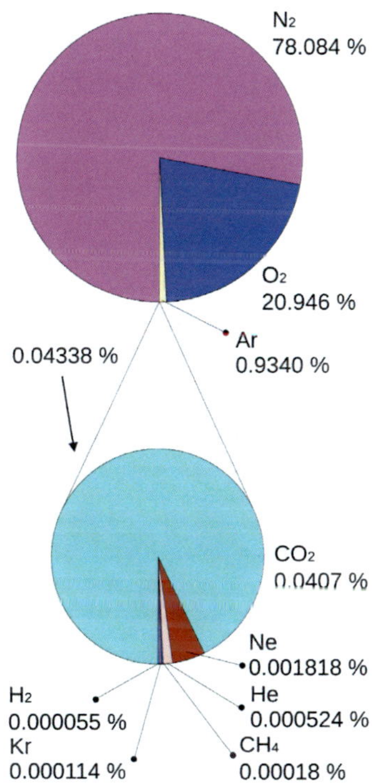

Figure 2.1. *Atmospheric gases and their proportions.*

Source: By Mysid – Vectorized version of w:Image:Atmosphere gas proportions.gif (originally by Brockert). I SVG'd it: (a) to make it more international (chemical symbols), (b) to make it (hopefully) clearer when rendered as a thumbnail, and (c) to make it easier to modify. Public Domain, https://commons.wikimedia.org/w/index.php?curid=823817.

2.1.1.2. Argon

Argon is the third-largest gas in the atmosphere in terms of percentage, making up 0.934 of the entire dry atmosphere (Figure 2.1). It is an inactive gas in terms of chemistry. It can also be found in seawater and the crust of the earth. It is a component of fluorescent and electric bulbs.

2.1.1.3. Carbon Dioxide

It is the atmosphere's fourth most prevalent gas. It is the densest gas and is located at the bottom. It is primarily found in the lower strata and can be found up to a height of around 30 km. Its proportion is extremely low, that is 0.04% (Figure 2.1), but the growth of the biosphere's vegetative life depends on it most. Although it is transparent to incoming solar radiation, it prevents some of it from escaping. It is therefore referred to as a greenhouse gas. It is extremely important in raising the global temperature.

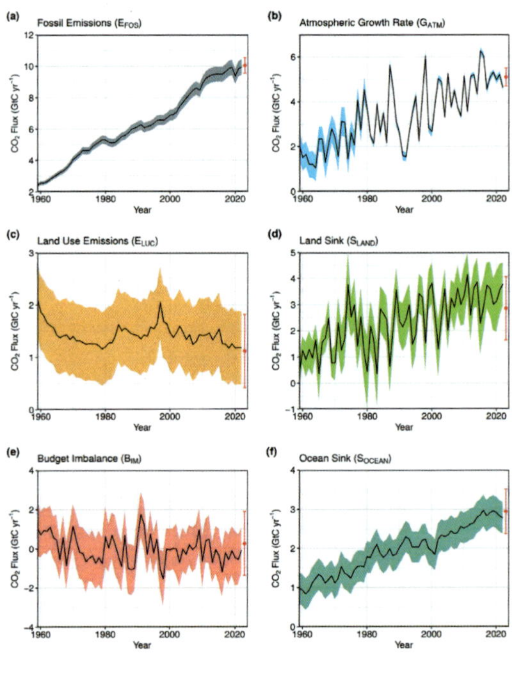

Figure 2.2. *Different sources of carbon dioxide.*

Source: By Pierre Friedlingstein, Matthew W. Jones, Michael O'Sullivan, et al. – https://essd.copernicus.org/articles/11/1783/2019/, CC BY 4.0, https://commons.wikimedia.org/w/index.php?curid=97409285.

Since its amount is influenced by combustion, human activity, and the planet's vegetative cover, it is also known as variable gas. Numerous human activities, including the use of energy, transportation, industry, agriculture, and waste generation, are causing CO_2 to enter the atmosphere. (Figure 2.2). Aside from these naturally occurring sources, man-made sources also exist. Examples of these include plant respiration and the natural process of denudation, which releases stored carbon in rocks into the atmosphere. These are all contributing to the atmosphere's rise. As society's economic development progresses, so does the amount of this gas. Figure 2.3 illustrates the large volume of CO_2 gases produced by economically developed nations and regions.

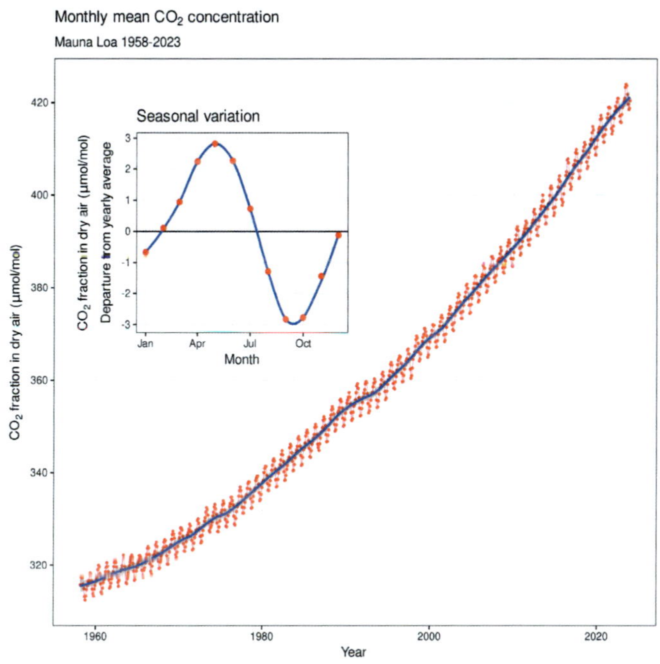

Figure 2.3. *Global carbon dioxide emission by region: upto 2023.*

Source: https://en.wikipedia.org/wiki/File:Mauna_Loa_CO$_2$_monthly_mean_concentration.svg.

2.1.1.4. Methane

Another greenhouse gas that raises air temperature is methane, which absorbs radiation. Methane production from paddy fields also enters the atmosphere. Additionally, wetlands, and soggy soils produce it, which is then released into the atmosphere. Another source of methane emissions into the atmosphere is fossil fuels. Its concentration in the atmosphere varies. One can observe the various methane sources from Figure 2.4.

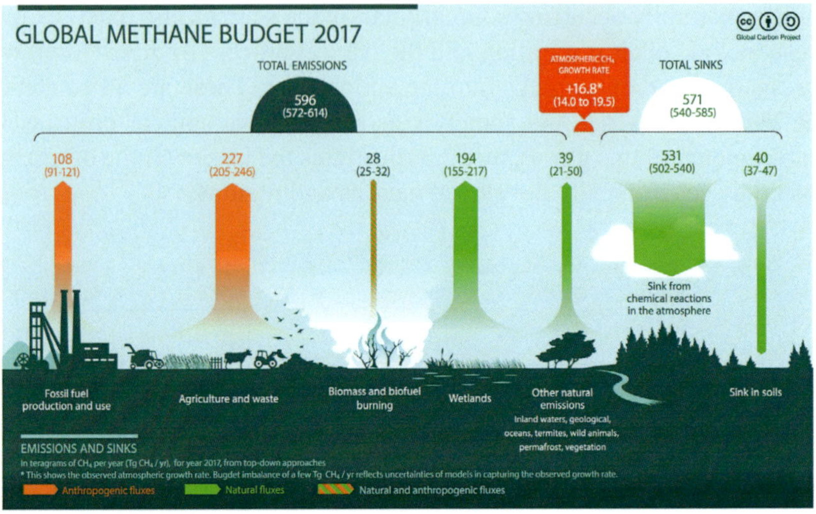

Figure 2.4. *Different sources of methane.*

Source: By The Global Carbon Project – "Increasing anthropogenic methane emissions arise equally from agricultural and fossil fuel sources," CC BY–SA 4.0, https://commons.wikimedia.org/w/index.php?curid=93291867.

2.2. STRUCTURE OF THE ATMOSPHERE

Structure is the grouping of various components into a single unit. To put it another way, it is the structure, organization, or anatomy of a whole as seen through the relationships between its constituent pieces. Given this context, the study of the various components of the atmosphere and their interactions is referred to as the atmosphere's structure. The atmosphere is split vertically into various layers and sections. As a result, the study of the various layers is referred to as atmosphere structure. Based on chemical composition, the atmosphere is classified into two. They are homosphere and heterosphere.

2.2.1. Homosphere

The region of the atmosphere known as the homosphere is where the chemical makeup of the air is consistent or comparable. Its chemical composition places it at the bottom of the chain. It reaches approximately 85 km from the surface of the earth or ocean (Figure 2.5). With the exception of water vapor, pollutants, ozone, and certain trace or very minor gases, the proportion of the major gases in the air remains constant throughout this layer despite the air's rapid density change with altitude.

Figure 2.5. *Homosphere and heterosphere.*

Source: By NASA Earth Observatory – http://eol.jsc.nasa.gov/scripts/sseop/photo.pl?mission=ISS013&roll=E&frame=54329, Public Domain, https://commons.wikimedia.org/w/index.php?curid=1722627.

The atmosphere is separated into five layers based on temperature variations (Figure 2.6), with three of the lower layers falling under the homosphere (i.e., inside an altitude of 85 km). The mesosphere, stratosphere, and troposphere are these.

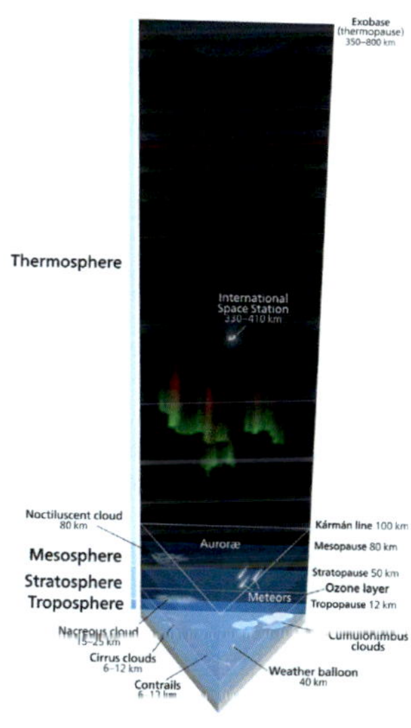

Figure 2.6. *Different layers of atmosphere.*

Source: By Kelvinsong – Own work, CC BY-SA 3.0, https://commons.wikimedia.org/w/index.php?curid=24006541.

2.2.1.1. Troposphere

It is the atmosphere's lowest and densest layer. It extends to a height of roughly 8 km over the pole and 18 km over the equator. This layer contains about 80% of the atmosphere's total mass. The temperature decreases as one rises in altitude and reaches the boundary of this stratum. The temperature drops by 6°C per km on average as one climbs higher. The upper boundary, which lies between 8 and 18 km, is referred to as the troposphere. The average temperature drops to between minus 50° and 60°C at this level (Figure 2.7). This layer contains a lot of water vapor; in fact, 99% of all atmospheric water vapor is concentrated here. However, there is a lot of variation in terms of height and longitude. Through the creation of the greenhouse effect, vapor plays a critical role in controlling the earth's temperature.

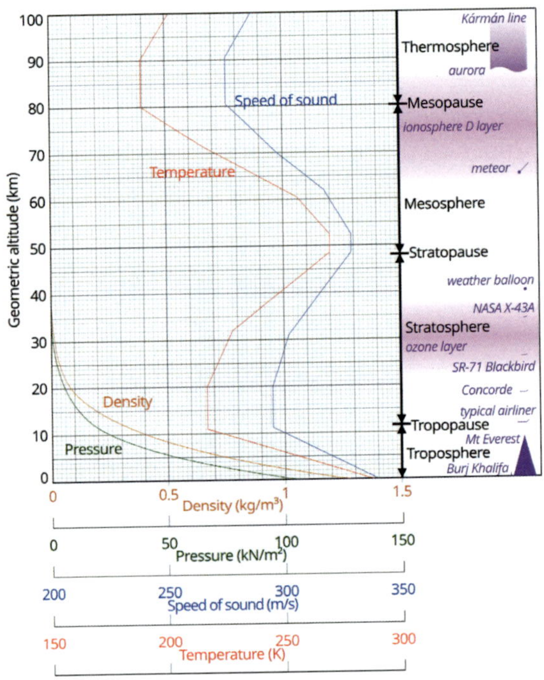

Figure 2.7. *State of temperature in different layers of atmosphere.*

Source: By Cmglee – Own work, CC BY-SA 3.0, https://commons.wikimedia.org/w/index.php?curid=17311330.

This layer is the only one where meteorological phenomena occur. All forms of clouds, atmospheric turbulence, and air mixing occur in the troposphere. Here, mixing is done both horizontally and vertically quite a bit. The Greek word "tropos," which means "turn," is actually where the word "troposphere" originates. The term "sphere" refers to a "ball" or a spherically-shaped object that achieves three dimensions. As a result, the troposphere is a three-dimensional object that has the ability to mix or turn. All forms of life are restricted to the biosphere, which consists of air, water, and land. The tropopause, a transition zone between the troposphere and the stratosphere, is the upper limit of the troposphere.

2.2.1.2. Stratosphere

The middle layer of the homosphere and its upward second layer is called the stratosphere. It begins at the tropopause and rises to about 50 km in altitude. Up to a height of 20 km, the temperature at the tropopause is essentially constant. Subsequently, it begins to rise and maintains this trend up to a height of 50 kilometers (Figure 2.7). The estimated temperature at this altitude is between minus 10° and 15¥. Despite the rising temperature, there isn't any turbulence in the atmosphere. There are no clouds or other meteorological phenomena present in this layer at all. It benefits from this property when supersonic jets or airplanes are flown through this layer over long distances. This layer's rise in temperature is a result of ozone's (O_3) absorption of solar radiation. This layer contains a lot of ozone, 90% of which is concentrated here, between 15 and 50 kilometers. The stratopause, a very thin transition zone strip that separates the upper limit from the mesosphere, is the upper limit.

Ozonosphere, a layer composed of ozone, is another name for the stratosphere. This layer shields the planet's life forms from the damaging effects of ultraviolet radiation. In 1985, scientists discovered the hole and ozone depletion over the Antarctic. Since then, we have noticed that the ozone layer in this layer has dangerously depleted. The primary sources of ozone depletion are emissions of carbon tetrachloride, hydrochlorofluorocarbons (HCFCs), and chlorofluorocarbons (CFCs). Contact between chlorine atom molecules and ozone (O_3) converts O_3 into O_2 or regular oxygen. More than a lakh ozone atoms can be destroyed by one chlorine atom. The chain of depletion is therefore extremely dangerous. Since O_3 was converted to O_2, an ordinary oxygen molecule cannot absorb UV radiation.

2.2.1.3. Mesosphere

The third and topmost layer of the homosphere is called the mesosphere. Heterosphere begins after this layer. The middle sphere is what the term "mesosphere" literally means. It is divided by the mesopause on top from the thermosphere and the tropopause below from the troposphere. It stretches from 50 to 85 kilometers away from the surface of the planet. There is very little air pressure. At the lowest limit, it is 1 millibar, and at the highest limit, it is 0.01 millibar. This layer has a decreasing temperature and is where the lowest/coldest recorded air temperature is found. The lowest recorded temperature in the vicinity of the mesosphere is approximately minus 130°C. The current temperature is below the lowest ever recorded over the Antarctic.

Noctilucent clouds, which are typically found during summer nights between 50° and 70° north and south latitudes, are located between 75 and 85 km from Earth. It means "night shining" in literal terms. Only when the sun is setting and the sunlight is still falling at that height can one see the deep twilight. Ice crystals make up the clouds that are visible. Meteoric dust particles, which fall and are created when a meteor burns due to friction, function as the building blocks for ice crystallization (Figure 2.8).

Figure 2.8. *Noctilucent clouds.*

Source: By ISS Expedition 23 crew – NASA Earth Observatory, Public Domain, https://commons.wikimedia.org/w/index.php?curid=10633526.

2.2.2. Heterosphere

Heterosphere refers to the atmosphere that lies outside of the homosphere. The name itself describes the region of the atmosphere where the air is not uniform, and it is self-explanatory. The air is scarce and the molecules are widely separated in this region of the atmosphere. Lighter gas molecules are compelled to be above, while relatively heavier gas molecules are concentrated in the lower region.

The composition of the atmosphere changes significantly with altitude above 85 km. It is impossible for the gases to mix because turbulence is not present. The nitrogen, oxygen, helium, and hydrogen gases are distinct layers that are clearly different from one another. Nonetheless, the thermosphere and exosphere are the two primary spheres that make up the heterosphere.

2.2.2.1. Thermosphere

This globe reaches from the mesopause that is between 85 and 650 kilometers away from Earth. The tiny amount of oxygen molecules in this layer is absorbing solar radiation, which is causing the temperature to rise. It is heavily reliant on solar activity. At an elevation of roughly 350 km, the temperature rises to over 1200°C, and by 650 km, it may even reach 2000°C. This extremely high temperature is mostly determined by the average molecular speed. This could lead to a high temperature. This temperature's efficacy is not very high. If astronauts emerge from the capsule, their contact with it has no effect whatsoever.

An obvious phenomenon seen in this layer is the aurora. They are a remarkable show of light that peaks in the magnetic polar regions of the northern and southern hemispheres between 10 p.m. and 2 a.m. In the northern hemisphere, it is referred to as Aurora Borealis (Figure 2.9), and in the southern hemisphere, Aurora Australis.

We call them the southern and northern lights. The upper atmosphere of the earth is reached by charged particles that originate from solar flares. The rare oxygen and nitrogen molecules there are collided with by these charged particles. Thc conspiracy produces countless tiny flashes of light. It is breathtaking to see how these lights are dispersed.

Figure 2.9. *Aurora borealis.*

Source: By Schnuffel2002 – Own work, CC BY-SA 4.0, https://commons.wikimedia.org/w/index.php?curid=50615000.

2.2.2.2. Exosphere

Exo is short for external. The exosphere, then, is the outermost or external layer of the atmosphere. Its lower boundary extends 650 kilometers from the thermopause to the point where the void space starts. About 10,000 kilometers are thought to be this limit.

This distance is slightly smaller than the earth's diameter. Indeed, the atmosphere's limit is incredibly large. Helium and hydrogen are very light gases that are traced in the exosphere. Their molecules are widely apart. When the exosphere reaches its upper limit, the surrounding area is regarded as empty.

2.3. THE RADIATION BALANCE

An accounting of the radiation's incoming and outgoing components is called the Earth system's radiation balance.

Over extended periods of time and throughout the entire Earth, these elements are in balance. The Earth would be continuously warming or cooling if they weren't. Radiant energy is unevenly distributed over the Earth in the short term, though (Figure 2.10).

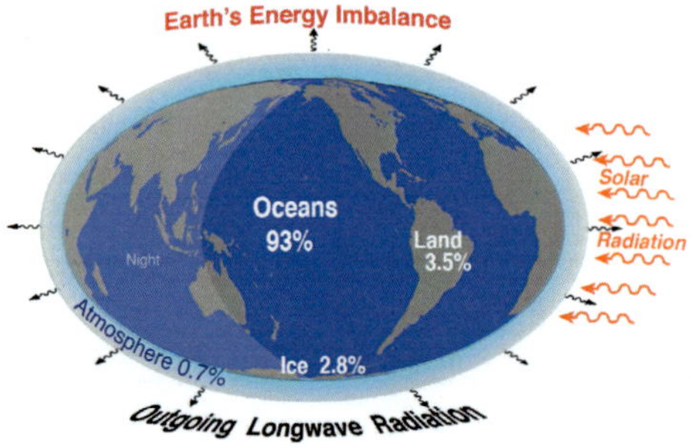

Figure 2.10. *Radiation balance of the earth.*

Source: By Kevin E Trenberth – Own work, CC BY-SA 4.0, https://commons.wikimedia.org/w/index.php?curid=138672738.

2.3.1. Shortwave Radiation

The vacuum of space allows shortwave radiation from the Sun to pass through to the outer edge of the atmosphere. Due to absorption and reflection, solar radiation starts to diminish as soon as it enters the atmosphere (Figure 2.11).

Figure 2.11. *Scattering by particles in the atmosphere causes a beam of light to be broken into several weaker beams of light.*

Source: By NASA/Scott Kelly – https://twitter.com/StationCDRKelly/status/675305899511128064 (direct link)https://www.flickr.com/photos/nasa2explore/25029799480/, Public Domain, https://commons.wikimedia.org/w/index.php?curid=45572985.

About 30% of the available solar radiation at the top of the atmosphere is reflected or scattered back to space by particulates and clouds before it reaches the ground. The gases of the atmosphere are relatively poor absorbers of solar radiation, absorbing only about 20% of what is available at the outer edge of the atmosphere. The remaining solar radiation makes its way to the surface as direct and diffuse solar radiation. *Direct solar radiation* (S) is shortwave radiation able to penetrate through the atmosphere without having been affected by constituents of the atmosphere in any way. *Diffuse radiation* (D) is shortwave radiation that has been scattered by gases in the atmosphere. Scattering is a process whereby a beam of radiation is broken down into many weaker rays redirected in other directions. Together, direct, and diffuse shortwave radiation accounts for the total incoming solar radiation or *insolation* (K↓). In equation form:

$$K\downarrow = S + D$$

The surface absorbs some of the incoming solar radiation, while some of it is also reflected away. The albedo (a) is the percentage of light reflected off a surface. The range of albedo values is 0 for no reflection and 1 for total reflection of light incident on the surface. To make albedo easier to understand, it can also be expressed as a percentage (albedo multiplied by 100). For example, the albedo of grass is approximately 0.23. This indicates that 23% of the solar radiation that strikes the grass is reflected back. Conversely, surfaces with high reflectivity, such as snow, have an albedo above.87, meaning that 87% of the sun's light is reflected back (Table 2.2).

Table 2.2. *Sample Albedos*

Surface	Typical Albedo
Fresh asphalt	0.04
Worn Asphalt	0.12
Bare Soil	0.17
Conifer forest (Summer)	0.08
Green Grass	0.25
New Concrete	0.55
Fresh Snow	0.80–0.90

Source: https://geo.libretexts.org/Bookshelves/Geography_(Physical)/The_Physical_Environment_(Ritter)/04%3A_Energy_and_Radiation/4.03%3A_Radiation_and_Energy_Balance_of_the_Earth_System/4.3.01%3A_The_Radiation_Balance.

The amount of reflection (K↑) is given by the following equation:

$K\uparrow = (S+D)a$

Darker surfaces reflect less than lighter ones, and the amount of reflection essentially depends on the color of the surface. The sun angle has an impact on albedo for some surfaces, such as water. If you've ever spent a daytime out on a lake, you may have noticed that when the sun is lower in the sky than it is when it is directly overhead, it reflects off the surface more.

Net shortwave radiation is the difference between incoming and outgoing shortwave radiation expressed as:

$K^{*} = (S+D) - (S+D)a$

During the day, K* is a positive value as incoming always exceeds outgoing shortwave radiation. At night, K* is equal to zero as the Sun is below the horizon. (No, moonlight doesn't count!)

2.3.2. Longwave Radiation

Terrestrial longwave radiation (L↑) is the energy that the Earth emits after it has absorbed it. The temperature of the surface largely determines how much energy is released. The surface will release more radiant energy the hotter it gets. Because the gases in the atmosphere are reasonably effective at absorbing longwave radiation, they also absorb the energy that the surface emits. Longwave atmospheric counter-radiation (L↓) is the downwardly directed portion of the absorbed radiation that is emitted in all directions.

The difference between incoming and outgoing longwave radiation is net longwave radiation expressed as:

$L^* = L\downarrow - L\uparrow$

Remember that, under typical circumstances, air temperature drops as altitude through the troposphere increases. This happens as a result of the earth serving as the air's primary source of heat. The surface is typically hotter than the air above because heat is transferred from warmer to cooler bodies. Since the Earth emits more longwave radiation than it receives from the atmosphere, net longwave radiation is typically negative. The earth's system loses some of the radiation that is released into space. The "greenhouse" effect is caused by material that is absorbed by the atmosphere and is not lost to space.

In other cases, the net longwave might be either positive or zero. If it were zero, the radiation released by the surface would be equivalent to the air's radiation. When the temperature of the surface and the air are equal, this is true. A positive value is present if the air is warmer than the ground. This might happen if there is a lot of cloud cover, as clouds are excellent longwave radiation radiators and absorbers, or if warm air moves across a colder surface.

2.3.3. Net Radiation

Net Radiation is the difference between incoming (i.e., S, and L↓) and outgoing components of radiation (S+D)a and L↑.

$Q^* = [(S+D) - (S+D)a] + L\downarrow - L\uparrow$

Net radiation may be zero, negative, or positive. When there is more radiation entering the system than leaving it, the net radiation is positive. This usually happens in the daytime when the air is at its warmest and the sun is out. Since there is no incoming solar radiation during the night, net longwave radiation is typically negative and is primarily determined by the outgoing terrestrial longwave flux. When the incoming and outgoing components are perfectly balanced, which doesn't happen very often, net radiation is zero. An alpine snow pack's incoming and outgoing radiation is measured by a net radiometer. Measurements like this one help determine how much snow melts and how much water is available to fill community reservoirs like the ones in the Boulder, Colorado area.

Many geographical patterns of the earth system's component parts can only be explained by having a basic understanding of net radiation and its distribution. The radiant energy that can be used to power the Earth system is known as net radiation. This activity takes many forms, such as warming the surface, evaporating water, and heating the air.

2.4. ATMOSPHERIC CIRCULATION

The movement of mass within the atmosphere creates atmospheric circulation. Similar to how warm air rises and becomes buoyant, this motion could be vertical. It can also be horizontal: air moves from high pressure, where it is compressed tightly, to low pressure, where it is less compressed. However, because of the earth's rotation, horizontal winds have curved trajectories. The air moves due to atmospheric forces, which alters the pressure differential. Pressure differences are indicated on a weather map by parallel lines known as isobars, which represent pressure changes, typically in steps of two to four millibars.

Sea breezes demonstrate how localized changes in pressure and temperature are caused by a combination of vertical and horizontal movements. Because land warms more quickly than water, coastal land regions get hotter during the day. This causes the air above the land to warm and rise, which raises the atmospheric pressure above the surface, where it begins to cool and form clouds. After that, it moves at an altitude from the high-pressure region over land to the low-pressure region over the sea. Air flows in from the sea to the land because there is more pressure at the sea's surface due to the increased mass over the sea and less mass over the land. The cycle is reversed at night when land cools faster than ocean (Figure 2.12).

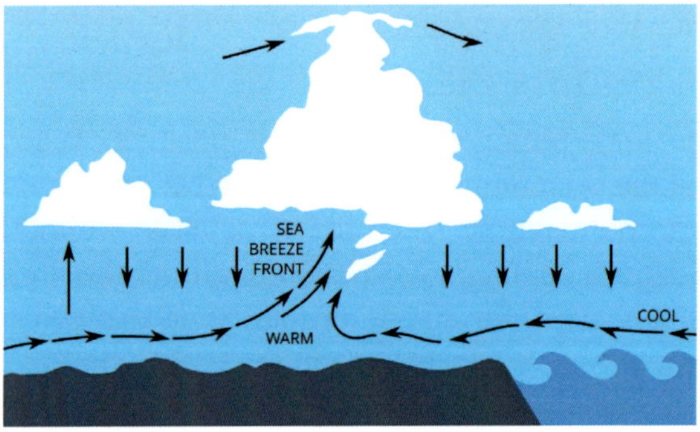

Figure 2.12. *Sea breeze.*

Source: By SeaBreeze.jpg: U.S. FAA AC 00-6A, Chapter 16, Figure 165.derivative work: Krzysztof Zajączkowski Malyszkz (talk) – SeaBreeze.jpg, Public Domain, https://commons.wikimedia.org/w/index.php?curid=17012202.

While many larger weather systems take less direct routes, in this example the sea breezes flow directly between two points. But their routes are not arbitrary. Long-distance winds appear to curve due to the Coriolis force, an apparent force brought on by Earth's rotation. Every 24 hours, all points on Earth's surface revolve once around its axis, causing this phenomenon. However, different points rotate at different speeds: air at the equator rotates at 1700 km/h, while air at a point at 60 degrees latitude, which is closer to the Earth's spin axis, rotates at 850 km/h.

Angular momentum, also known as energy of motion, is the property of objects on Earth's surface that describes how they rotate around a reference point. The product of an object's mass, velocity, and radius—the distance from the reference point—determines its angular momentum. Since an object's angular momentum is conserved when it moves on Earth, its velocity must increase if its radius of spin decreases (for example, when it moves from a low latitude to a high latitude). Figure skaters rotate more quickly during spins because of this relationship when they draw their arms in close to their bodies. A piece of air traveling north from the Equator toward the pole experiences the same process: as it gets closer to the Earth's axis of rotation, its radius of spin around the planet decreases and its rate of spin increases. The parcel deflects to the right of its original trajectory with respect to the planet's surface because its angular velocity is greater than the angular velocity of the Earth's surface at the higher latitude (Figure 2.13). The parcel would appear to deflect to the left in the Southern Hemisphere.

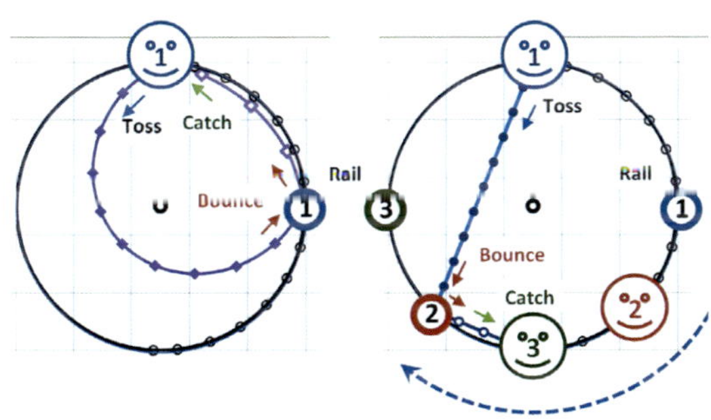

Figure 2.13. *Coriolis force.*

Source: By Brews ohare – Own work, CC BY-SA 3.0, https://commons.wikimedia.org/w/index.php?curid=4237726.

Gustave-Gaspard Coriolis, a French scientist, discovered this effect while attempting to determine why bullets fired from long-range cannons were missing their targets by a wide margin to the right. Sea breezes and other local weather patterns are not affected by the Coriolis force because it only affects masses that travel over long distances. Additionally, it doesn't cause water draining from a sink or toilet to rotate in one direction in the Northern Hemisphere and another direction in the Southern Hemisphere, despite a common misconception.

However, winds appear to blow almost parallel to isobars rather than straight across them from high to low pressure due to the Coriolis force.

In low-pressure weather systems like hurricanes, the Coriolis force causes the winds to rotate, spiraling into a clockwise direction in the Northern Hemisphere and a counterclockwise direction in the Southern Hemisphere. Pressure gradients that force air to move from high pressure to low pressure regions are what cause it to initially begin moving through the atmosphere. The Coriolis force begins to bend its path as it moves. The motion is directed toward a state known as geostrophic flow, in which the Coriolis force and the gradient force of pressure perfectly balance each other. The air parcel is no longer traveling from a zone of high pressure to one of low pressure at this point. Rather, it travels in a direction that is parallel to the isobars. The air parcel at point A_3 in Figure 2.14 is in geostrophic flow.

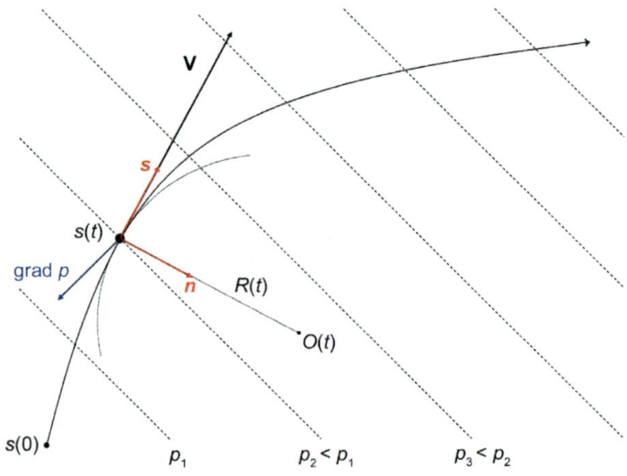

Figure 2.14. *Geostrophic flow.*

Source: By Xavierstuvw – Own work, CC BY-SA 3.0, https://commons.wikimedia.org/w/index.php?curid=3669147.

Air from the outside is forced toward a low-pressure area when one forms in the Northern Hemisphere. When air enters the system in reaction to this force, it is deflected to the right and revolves counter-clockwise. A high-pressure area, on the other hand, generates a pressure force that is directed away from the high. In the Northern Hemisphere, air that begins to move in reaction to this force is deflected to the right, creating a clockwise circulation pattern around an area of high pressure (Figure 2.15).

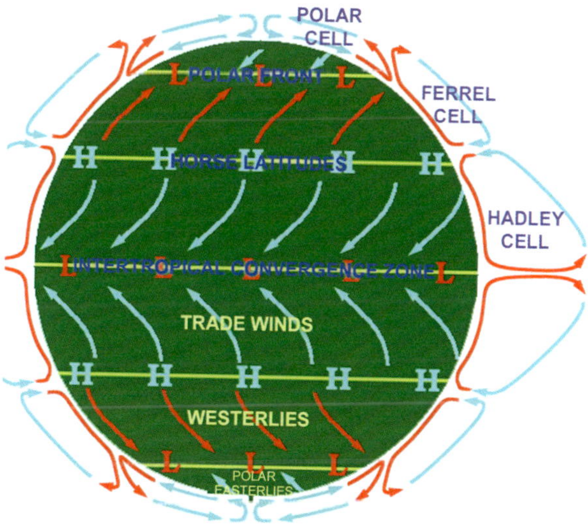

Figure 2.15. *Circulation of air around regions of high and low pressure in the northern hemisphere*

Source: By DWindrim – Own work, CC BY-SA 3.0, https://commons.wikimedia.org/w/index.php?curid=849141.

At altitudes less than one kilometer, this pattern changes as winding with ground based objects slows down. The Coriolis force decreases with wind speed, but pressure gradient forces do not change. Winds close to the ground are therefore diverted toward regions of low pressure. Air parcels will rise toward the center of low-pressure areas after spiraling into low-pressure zones close to the surface.

The air cools as it rises, resulting in rain, clouds, and condensation. Air parcels, on the other hand, will spiral away from surface-level high pressure areas and toward low-pressure areas. Air will descend from above to preserve barometric equilibrium. As a result, the descending air will warm and become less humid, which will typically result in sunny weather (Figure 2.16).

English meteorologist George Hadley made the first attempt to demonstrate in 1735 how weather patterns combined to create a general circulation of the atmosphere. Hadley sought an explanation for why sailors encountered "trade winds" from the east near the equator and westerly winds at midlatitudes. He concluded that the rotation of the Earth was the cause of this trend.

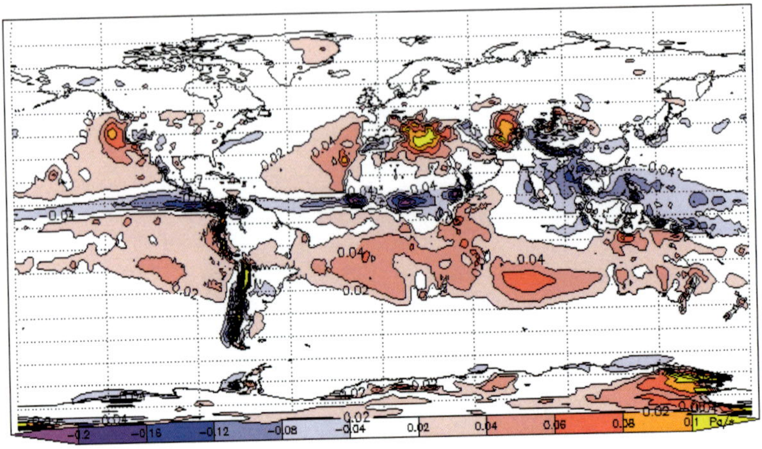

Figure 2.16. *Winds around highs and lows.*

Source: CC BY-SA 3.0, https://commons.wikimedia.org/w/index.php?curid=849148.

In many ways, Hadley's model was correct. Pressure gradients are produced when mass in the atmosphere is moved upward and poleward due to differential heating of the Earth, which produces more warmth near the equator and cooling due to radiation from space.

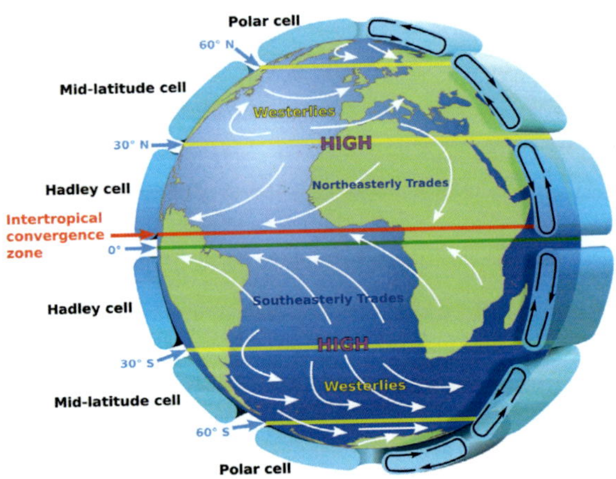

Figure 2.17. *General circulation of the atmosphere.*

Source: By Kaidor – Own work based on File:NASA depiction of earth global atmospheric circulation.jpg, CC BY-SA 3.0, https://commons.wikimedia.org/w/index.php?curid=23902538.

The Hadley circulation actually ends at a latitude of about 30°, despite the atmosphere's best efforts to create a straightforward circulation that would upwell near the equator and descend in polar regions. The Coriolis force, which generates easterly winds near the surface at low latitudes (known as "trade winds") and westerly winds at high latitudes, causes the air to sink to the ground and flow back toward the tropics at this point. Two additional sets of circulation zones, or "cells," between the tropics and the poles further to the north and south repeat this pattern (Figure 2.17). The friction that slows down the winds and motions brought on by differential heating combine to control the strength of the atmospheric circulation.

REFERENCES

1. Adonis, M., & Gil, L. (2001). Indoor air pollution in a zone of extreme poverty of metropolitan Santiago, Chile. *Indoor and Built Environment, 10*(3, 4), 138–146.
2. Alberini, A., & Krupnick, A. (1997). Air pollution and acute respiratory illness: Evidence from Taiwan and Los Angeles. *American Journal of Agricultural Economics, 79*(5), 1620–1624.
3. Lide, D. R. (1996). *Handbook of Chemistry and Physics* (pp. 14–17). Boca Raton, FL: CRC Press.
4. US Environmental Protection Agency. (1985). *Polymer Manufacturing Industry—Background Information for Proposed Standard* (EPA 450/3-85-019a; NTIS PB88-114996). Washington, DC: US Environmental Protection Agency.
5. US Environmental Protection Agency. (1986). *Handbook: Control Techniques for Hazardous Air Pollutants* (EPA 625/9-86-014; NTIS PB91-228809). Washington, DC: US Environmental Protection Agency.
6. US Environmental Protection Agency. (1989). *Evaluation of Continuous Compliance Monitoring Requirements for VOC Add-on Control Equipment* (US EPA Contract No. 68-02-4464, WA 60). Washington, DC: US Environmental Protection Agency.

CHAPTER 3

CYCLING OF PRIMARY GASEOUS POLLUTANTS

Contents

INTRODUCTION

Primary gaseous pollutants, such as carbon monoxide (CO), sulfur dioxide (SO_2), nitrogen oxides (NOx), and volatile organic compounds (VOCs), play a critical role in atmospheric chemistry and environmental health. These pollutants are directly emitted into the atmosphere from various sources, including industrial activities, vehicle exhaust, and natural processes like wildfires and volcanic eruptions. Understanding the cycling of these pollutants is essential for comprehending their environmental impacts, including their contribution to air quality degradation, climate change, and human health issues.

Once released into the atmosphere, primary gaseous pollutants undergo various physical and chemical transformations. For instance, nitrogen oxides can react with VOCs in the presence of sunlight to form ground-level ozone (O_3), a secondary pollutant that significantly affects air quality and respiratory health. Sulfur dioxide can be oxidized to form sulfuric acid (H_2SO_4), leading to acid rain, which has detrimental effects on aquatic ecosystems, soil quality, and built environments. Carbon monoxide, a byproduct of incomplete combustion, can interfere with the blood's oxygen-carrying capacity in humans and animals, posing severe health risks.

The cycling of these pollutants involves their dispersion, deposition, and chemical reactions in the atmosphere. Meteorological factors such as wind speed, temperature, and humidity influence their transport and dispersion, affecting their concentration levels and distribution. Wet and dry deposition processes remove pollutants from the atmosphere, depositing them onto surfaces like soil, vegetation, and water bodies. Furthermore, atmospheric chemistry plays a crucial role in transforming these primary pollutants into secondary pollutants, which can have more persistent and widespread environmental and health impacts.

Monitoring and managing the cycling of primary gaseous pollutants are vital for mitigating their adverse effects. Regulatory measures, such as emission standards and air quality guidelines, aim to reduce the release of these pollutants from anthropogenic sources. Technological advancements in emission control and pollution monitoring provide essential tools for tracking pollutant levels and implementing effective mitigation strategies.

The cycling of primary gaseous pollutants encompasses their emission, transformation, and removal processes, significantly influencing environmental quality and public health. A comprehensive understanding of these processes is crucial for developing effective policies and practices to address air pollution and its associated impacts.

3.1. PRIMARY POLLUTANTS

Primary pollutants are those kinds of pollutants that come straight from the source and into the environment. Primary pollutants are different from secondary pollutants in that the former does not require the presence of an atmosphere to be produced. Primary pollutants can also be produced by a wide range of other sources, such as open burning, vehicles, coal-fired power plants, natural gas-fired power plants, burning biofuel, volcanoes, and many more. Primary pollutants can have detrimental effects on humans, animals, and plants, putting all three at risk. Furthermore, their alarming role in the development of secondary pollutants is responsible for the formation of various forms of smog, including dangerous ground-level ozone, especially in densely populated areas like Los Angeles. Recent years have seen a notable decrease in the emissions of primary pollutants, primarily due to stricter regulations, advancements in technology, and changes in the economy.

Primary pollutants are substances that are deemed hazardous to the environment and human health because they create pollution levels in the state in which they are released at the source. Secondary pollutants are new substances that are created through chemical reactions between specific primary pollutants and other elements or compounds. Since primary pollutants are released into the ecosystem directly, they pollute the surrounding area. The material released may have been present in some previous amounts, but if the additional release raises the total physical quantity to levels that are deemed hazardous to human health, it is considered the primary pollutant (Ayres and Maynard, 2006; Vallero, 2014) (Figure 3.1).

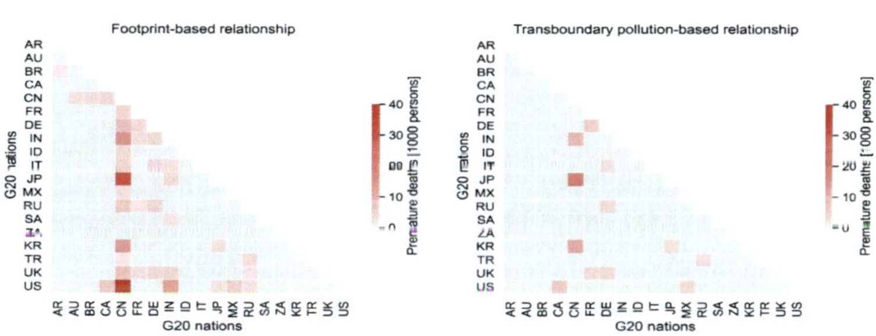

Figure 3.1. *Primary pollutants.*

Source. By Authors of the study: Keisuke Nansai, Susumu Tohno, Satoru Chatani, Keiichiro Kanemoto, Shigemi Kagawa, Yasushi Kondo, Wataru Takayanagi and Manfred Lenzen – https://www.nature.com/articles/s41467-021-26348-y, CC BY 4.0, https://commons.wikimedia.org/w/index.php?curid=113340923.

3.1.1. Examples of Primary Pollutants

Primary pollutants are one kind of air pollutant that comes straight from the source. CO, NOx, NO, SO_2, SOx, VOCs, and aerosols (dust, ash, salt particles) are a few examples of primary pollutants.

3.1.2. Criteria of Primary Pollutants

The six criteria pollutants that must be taken into account are carbon monoxide, ground-level ozone, lead, nitrogen dioxide, particulate matter, and sulfur dioxide.

3.1.2.1. Nitrogen Oxides Are a Type of Gas (NOx)

Nitrogen oxides are primarily produced by burning fossil fuels, such as gasoline and diesel engines, in urban areas. In contrast, farming practices, including the use of synthetic fertilizers, and the activity of microorganisms in compost, are the main causes of nitrogen oxides in rural areas. The respiratory system may become inflamed as a result of nitric oxide exposure.

3.1.2.2. Carbon Monoxide Is a Gas That Exists Naturally (CO)

An odorless and colorless gas known as CO is created when fuel is burned improperly. The main sources of CO in the atmosphere are burning biomass and gasoline or diesel-powered vehicles. An increased risk of cardiovascular disease has been linked to CO, a highly toxic gas. Extreme CO exposure has the potential to cause unconsciousness or even death.

3.1.2.3. Sulfur Oxides Are Substances That Contain Sulfur (SOx)

It's a strong-smelling gas that's hidden. Sulfur dioxide is an anthropogenic gas that is mostly produced by burning fossil fuels and handling sulfur-containing mineral ores. Exposure to sulfur dioxide can lead to serious respiratory issues in both humans and animals. When sulfur dioxide and water are combined in the atmosphere, dangerous acid rain can result.

3.1.2.4. Particulate Matter Is a Type of Substance That Is Made Up of Small Particles (PM)

Particulate matter is the term used to describe the solid and liquid airborne particles that make up the atmosphere. There may be natural primary particles that come from these sources, such as soil particles and sea spray. When their references include, among other things, tire breaks, exhausts, and metallic materials processes, they can also be connected to the industrial and transportation sectors.

3.1.2.5. Ozone

The gas O_3 is not directly released by human processes. However, it is also produced as a result of the interaction between volatile organic compounds and nitrogen oxides, two common sources of industrial pollution. Exposure to ozone can cause breathing problems and damage to the airways. As with other air pollutants, breathing in O_3 can aggravate lung diseases and other breathing difficulties.

3.1.2.6. Lead

Common sources of this primary air pollutant include waste burning, the production of lead-acid chemicals, and the raw material and metal production process. One of the main causes of lead in the atmosphere has historically been the burning of leaded gas by cars. Lead is a likely carcinogen that can cause cancer in people. It may also have negative effects on a child's development, such as a decline in IQ.

Many sources, such as dust from mining operations, emissions from automobile and industrial exhausts, and pesticide dust from heavy agricultural spraying, can produce an environmental primary pollutant. A small number of primary pollutants consist of fine particles that are challenging for the environment to disperse. The burning of fossil fuels or even other agricultural and industrial processes releases smoke, coal dust, dust, and liquid droplets into the atmosphere, which are among the main pollutants. First-party pollutants are also generated by natural processes. This category includes the ash and dust produced by volcanic eruptions, as well as the grit and dust from volcanic explosions. Primary pollutants include gaseous pollution from wetlands and marshes, as well as other decomposing matter, and salt dust that is carried away from the coast by strong oceanic winds. Rain and wind-borne pollutants are examples of secondary pollutants (Ayres and Maynard, 2006; Vallero, 2014).

3.2. GASEOUS AIR POLLUTION

Numerous natural phenomena, including volcanoes and forest fires, release gaseous air pollutants into the atmosphere. However, due to industrialization and population growth, anthropogenic emissions of some gases may surpass those of natural gases.

There was little pollution in the past. Still, there would have been local issues even then. For example, smoke from cooking and heating wood fires would have led to poor air quality inside poorly ventilated homes. Burns from hunting societies would have produced significant emissions of particulate carbon (soot), carbon dioxide, and other gases that would have momentarily harmed the quality of the surrounding air. Hunting cultures frequently used fire to drive game animals and enhance local forage. But all in all, these effects were not that significant.

Naturally, air pollution grew more and more of a concern as the world's population increased and became more industrialized. Since coal swiftly replaced other fuels as the primary source of heat and energy for machinery at the start of the Industrial Revolution (about 1750), air pollution has gotten worse. In the industrial towns and cities of Europe and the Americas, the extensive use of coal resulted in serious pollution from soot and sulfur dioxide (SO_2). Pollutant emissions have increased since 1900 as a result of growing industrialization and new technologies like power plants and automobiles.

When conditions are stable and calm in the lower atmosphere, air pollution can be particularly bad. These circumstances frequently arise beneath an atmospheric phenomenon known as an inversion, which is characterized by the trapping of a layer of cooler air beneath a higher layer of warmer air (temperature typically decreases with altitude). A comparatively stable situation known as an atmospheric inversion keeps cleaner air from higher altitudes from mingling with contaminated ground-level air. Smog, a word derived from the words "smoke" and "fog," is the term for pollution that occurs when an atmospheric inversion is accompanied by fog. Even as recently as the 1950s, thousands of urban residents perished from respiratory distress and thousands more were killed by sporadic "killer smogs" (also known as "reducing smog") rich in SO_2 and soot. The most well-known killer smogs happened in Pittsburgh, Pennsylvania, London, Glasgow, and a few other industrial cities in Europe.

Governments started enacting laws to reduce emissions of air pollution once scientists realized the serious harm it was causing. The discovery by medical researchers of a direct correlation between human health and death and air quality led to an increased focus on pollution control. Since the killer smogs were most common in urban areas, special attention was given to air pollution there. Governments typically responded with important control actions, such as the following:

1. Switching from coal, which is a relatively "dirty" fossil fuel, to "cleaner" ones such as natural gas or oil, or to alternative energy technologies such as nuclear power and hydroelectricity.
2. Constructing tall smokestacks to spread emissions over a much wider area so that ground-level exposures become less common and less intense – this tactic is the "dilution solution to pollution."

3. Centralizing energy production in large power plants to replace much of the relatively dirty and inefficient burning of coal in home fireplaces and furnaces, thereby permitting better control of emissions.
4. Treating waste gases to remove some of their pollutant content, thereby reducing emissions to the atmosphere.

In many countries, lowering smog levels has become much less important as a result of these beneficial regulatory actions. But in many areas, so-called oxidizing smog is now a more significant source of damage. Through intricate photochemical processes that convert hydrocarbons and nitrogen oxides into ozone (O_3) and other gases, oxidizing smog arises in the atmosphere. In addition to damaging plants, ozone, and other oxidizing gases irritate people's eyes and respiratory systems. When there are hydrocarbons and nitrogen oxides from industrial and vehicle emissions combined with sunshine and an atmospheric inversion that hinders the dispersion of the contaminated air mass, oxidizing smog can occur.

Generally speaking, as one ascends higher into the atmosphere, the temperature drops. An atmospheric inversion, also referred to as a temperature inversion, is a phenomenon that occurs when a layer of warmer air becomes trapped beneath a layer of relatively cool air.

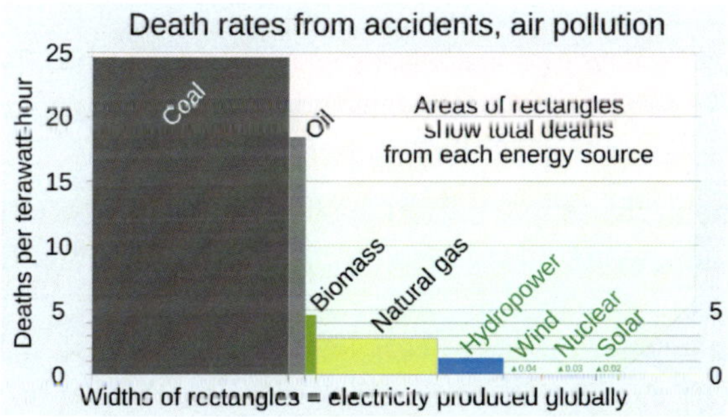

Figure 3.2. *Chart showing relation of death rates per unit of energy generated, total death rates, and energy produced globally, each as a function of energy type.*

Source: What are the safest and cleanest sources of energy? Our World in Data (2021).

Source: By RCraig09 – Own work, CC BY-SA 4.0, https://commons.wikimedia.org/w/index.php?curid=144059308.

A clear, cloudless night may lead to the development of an atmospheric inversion. In these circumstances, heat absorbed during the day is radiated outward, causing the ground to cool rapidly. As a result, there may be a layer of cold air below a warmer layer above. A developing atmospheric inversion can be especially dangerous in hilly terrain because relatively dense, cool air can flow down from hilltops and collect in valleys. A still-mixing air mass at ground level may occasionally be capped by stable atmospheric conditions that form at higher altitudes during the summer.

A stable atmospheric inversion is possible. It has the ability to trap and hold onto air pollutants released during the inversion event until they are dispersed by strong winds. The development and maintenance of stable inversions for several days can lead to severe episodes of air pollution.

Some locations, like Los Angeles, Mexico City, and Greater Vancouver, experience less persistent inversions on a regular basis. The inversions form in the morning and usually disperse in the afternoon in these locations. However, oxidizing air pollutants like ozone can build up and create photochemical smog in the interim (Ayres and Maynard, 2006; Vallero, 2014).

3.3. SULFUR GASES

3.3.1. Emissions and Transformations

Primarily, SO_2 is a significant gaseous air pollutant. SO_2 is a colorless gas with a strong smell. Only at concentrations of 0–1 ppm (parts per million; 1 ppm = 2–6 mg/m^3) can humans detect its bitter taste. Another sulfur gas, hydrogen sulfide (H_2S), is odorous and tastes like rotten eggs when present in quantities less than 1 ppb (parts per billion; 1 ppm = 1.4 mg/m^3).

Following their release into the atmosphere, SO_2 and H_2S oxidize to produce various compounds, which in turn lead to the formation of sulfate (SO_4^{2-}). The sulfate ion is an anion (any ion, atom, or molecule with a negative charge) because it has a negative charge. H_2S has an atmospheric residence time of less than a day due to its rapid oxidation to SO_2 (the time it takes for an initial amount to completely disappear). More slowly, at a rate of less than 1 to 5% per hour, SO_2 oxidizes in response to factors like humidity, sunshine, strong oxidants like ozone, and the presence of particulates containing metals that may serve as catalysts. About four days is the average residence time of SO_2 in the atmosphere. As a result, SO_2 may spread out from its emission point over a considerable amount of space before oxidizing or depositing on a surface, either land or water. Long-range transportation of air pollution is the term used to describe this type of dispersal (LRTAP).

Atmospheric sulfate (which is created when SO_2 or H_2S oxidizes), it can react with positively charged ions (cations) to generate a variety of compounds. Particularly ammonium sulfate ($(NH_4)_2$ SO_4), the majority of atmospheric sulfate exists as minute particles. This is the main cause of the fine particulate haze that frequently reduces visibility in cities, along with ammonium nitrate (NH_4NO_3). In certain rural areas, haze is also a result of pollutants imported by LRTAP from other emission sources. Additional cations that react with sulfate are sodium (Na^+), calcium (Ca^{2+}), and magnesium (Mg^{2+}). Frequently, though, there aren't enough of these cations to counteract all of the sulfate's (SO_4^{2-}) negative charges. In these circumstances, some of the negative charges are balanced by hydrogen ions (H^+), which produce an aerosol that contains sulfuric acid (H_2SO_4), the main ingredient in acidic precipitation.

3.3.2. Emissions of Sulfur Gases

Sulfur gases are naturally emitted by volcanoes. 90% of the 12 million tons of sulfur gases that volcanoes emit annually are SO_2 and the remaining 10% are H_2S. But during the massive 1991 eruption of Mount Pinatubo in the Philippines, 7–10 million tons of SO_2 (also known as SO_2-S, or sulfur content) were released into the atmosphere. The considerably smaller Mount St. eruption. Helens in Washington State released approximately 0.2 million tons of SO_2-S in 1980. Naturally occurring SO_2 emissions also happen during wildfires.

An estimated 150 million tons of SO_2 are released annually due to human activity, which is 3.8 times more than what occurs naturally. About half of the emissions caused by humans come from the burning of fossil fuels. Both organic and mineral sulfur compounds, including pyrite (FeS_2), are found in coal and petroleum. Burning these fuels converts the sulfur to gaseous SO_2, which can be released into the atmosphere. The sulfur content of hard coals mined in eastern North America ranges from 1 to 12%; softer coals from western regions have <0 to 10.5%; crude oil has 0 to 10.0%; residual fuel oils, like bunker-C, have 0 to 3%; and motor fuels, like gasoline and diesel, have 0 to 4%. The smelting of metal ores (7%) and the burning of natural habitats during agricultural conversions (the majority of the remaining emissions) are significant additional sources of SO_2 emissions. Manufacturing processes account for 23% of global emissions (Table 3.1).

Table 3.1. *Air Pollution Effects in Animal Models of Neurological and Mental Disorders*

Studies and Major Outcomes
Stroke (cerebral ischemia)
Seasonal variation in air particulate matter (PM10) exposure-induced ischemia-like injuries in the rat brain was attributed to varying toxin (PAHs) loading of the particles

SO_2 inhalation contributes to the development and progression of ischemic stroke in the rat brain by enhanced endothelin-1 activity and inflammation (iNOS, COX-2, and ICAM-1 mRNA) followed by activation of caspase-3 and higher cerebral infarct volume
Air pollutants PM generated by different engines and aluminum sulfate aerosols) caused cortical selective neuronal loss, nuclear pyknosis, karyolysis, and karyorrhexis as well as activation of microglia and astrocytes (also features of stroke and other neurological disease) as revealed by magnetic resonance imaging
Stroke damage is aggravated by nano-size particulate matter in mice, secondary to more pronounced DNA damage (8-hydroxyguanosine) and oxidative stress (gp91phox, p47phox) as well as higher number of inflammatory cells (CD68 and Ly6g positive)
Glutamatergic neurons in rodent models respond to nanoscale particulate urban air pollutants ($PM_{0.2}$) in mice, suggesting additive effects of air pollution and ischemic stroke on cerebral damage
Astrocyte activation plays a role in fine particulate matter (PM2.5)-dependent aggravation of ischemic stroke in male rats
Dementia (Alzheimer's disease)
Diesel engine exhaust accelerates amyloid β42 plaque formation in the 5X Familial AD mouse model of Alzheimer's disease, although no additive effects on spatial working memory deficits (assessed by Y-maze and X-maze tests) and markers of inflammation (IL-1β, RANTES, and MCP-1) were observed
Central role of Toll-like receptor 4 for glial inflammatory responses (higher TLR4, MyD88, TNFα, and TNFR2 mRNA) to air pollution ($PM_{0.2}$) in rats leading to a neuroinflammatory, accelerated cognitive aging and dementia-like phenotype
NO_2 inhalation promotes Alzheimer's disease-like progression via cyclooxygenase-2-derived prostaglandin E2 modulation, altered astrocyte and microglia function, all of which leading to deterioration of spatial learning and memory as well as aggravated amyloid β42 accumulation in wildtype C57BL/6J or Alzheimer's disease-prone APP/PS1 mice
Neurotoxicity of diesel exhaust nanoparticles in the rat brain is associated with increased levels of pro-inflammatory cytokines, amyloid β42, reactive oxygen species, hydrogen peroxide, nitrogen oxide metabolites and apurinic/apyrimidinic sites (DNA damage)

Exposure of mice to particulate urban air pollution reduced the repressive epigenetic marks (H3K9me2/me3) and increased DNA damage (γ-H2AX) as well as Alzheimer's disease hallmarks (hyperphosphorylated tau and amyloid-β plaques) in the brain
Traffic-related air pollutants (nano-sized PM) promote neuronal amyloid genesis (amyloid-β deposition) through oxidative damage (4-HNE, 3NT) in lipid rafts of mice
Parkinson's disease
Developmental exposure to concentrated ambient ultrafine particle air pollution (similar to the paraquat and maneb model) cause a Parkinson's disease phenotype in male mice with locomotor dysfunction and dopaminergic and glutamatergic changes
Cognitive and memory impairment
Selective memory and behavioral alterations after ambient ultrafine particulate matter exposure (using the Harvard ultrafine concentrated ambient particle system) in aged 3xTgAD Alzheimer's disease mice
Developmental exposure to low level concentrated ambient ultrafine particle air pollution and cognitive dysfunction in mice revealed by complementary learning (repeated learning), memory (novel object recognition, NOR), impulsive-like behavior (differential reinforcement of low rate (DRL), schedule of reward and delay of reward (DOR)), motor activity (locomotor behavior) and motivation (progressive ratio schedule) assessment assays
Activation of NLRP3 in microglia exacerbates diesel exhaust particles-induced impairment in learning and memory in mice
Impairment of learning and memory, induction of oxidative stress and dysregulation of monoamine neurotransmitters in the brains of mice by exposure to volatile organic compounds and carbon monoxide mixtures
PM2.5, SO_2 and NO_2 co-exposure impairs neurobehavior and induces mitochondrial injuries in the mouse brain
Effects of diesel engine exhaust origin secondary organic aerosols on novel object recognition ability and maternal behavior in mice
Exposure to ambient dusty particulate matter impairs spatial memory and hippocampal long-term potentiation by increasing brain inflammation and oxidative stress in rats
Mental disorders

Involvement of oxidative stress and mitochondrial mechanisms in air pollution (simulated vehicle exhaust)-related neurobiological impairments in rats leading to anxiety- and depression-like behavior
Ambient PM2.5 exposure caused depressive-like responses in mice through Nrf2/NLRP3 signaling pathway and altered inflammation
Psychological impact of vehicle exhaust exposure (CO, CO_2, NO_2) as revealed by anxiety- and depression-like behavior as well as impaired memory in rats
Ozone exposure of rats (Flinders Sensitive Line translational model) caused neurobiological oxidative stress and a depression-like phenotype
Other adverse effects on the brain (e.g., impaired BBB)
Early postnatal exposure to ultrafine particulate matter air pollution leads to dysregulated CNS neurotransmitters, cytokines, and glial activation preferentially in male mice. In addition, lateral ventricle dilation (=ventriculomegaly) was observed in exposed male mice, which is associated with poor neurodevelopmental outcome, autism, and schizophrenia
Exposure to traffic-generated air pollutants mediates alterations in brain microvascular integrity (disrupted blood-brain barrier) and enhanced oxidized low density lipoprotein signaling in wildtype mice on a high-fat diet, indicating additive adverse effects of obesity and air pollution on brain function

Source: Hahad O, Lelieveld J, Birklein F, Lieb K, Daiber A, Münzel T. Ambient Air Pollution Increases the Risk of Cerebrovascular and Neuropsychiatric Disorders through Induction of Inflammation and Oxidative Stress. International Journal of Molecular Sciences. 2020; 21(12):4306. https://doi.org/10.3390/ijms21124306.

Since the start of the Industrial Revolution, anthropogenic SO_2 emissions have significantly increased. In 2000, there were roughly 150 million tons of emissions, compared to approximately 5 million tons in 1860. Since then, the majority of the wealthier nations have made significant investments in clean-air technologies for power plants and other industries that use coal and oil in order to lower the amount of harmful SO_2 emissions into the atmosphere. The clean-air actions include the following:

1. The installation of technologies to capture SO_2 from post-combustion waste gases (this is known as flue-gas desulfurization or scrubbing).
2. The removal of some of the sulfur content of fuels (known as fuel desulfurization or coal washing).
3. The installation of particulate-control devices (such as electrostatic precipitators) to greatly reduce the emissions of particulates (although this has little effect on SO_2 emissions).

4. Switching to low- or no-sulfur fuels such as natural gas, or to no-sulfur energy technologies such as hydroelectricity and nuclear power (known as fuel-switching).
5. Energy conservation to reduce the overall demand for fuel and associated emissions of pollutants.
6. Building taller smokestacks to disperse emissions of SO_2 more widely, which helps to decrease the local ground-level pollution.

Future global emissions will inevitably rise, though. Burning coal and petroleum fuels, China, India, and other rapidly industrializing nations provide for a large portion of their growing energy needs. Coal is the main industrial energy source in China, providing 68% of the country's energy supply (in 2013). China's SO_2 emissions rose from 10 million tons in 1980 to 34 million tons in 2000 as a result of growing industrialization; this represents a 384-fold increase in just 20 years.

Depending on factors like population density, level of industrialization, and primary fuel consumption, countries emit varying amounts of SO_2. For example, SO_2 emissions from Canada are only about 22% of U.S. emissions. That being said, Canada's population is only 11% that of the U.S.. As a result, Canada's per-capita SO_2 emissions are roughly double those of the U.S. (Table 3.2).

Table 3.2. *Summary of Original Research on Cognitive Impact of Air Pollution Across Life Course*

S. No.	Author (Year)	Study Site	Study Population	Exposure Studied	Outcome Variable Studied	Cognitive Impact
A	Studies from High Income Settings					
1	Lee et al. (2019)	South-eastern part of the United States	94 million follow-up records from fee-for-service Medicare records for 13 million Medicare beneficiaries of fee for service (FFS) residing in the southeastern United States (U.S.) from 2000 to 2013.	Spatially and temporally continuous $PM_{2.5}$ exposure data	Hospitalization rates for dementia	Long-term exposure to a high $PM_{2.5}$ levels associated with increased hospitalization with dementia per 1 μg/m³ increase in annual $PM_{2.5}$, with higher risk for vascular dementia. Did not report analysis stratified by sex

2	Lo et al. (2019)	Taiwan	2241 community-dwelling, free-living elderly population with mean age at the time of recruitment 73.62 years; M:F = 57.5:42.5) followed from 1996 to 2007	PM_{10} Ozone	Short Portable Mental Status Questionnaire	Long-term exposure to PM_{10} and O_3 associated with cognitive impairment with greater impact of the joint effect of exposure to PM_{10} and O_3 Did not report analysis stratified by sex
3	Molina-Sotomayor et al. (2019)	Chile	181 older women, patients of the "La Estrella" Health Center of the Pudahuel commune, Metropolitan Region of Santiago de Chile, and patients of the Senior Centers of Viña del Mar City-Chile	Average PM_{10} $PM_{2.5}$ NO_2 SO_2 Ozone Concentration between 2012–2014 Cardio-respiratory Aerobic exercise indicators: Maximum oxygen uptake (VO_{2max}), estimated by the Six-Minute Walk Test (6mWT); heart rate (HR); and oxygen saturation (SpO_2).	Mini Mental State Examination (MMSE)	Significant differences ($p < 0.05$) between the Active Group residing in relatively less polluted areas versus sedentary group residing in more polluted areas on all the MMSE dimensions except "Registration," and in all the physiological variables (VO_{2max}, SpO_2, HR). Aerobic exercise may be a protective factor adverse cognitive effects of air pollution have on cognition probably due to improvement in the mechanisms of oxygen transport.
4	Andersson et al. (2018)	Umea, Sweden	Data of 1721 participants aged 55–85 years at baseline (Male: Female = 985:736 of Betula project, a longitudinal study of health and ageing aged 55–85 years at baseline	Estimates of annual mean levels of nitrogen oxides (NO_x) at the participants' residential address using a land-use regression model. Modelled data for road traffic noise levels at the participants' residential address	Dementia incidence	302 of 1721 participants at baseline, 302 developed dementia during the follow up period. Residing in the two highest quartiles of NO_x exposure was associated with an increased risk of dementia which was not modified by adjusting for noise. Did not report analysis stratified by sex
5	Berghuis et al. (2018)	Netherlands	101 children aged 13–15 years (M:F 55:46) participating in Development at Adolescence and Chemical Exposure (DACE)-study a follow-up of two Dutch birth cohorts.	Maternal pregnancy serum levels of PCB-153 and three OH-PCBs, 9 PCBs, 5 polybrominated diphenyl ethers (PBDEs), dichloroethane (DDE), pentachlorophenol (PCP) and hexabromocyclododecane (HBCDD) in different parts of the cohort Maternal smoking and alcohol use	Wechsler Intelligence Scale for Children, third edition, Dutch version (WISC-III-NL) Dutch version of the Rey's Auditory Verbal Learning Test (AVLT) Movement Assessment Battery for Children (Movement-ABC)	Significant association between PCB-183 and lower total intelligence, HBCDD with lower performance intelligence and PBDEs with lower verbal memory Positive trends between OH-PCBs and verbal intelligence, and a negative trend between BDE-153 and fine motor skills was observed Several OH-PCBs associated with more optimal sustained attention and balance CONTD ON NEXT PAGE Boys had poorer outcomes on fine motor skills and better outcomes on ball skills with positive trend was seen between 4-OH-PCB-107 and fine motor skills In girls, Prenatal DDE levels were associated with (sub)clinical motor performance

6	Carey et al. (2018)	London, UK	Retrospective cohort of 130,978 adults aged 50–79 years	Average annual concentrations of NO_2, $PM_{2.5}$, ozone (O_3), traffic intensity, distance from major road and night-time noise levels	Clinical diagnosis of Dementia	Positive exposure-response relationship between dementia and all measures of air pollution except O_3. Did not report gender based findings
7	Cullen et al. (2018)	UK	86,759 middle- to older-aged adults from the UK Biobank	Cumulative impact of outdoor air pollution exposure (PM_{10}, $PM_{2.5}$, NO_2, NO_x)	Cognitive function including reasoning test, pairs matching test, reaction time, prospective memory, visuospatial memory, numeric memory	Weak association between air pollutant exposure and cognitive performance at baseline (dose-dependent lower reaction time, higher error rate on a visuospatial memory test and lower numeric memory scores) with no such association at 2.8 years follow-up. Did not report analysis stratified by sex
8	Guxens et al. (2018)	Netherlands	Data from population-based birth cohorts GENERATION R (The Netherlands) (2002–2006) that recruited mother infant dyads including 8879 pregnant women and 1932 children born between April 2002 and January 2006 were taken of which 783 children between ages of 6–10 years participated in MRI sub-study	Prenatal exposure to air pollutants such as (NO_2, NO_x) in all regions and PM ($PM_{2.5}$, PM_{10} and PM_{coarse}) and $PM_{2.5}$ absorbance in a subgroup using land-use regression models for the period 2008 to 2011 and then modelled for air pollutant profile for the exact pregnancy periods using background monitoring sites.	MRI Brain Cognitive and psychomotor development	Air pollution exposure was not associated with global brain volumes Higher prenatal PM exposure per 5 μg/m^3 had a dose-dependent thinning of cortex increase in several brain regions of both hemispheres (e.g., cerebral cortex of the precuneus in Right hemisphere). Reduced cerebral cortex in precuneus and rostral middle frontal regions partially mediated the association between PM exposure and impaired inhibitory control. Did not report analysis stratified by sex
9	Kerin et al. (2018)	California, USA	327 children with Autism Spectrum Disorder (ASD) from the Childhood Autism Risks from Genetics and the Environ ment study	Exposure to NO_2, $PM_{2.5}$ and PM_{10}, ozone, and near-roadway air pollution in each trimester of pregnancy and first year of life.	Mullen Scales of Early Learning (MSEL), the Vineland Adaptive Behavior Scales (VAB3), and the Autism Diagnostic Observation Schedule calibrated severity score.	ASD severity not associated with any air pollutant exposure. Prenatal and First year exposure to NO_2 associated with impaired cognitive performance, adaptive functioning and behavioral indices but not severity of ASD 3rd Trimester PM_{10} exposure was paradoxically associated with improved behavioral performance Did not report analysis stratified by sex

10	Oudin et al. (2018)	Umea, Northern Sweden	1806 participants from Betula project from Umeå, Northern Sweden enrolled between (1993–1995) and followed upto 2010	Modelled levels of source-specific residential fine PM exposure to wood stoves or wood boilers and traffic	Validated data on dementia diagnosis	Increased dose-dependent risk for incident dementia (Vascular Dementia and Alzheimer's Disease) with local residential wood burning and traffic exhaust Did not report analysis stratified by sex
11	Ailshire et al. (2017)	USA	779 U.S. adults age ≥ 55 years from the 2001/2002 wave of the Americans' Changing Lives study	Annual average $PM_{2.5}$ concentration in 2001 in the area of residence by linking respondents with EPA air monitoring data using census tract identifiers. Exposure to neighborhood social stressors using perceptions of disorder and decay including subjective evaluations of neighborhood upkeep, presence of deteriorating/abandoned buildings, trash, and empty lots.	Error rate on Short Portable Mental Status Questionnaire (SPMSQ).	Association between higher rates of cognitive errors with high concentrations of $PM_{2.5}$ which was stronger in high stress neighborhoods indicating towards a possible role of social stressors and environmental hazards No statistically significant gender based differences found
12	Alvarez-Pedrerol (2017)	Barcelona, Spain	1234 children aged 7–10 years from 39 schools who commuted to school by foot	TRAP exposure (Average $PM_{2.5}$, Black Carbon (BC) and NO_2 concentrations) for the shortest walking route to school	Working Memory (the three-back numbers test) and inattentiveness (hit reaction time standard error of the Attention Network Test)	$PM_{2.5}$ and Black Carbon were associated with a reduction in the growth of working memory with no significant association of working memory with NO_2 Associations of TRAP concentrations (BC and $PM_{2.5}$) and working memory were stronger for males than females
13	Cacciottolo et al. (2017)	USA	3647 women aged 65 to 79 years from Women's Health Initiative Memory Study (WHIMS)	$PM_{2.5}$	Global cognitive decline and all-cause dementia APOE 4/4 status	Residence in places with high $PM_{2.5}$ was associated with an increased risk for global cognitive decline and all-cause dementia by 81% and 92% respectively, with risk exacerbated by APOE 4/4 allele status
14	Chen et al. (2017)	Ontario, Canada	All Ontario residents who were 55–85 years old on 1 April 2001, Canadian-born, and free of physician-diagnosed dementia who were followed up to 2013	Long-term average residential exposure to Air Pollutants including $PM_{2.5}$, NO_s, and ozone	Dementia incidence	Every interquartile-range increase in exposure to $M_{2.5}$ and NO_2 even at low levels of air pollution associated with higher incidence of dementia No statistically significant gender based differences

15	Colicino et al. (2017)	USA	428 older men in the Veterans Affairs (VA) Normative Aging Study	Black carbon	MMSE Score Telomere length C Reactive Protein	Black Carbon associated with lower cognition. Each doubling in BC level associated with 1.57 (95% CI: 1.20, 2.05) times higher odds of low MMSE scores in individuals with longer blood Telomere length (OR = 3.23; 95% CI: 1.37, 7.59; $p = 0.04$ for BC-by-TL-interaction).
16	Forns et al. (2017)	Barcelona, Spain	1439 of 2897 children recruited from 39 schools across Barcelona who participated in the BREATHE project 2012/2013	Composite exposure to indoor and outdoor levels of various TRAPs such as elemental carbon (EC), nitrogen dioxide (NO_2), $PM_{2.5}$ and ultrafine particles (UFP) at school	Working Memory	Slower development of working memory in children over 3.5 years period associated with higher schools based exposure to air pollution Did not report analysis stratified by sex
17	Lett et al. (2017)	USA	Sub population of Early Childhood Longitudinal Study, Birth Cohort ($n = 4050$)	Isophorone exposure using 2002 National Air Toxics Assessment levels Home learning environment assessed with a modified Home Observation for Measurement of the Environment (HOME) Inventory	Standardized math assessment scores as a measure of early cognitive skills.	High isophorone levels (>0.49 ng/m^3) and low HOME score were associated low math scale score. Decrement in math scale score was more than the additive effect of each exposure especially in male children.
18	Oudin et al. (2017)	Umeå, Northern Sweden	1469 participants aged 60 to 85 years from Betula project followed up every five years from 1988 to 2010	Cumulative annual residential mean of NO_x (marker of long-term exposure to TRAP)	Episodic memory	No association between long-term exposure to air pollution especially TRAP and episodic memory. Did not report analysis stratified by sex
19	Stingone et al. (2017)	USA	6900 children enrolled in the Early Childhood Longitudinal Study Birth Cohort	Residential concentrations of 104 ambient air toxins (including trichloroethylene, isophorone) from the National Air Toxics Assessment (2002) at age 9 months as per ZIP codes	Mathematics Tasks score	High isophorone levels (>0.49 ng/m^3) were associated with low mathematics task scores in urban and highly populated urban areas Did not report analysis stratified by sex

20	Sunyer et al. (2017)	Spain	2687 school children from 265 classrooms in 39 schools across Barcelona, Spain over one year from January 2012 to March 2013	Ambient TRAP exposure with daily levels of nitrogen dioxide (NO_2) and elemental carbon (EC) in $PM_{2.5}$ measured from $PM_{2.5}$ filters at a fixed air quality background monitoring station and in schools.	Computerized child Attention Network test (ANT)	TRAP was associated with attentional impairment Daily ambient levels of NO_2 and $PM_{2.5}$ contributed to an estimated equivalent to a 1.1 month delay in age appropriate improvement in response speed as part of natural developmental trajectory. Boys had worse cognitive performance on Hit Reaction Time with incremental increase of both Elemental Carbon and NO_2
21	Tallon et al. (2017)	USA	3377 participants aged 57 to 85 years (from Wave 2, August 2010 to May 2011 in National Social Life, Health, and Aging Project (NSHAP) cohort study	$PM_{2.5}$ exposure (estimated using GIS-based spatio-temporal models) and nitrogen dioxide (NO_2) exposures (obtained from EPA monitors).	Chicago Cognitive Function Measure (CCFM)	High $PM_{2.5}$ exposures associated with decrease in Chicago Cognitive Function Measure scores equivalent to aging by 1.6 years for $PM_{2.5}$ and 1.9 years for NO_2 exposure. Cognitive impact of $PM_{2.5}$ was modified by stroke, anxiety, stress, and mediated by depression. Cognitive impact of NO_2 were mediated by stress with effect modification by impaired activities of daily living. Did not report analysis stratified by sex
22	Tzivian et al. (2017)	Bochum, Essen, and Mülhei, Germany	Heinz Nixdorf Recall population based cohort study with 4086 participants	Land use regression was used to assess long-term residential concentrations for size-fractioned PM and nitrogen oxides. Assessment of road traffic noise	Cognitive assessment using five neuropsychological subtests and an additively calculated global cognitive score	Association of air pollutants with cognitive dysfunction was amplified by higher noise exposure at high levels of exposure. Did not report analysis stratified by sex
23	Best et al. (2016)	USA	($n = 454$; age $\geq$ 60 years from the 2001–2002 National Health and Nutrition Examination Survey).	Urinary 1-hydroxypyrene (indicator of PAH exposure)	Digit Symbol Substitution Test (DSST)	Dose-dependent 1% increase in urinary 1-hydroxypyrene associated with 1.8% poorer performance on DSST
24	Oudin et al. (2016)	Sweden	1469 persons aged 60 to 85 years at inclusion in the Betula project and followed up to 22 years, five years apart between 1988 and 2010	Exposure to traffic-related air pollution	Dementia incidence	Participants in the group with the highest exposure to TRAP more likely to be diagnosed with dementia. Did not report analysis stratified by sex

25	Porta et al. (2016)	Rome, Italy	474 children from a birth cohort of 719 newborns enrolled in 2003–2004 as part of the GASPII project evaluated at the age of 7 years	Air pollutants (NO_2, PM_{coarse}, $PM_{2.5}$, $PM_{2.5}$ absorbance) during pregnancy and at birth Maternal smoking	IQ assessed with Wechsler Intelligence Scale for Children-III	Traffic intensity in a 100 m buffer around home and an incremental 10 μg/m³ higher exposure of NO_2 exposure in intra-uterine period was associated reduced verbal IQ and verbal comprehension IQ. Other pollutants also showed negative associations with much larger confidence intervals. Did not report analysis stratified by sex
26	Tziv-ian et al. (2016)	Bochum, Essen, and Mülhei, Germany	Heinz Nixdorf Recall population based cohort study with 4086 participants aged 50–80 years old	Land use regression was used to assess long-term residential concentrations for size-fractioned PM and nitrogen oxides. Assessment of road traffic noise	Cognitive assessment using five neuropsychological subtests and an additively calculated global cognitive score (GCS)	Long-term exposures to AP and traffic noise are negatively correlated with four subtests including memory and executive functions and GCS in dose-dependent relationship independent of noise exposure e.g., an interquartile range rise in $PM_{2.5}$ correlated with verbal fluency, labyrinth test, and immediate and delayed verbal recall. Did not report analysis stratified by sex
27	Tziv-ian et al. (2016)	Vide supra	Vide supra	Vide supra	Diagnosis of Mild Cognitive Impairment based on five neuropsychological tests (Vide supra) and subjective memory complaint	Positive dose-dependent association between long-term $PM_{2.5}$ exposure and mild cognitive impairment, mainly amnesic subtype (aMCI) Did not report analysis stratified by sex
28	Chen et al. (2015)	USA	1403 community-dwelling older women aged 65–80 years without enrolled in the Women's Health Initiative Memory Study (WHIMS), 1996–1998	Cumulative $PM_{2.5}$ exposure in 1999–2006 Based on given residential histories and air monitoring data	MRI Brain	Greater $PM_{2.5}$ exposures associated with significantly smaller white matter (WM) volumes in frontal and temporal lobes and corpus callosum (equivalent to 1–2 years of brain ageing), but not of gray matter or hippocampus
29	Harris et al. (2015)	Eastern Massachusetts, USA	1109 mother-child pairs in Project Viva, a prospective birth cohort study in eastern Massachusetts (USA)	Prenatal and childhood exposure to TRAPs including black carbon (BC) and $PM_{2.5}$ assessed by distance of residence from roadways and traffic density	Verbal and nonverbal intelligence, visual motor abilities, and visual memory assessed at mean age of 8 years	Children with a residence less than 50 m away from major highway had lower nonverbal IQ and lower verbal IQ and visual-motor abilities Equivocal findings for cross-sectional associations with major roadway proximity, prenatal, and childhood exposure to traffic density and $PM_{2.5}$, third-trimester, and childhood BC exposures Did not report analysis stratified by sex

30	Peterson et al. (2015)	New York, USA	40 children aged 7 to 9 years born to any of the 665 urban Latina (Dominican) or African American women 18–35 years old who were not cigarette smokers or users of other tobacco products or illicit drugs, with initial prenatal care by the 20th week of pregnancy, and who were free of diabetes mellitus, hypertension, and known human immunodeficiency virus recruited between 1998 and 2006 through the local prenatal care clinics who had completed survey and who had a full range of prenatal PAH exposure levels; no or very low prenatal exposure to environmental tobacco smoke (classified by maternal report validated by cotinine levels of less than 15 µg/L in umbilical cord blood) and low chlorpyrifos exposure (below 4.39 pg/g) 20 children were randomly selected from above the median PAH level group and 20 from below the median PAH level group	Prenatal airborne PAH exposure by the sum of 8 Nonvolatile PAH: benzo[a]anthracene, chrysene/iso-chrysene, benzo[b]fluoranthene, benzo[k] fluoranthene, benzo[a]pyrene, indeno[1,2,3c,d] pyrene, dibenzo[a,h] anthracene, and benzo[g,h,i]perylene. measured with personal air monitoring of the mothers over a 48-h period in the third trimester of pregnancy but assumed to be index exposure for the entire gestation Prenatal PAH exposures validated against concurrent 2-week monitoring of residential air samples in the final trimester of pregnancy Postnatal PAH exposure by spot urine sample at age of 5 years for 9 metabolites (1-Hydroxynaphthalene, 2 Hydroxynaphthalene, 2-Hydroxyfluorene, 3-Hydroxyfluorene, 9-Hydroxyfluorene, 1 Hydroxy phenanthrene, 2-Hydroxyphenanthrene, 3-Hydroxyphenanthrene, and 4 Hydroxyphenanthren.)	CBCL WISC-IV MRI Brain	A dose-response inverse relationship between prenatal PAH exposure and reductions of the white matter surface in most of the frontal, superior temporal, and parietal lobes, as well as the entire rostrocaudal extent of the mesial surface, in the left but not the right hemisphere of the brain and reduced white matter in later childhood associated with slower information processing speed during intelligence testing and more severe externalizing behavioral problems such as ADHD and Conduct disorder. Postnatal PAH exposure correlated with white matter surface measures in bilateral dorsal prefrontal cortex bilaterally independent of prenatal PAH exposure Stronger inverse associations of prenatal PAH exposure with white matter surface measures in girls as compared to boys

31	Schikowski et al. (2015)	Ruhr Area and Borken, Germany	4874 women from the SALIA cohort (aged 55 years at baseline) enrolled between 1985 and 1994 and followed up in 2006 (*n* = 2116) and 2008 (*n* = 834) of which complete information available (*n* = 789)	Particulate matter (PM) size fractions and nitrogen oxides (NO_x) Traffic load	CERAD-Plus test Apolipoprotein E (ApoE) alleles	Air pollution was inversely related to visuospatial abilities on cognitive assessment with significant adverse association of traffic load in carriers of ApoE 4 risk alleles.
32	Sunyer et al. (2015)	Barcelona, Spain	2715 children aged 7 to 10 years from 39 schools in Barcelona	Chronic traffic air pollution [elemental carbon [(EC), nitrogen dioxide (NO_2), and ultrafine particle number (UFP; 10–700 nm)] measured twice during 1-wk campaigns both in the courtyard (outdoor) and inside the classroom (indoor)	*n*-back and the attentional network tests	Detrimental associations between Traffic related air pollution and cognitive performance were stronger in boys than in girls
33	Ailshire and Clarke (2014)	USA	Cross sectional data of 780 non-Hispanic black and white men and women aged ≥ 55 years from the 2001/2002 Americans' Changing Lives Study	$PM_{2.5}$ using PA air monitoring data linked to respondents using census tract identifiers.	Tests of working memory and orientation	Exposure to high $PM_{2.5}$ concentrations associated with 1.5 times greater error rate Male gender was associated with more cognitive errors
34	Ailshire and Crimmins (2014)	USA	13,996 men and women aged 50 years or older from the 2004 HRS survey	Residence in areas with higher $PM_{2.5}$ concentrations	Cognitive Function	Living in areas with higher $PM_{2.5}$ concentrations was associated with worse cognitive function especially episodic memory No statistically significant gender based differences
35	Gatto et al. (2014)	Los Angeles Basin, USA	1496 individuals (mean age of 60.5 years)	Air pollutants [O_3, $PM_{2.5}$ and nitrogen dioxide (NO_2)]	Six domains of cognitive function and global cognition	Increased exposure to $PM_{2.5}$ associated with lower verbal learning Exposure to NO_2 > 20 ppb associated with lower logical memory Exposure to O_3 above 49 ppb associated with lower executive function No air pollutant significantly associated with global cognition. No statistically significant gender based differences

36	Guxens et al. (2014)	Europe (Netherlands, Germany, France, Italy, Greece, Spain)	Mother-, infant pairs recruited between 1997 to 2008 yielding a total sample of 9482 children from 6 European population-based birth cohorts—GENERATION R (The Netherlands), DUISBURG (Germany), EDEN (France), GASPII (Italy), RHEA (Greece), and INMA (Spain)	Prenatal exposure to air pollutants such as (NO_2, NO_x) in all regions and PM ($PM_{2.5}$, PM_{10} and PM_{coarse}) and $PM_{2.5}$ absorbance in a subgroup using land-use regression models for the period 2008 to 2011 and then modelled for air pollutant profile for the exact pregnancy periods using background monitoring sites.	Assessment for cognitive and psychomotor development at 1 and 6 years of age.	Prenatal air pollution exposure during pregnancy, particularly NO_2, was associated with dose-dependent reduction in psychomotor development but not cognitive development. Did not report analysis stratified by sex
37	Tonne et al. (2014)	London, UK	2867 white men retired from work (mean age 66 years) from Whitehall II cohort	Particulate matter from traffic exhaust	Alice Heim 4-I test, 20-word free-recall test, semantic, and phonemic verbal fluency	Higher $PM_{2.5}$ of 1.1 μg/m^3 was negatively associated with reasoning and memory but not verbal fluency and significant 5-year decline in standardized memory score.
38	Loop et al. (2013)	USA	20,150 men and women enrolled in the Reasons for Geographic And Racial Differences1 in Stroke (REGARDS) cohort	Satellite-derived estimate of $PM_{2.5}$ concentration map	Cognition	No consistent increase in odds of cognitive impairment with every 10 μg/m^3 increase in $PM_{2.5}$ concentration. No statistically significant gender based differences
39	Power et al. (2013)	USA	628 men (mean age of 70 years) from the VA Normative Aging Study	TRAP exposure on cognitive function.	HFE C282Y variant (hemochromatosis gene polymorphisms) Cognitive function	Older adults lacking HFE C282Y variant had greater adverse cognitive impact of TRAP exposure
40	Weuve et al. (2012)	USA	19,409 U.S. women aged 70 to 81 years from Nurses' Health Study Cognitive Cohort	Long-term exposure to higher levels of both $PM_{2.5-10}$ and $PM_{2.5}$	Cognition	Dose-dependent association between Long-term exposure to higher levels of both $PM_{2.5-10}$ and $PM_{2.5}$ and significantly faster cognitive decline at 2 years. Every 10 μg/m^3 increment in long-term (2 years) exposure to $PM_{2.5-10}$ and $PM_{2.5}$ cognitively equivalent to aging by approximately 2 years.

41	Power et al. (2011)	USA	680 men (mean ± SD, 71 ± 7 years of age) from the U.S. Department of Veterans Affairs Normative Aging Study	Traffic-related air pollution including Black carbon (BC) exposure	Mini Mental Status Examination (MMSE)	Significant association between Black carbon (BC) exposure and lower MMSE score
42	Edwards et al. (2010)	Krakow Poland	214 offspring of a cohort of pregnant, healthy, nonsmoking women of Krakow, Poland, between 2001 and 2006	Maternal 48-hr personal air monitoring to estimate fetal Polyaromatic Hydrocarbon air pollutant exposure and PAH estimation from maternal blood sample and/or a cord blood sample at the time of delivery	Raven's Colored Progressive Matrices (RCPM) at 5 years of age	Higher (above the median of 17.96 ng/m^3) prenatal exposure to airborne PAHs associated with decreased scores on RCPM corresponding to an estimated average decrease of 3.8 IQ points. Did not report gender-based differences
43	Freire et al. (2010)	Spain	210 boys from a population-based birth cohort from southern Spain	NO_2	Cognitive development	No significant association between NO_2 and cognitive development
44	Chen and Schwartz (2009)	USA	1764 adult participants (aged 37.5 ± 10.9 years) of the Third National Health and Nutrition Examination Survey in 1988–1991	Geocoded Residential Ambient annual PM_{10} and ozone	Neurobehavioral Evaluation System-2 (NES2) data (including a simple reaction time test [SRTT] measuring motor response speed to a visual stimulus; a symbol-digit substitution test [SDST] for coding ability; and a serial-digit learning test [SDLT] for attention and short-term memory	Consistent associations between ozone and not PM_{10} and reduced performance on symbol-digit substitution test and a serial-digit learning test but not in simple reaction time test. Each 10-ppb increase in annual ozone was associated with equivalent to 3.5 and 5.3 years of aging-related decline in cognitive performance. Did not report analysis stratified by sex

45	Perera et al. (2009)	New York, USA	249 Children born to non-smoking black or Dominican-American women residing in New York City who had undergone prenatal ambient personal PAH monitoring	PAH exposure	Wechsler Preschool and Primary Scale of Intelligence-Revised (WPPSI-R)	High Prenatal Polyaromatic Hydrocarbons (PAH) levels (above the median of 2.26 ng/m^3) inversely associated with full-scale IQ and verbal IQ with a decrement of 4.31 to 4.67 points of IQ at age of 5 years. Did not report analysis stratified by sex
46	Ranft et al. (2009)	Germany	399 women aged 68–79 years	Traffic-related fine PM	Mild Cognitive Impairment (MCI)	Exposure to traffic-related fine PM consistent and significant risk factor for MCI
47	Lee et al. (2007)	USA	278 children aged 12–15 years included in the National Health and Nutrition Examination Survey 1999–2000	POPs such as 3,3,'4,4,'5-pentachlorobiphenyl, 1,2,3,4,6,7,8-heptachlorodibenzo-p-dioxin (HPCDD), 1,2,3,4,6,7,8,9-octachlorodibenzo-p-dioxin (OCDD), 1,2,3,4,6,7,8-heptachlorodibenzofuran (HPCDF), beta-hexachlorocyclohexane, p,p'-dichlorodiphenyltrichloroethane, and trans-nonachlor.	Prevalence rates of learning disability (LD) and attention deficit hyperactivity disorder (ADHD), both of which are characterized by cognitive impairment	Direct association between POPs and LD/ADHD. No statistically significant gender based differences found
B	Studies from Upper and Middle Income Settings					
1	Saenz et al. (2018)	Mexico	13,023 Mexican adults over age 50 participating in 2012 Wave of the Mexican Health and Aging Study	Indoor air pollution (Use of wood or coal as primary cooking fuel)	Verbal learning, verbal recall, attention, orientation, and verbal fluency	Exposure to indoor air pollution associated with poorer cognitive performance Did not report analysis stratified by sex
2	Zhang et al. (2018)	China	25,486 individual respondents over age 10 in 2010 and 2014, from China Family Panel Studies (CFPS)	Cumulative and transitory exposures to air pollution	Verbal and Math tests	Adverse cognitive impact of air pollution on performance in verbal and math tests with greater deficits on verbal tasks as people aged. Significantly worse verbal tests performance in males on verbal tasks with short-term and cumulative air pollution exposure
3	Calderón-Garcidueñas et al. (2012)	Mexico	30 children (20 from Southwest Mexico City (SWMC) and 10 from Polotitlan)	High pollution versus low pollution areas	Frontal tau hyperphosphorylation with pre-tangle material amyloid-beta diffuse plaques	Nearly 40% of highly exposed children and young adults had frontal tau hyperphosphorylation with pre-tangle material and 51% had amyloid-beta diffuse plaques compared versus 0% of controls living in low pollution areas. Did not report analysis stratified by sex

5	Calde-rón-Garcidueñas et al. (2011)	Mexico	20 children from Mexico City (Mean age = 7.1 years, SD = 0.69) and 10 children from Polotitlán (Mean age = 6.8 years, SD = 0.66)	High pollution Areas versus Low pollution areas	MRI Brain Wechsler Intelligence Scale for Children-Revised (WISC-R) Inflammatory markers interleukin 12, cytokines, and chemokines	Complex modulation of cytokines and chemokines influences children's central nervous system structural and volumetric responses and cognitive correlates resulting from environmental pollution exposures Did not report analysis stratified by sex Exposure high pollution associated with low white matter volumes especially in the right parietal and bilateral temporal areas irrespective of white matter abnormalities as well as progressive deficits on WISC-R Vocabulary and Digit Span subtests
5	Calde-rón-Garcidueñas et al. (2011)					White matter hyperintensities associated with evidence of inflammation, immunoregulation, and tissue remodeling on MRI. Children without white matter hyperintensities had raised levels of interleukin 12, pro-inflammatory cytokines, and low levels of neuroprotective cytokines and chemokines. Did not report analysis stratified by sex
6	Calde-rón-Garcidueñas et al. (2008)	Mexico	55 Children (mean age: 9.2 years) from Mexico City with high air pollution and 18 children (mean age: 10.5 years) from Polotitlán with low air pollution	Air Quality	Psychometric testing MRI Brain	Residence in high air pollution area associated with deficits in a combination of fluid and crystallized cognition tasks, high rates of prefrontal white matter hyperintense lesions. Did not report analysis stratified by sex

Sources: Chandra M, Rai CB, Kumari N, Sandhu VK, Chandra K, Krishna M, Kota SH, Anand KS, Oudin A. Air Pollution and Cognitive Impairment across the Life Course in Humans: A Systematic Review with Specific Focus on Income Level of Study Area. International Journal of Environmental Research and Public Health. 2022; 19(3):1405. https://doi.org/10.3390/ijerph19031405.

Large industrial sources, especially metal smelters, account for about 66% of Canada's SO_2 emissions, with "fuel combustion," or primarily fossil fuel-fueled power plants, accounting for 24%. As an illustration, 84% of U.S. Power plants account for 90% of emissions, with 10% coming from industrial sources. The majority of this discrepancy can be attributed to two things: metal smelting is a significant industry in Canada, and nuclear and hydropower technologies account for a sizable amount of the country's electricity generation. These technologies do not release SO_2.

ost developed countries have taken action to lower their emissions due to the harm that SO_2 and other air pollutants can do to human health and the environment. In comparison to the United States, where SO_2 emissions decreased by 32% between 1985 and 2012, Canada saw a reduction of roughly 41%. In both countries the reductions were achieved by several methods:

- Switching to low- or no-sulfur fuels for some major uses, especially for electricity generation.
- Removing sulfur from some fuels prior to combustion, mostly by coal washing.
- Reducing energy demands through conservation.
- Installing scrubbers to remove SO_2 from post-combustion waste gases before they are vented to the atmosphere.

On the other hand, as Figure 3.2 illustrates, emissions reductions from stationary sources—primarily coal-fueled power plants—and industrial sources—particularly metal smelters—have been particularly significant. The significant reductions in SO_2 that have been attained in North America over the last few decades have come at a significant financial cost and ought to be recognized as a significant "success story" in the field of pollution control.

Unlike SO_2, the majority of the world's emissions of sulfide gases come from natural sources. The two main sources are dimethyl sulfide ($(CH_3)_2S$), which is produced by phytoplankton and outgassed from oceanic waters, and H_2S, which is released from anoxic sediment in shallow marine and inland waters. About 100 million tons of H_2S and 15 million tons of dimethyl sulfide are naturally emitted annually; these emissions are expressed as tons of S, or sulfur equivalent. Three million tons of H_2S are emitted annually due to anthropogenic sources, primarily the chemical industry, sewage treatment facilities, and animal manure. Almost 300 million tons of sulfur are emitted annually worldwide from all gases containing sulfur. Anthropogenic sources account for about half of the world's emissions. SO_2 and H_2S concentrations in clean air are normally less than 0.2 ppb. The amounts of SO_2 and H_2S in air that have been contaminated by emissions can vary greatly. In urban atmospheres, they are usually around 0–2 ppm, but they can rise to 3 ppm in close proximity to significant emission sources (Ayres and Maynard, 2006; Vallero, 2014).

3.3.3. Toxicity of Sulfur Gases

Seldom are environmental H_2S concentrations high enough to be harmful to plants. On the other hand, unintentional emissions from sour-gas plants—where

H_2S is extracted from natural gas—may harm nearby vegetation. On the other hand, SO_2 concentrations in urban areas and close to industrial sources are frequently high enough to harm both domesticated and wild plants. As we look at later, there has been significant damage to the vegetation near some metal smelters.

Most plant species will experience acute damage after being exposed to 0 ppm SO_2 for an hour or 0 ppm SO_2 for eight hours. However, it's crucial to remember that some species—known as hypersensitive species—can experience acute injuries at concentrations lower than those mentioned above due to their extreme sensitivity to SO_2 exposure.

Furthermore, when exposed to concentrations of SO_2 lower than those necessary to cause an acute injury, plants frequently show a reduction in yield. This kind of reaction is known as "hidden injury" since it takes place in the absence of signs of acute tissue damage. Plants in experimental chambers are enclosed and exposed to either ambient level SO_2 air or air that has been filtered through charcoal to remove that gas in order to measure hidden injuries to wild and agricultural vegetation. If output is higher in the filtered air, then there was a hidden harm being caused by the ambient SO_2. Research on pasture grasses has shown that yield reductions due to hidden injuries would result from exposure to SO_2 concentrations as low as 0.04 parts per million on average.

A maximum of 450 $\mu g/m^3$ (0.17 ppm) of ambient SO_2 in the atmosphere during an hour of exposure is the air-quality guideline for Canada. This recommendation is based on the concentration at which the majority of agricultural plants experience acute foliar (leaf) injuries. While areas that fall within this guideline might not exhibit severe acute damage to vegetation, more vulnerable species could be impacted by covert injuries, which could lead to large yield losses.

The majority of other animals, including humans, are far less sensitive to SO_2 than are plants. The fact that longer-term exposures to low concentrations can be just as important as higher acute exposures in terms of dosage received is taken into account in guidelines for permissible exposures of humans to SO_2 and other potentially toxic gases.

Air pollution guidelines for occupational exposures are often higher than ambient guidelines. It is advised that occupational SO_2 exposures in North America be kept to a maximum of 2 parts per million (5 ppm) for extended periods of time and 5 parts per million (13 mg/m^3) for shorter exposures. However, some individuals are more sensitive to SO_2, and they may experience asthma or other respiratory distress at concentrations below 1 ppm (2.06 mg/m^3). Furthermore, a few studies have hypothesized that prolonged exposure to sulfate particulate aerosols in urban areas, which are ultimately derived from gaseous SO_2, may cause slight increases in the incidence of respiratory and circulatory disorders, most likely in hypersensitive individuals.

1. **Air Pollution Chemistry:** Specific chemicals that are released into the atmosphere have the potential to react with other molecules to form new ones.

   Several examples follow:

i. Oxidation of SO_2

$SO_2 + OH \rightarrow HO{\bullet}SO_2$ (3.1)

$HO{\bullet}SO_2 + O_2 \rightarrow HO_2 + SO_3$ (3.2)

$SO_3 + H_2O \rightarrow H_2SO_4$ (3.3)

$H_2SO_4 \rightarrow 2H^+ + SO_4^{2-}$ (in aqueous solution) (3.4)

Note that the SO_4^{2-} produced is important as a constituent of acid rain and as particulate ammonium sulfate ($(NH_4)_2SO_4$), which are important pollutants.

ii. **Oxidation of NO**

$NO + HO_2 \rightarrow NO_2 + OH$ (3.5)

$NO_2 + OH \rightarrow HNO_3$ (3.6)

$HNO_3 \rightarrow H^+ + NO_3-$ (in aqueous solution) (3.7)

Similarly, the NO_3- produced is important in acid rain and as particulate NH_4NO_3.

Formation and Destruction of Stratospheric O_3

O_2 + UV radiation $\rightarrow O + O$ (3.8)

$O + O \rightarrow O_2$ (3.9)

$O + O_2 \rightarrow O_3$ (3.10)

O_3 + UV radiation $\rightarrow O_2 + O$ (3.11)

Reaction 11 is an ultraviolet photodissociation of O_3. Ozone can also be consumed by reactions with NO, NO_2, N_2O, and with ions or simple molecules of chlorine (especially ClO), bromine, and fluorine. These reactions are too complex to describe here.

Formation and Destruction of Ground-Level O_3

$O + O_2 \rightarrow O_3$ (3.10)

NO_2 + UV radiation $\rightarrow NO + O$ (3.12)

$NO + O_3 \rightarrow NO_2 + O_2$ (3.13)

$NO + RO_2 \rightarrow NO_2 + RO$ (3.14)

Atomic O, created by the photodissociation of NO_2, is necessary for the synthesis of O_3 (reaction 3.10). (reaction 3.12). Reaction 3.13: Ozone can be consumed by reacting with NO, an expelled gas, to produce NO_2. However, because these operate in competition with reaction 3.13 for NO, atmospheric O_3 can accumulate if other reactions (like reaction 3.14) convert NO to NO_2. (The species RO_2 is made up of a variety of substances referred to as peroxy radicals, which are created when organic molecules are broken down by hydroxyl radicals and then molecular O_2 is added. The chemically reduced form of RO_2 is called RO (Ayres and Maynard, 2006; Vallero, 2014).

3.4. NITROGEN GASES

3.4.1. Emissions and Transformations

Nitrous oxide (N_2O), ammonia (NH_3), nitrogen dioxide (NO_2), and nitric oxide (NO) are the most significant nitrogen-containing gases. NO and NO_2 are frequently regarded as a complex, or NOx, when combined.

The colorless gas NH_3 is mostly released by wetlands, where it is created by the anaerobic breakdown of dead biomass. The annual natural emissions of NH_3 are approximately 1.2 trillion tons. Fossil fuel combustion (4 million tons annually) and animal husbandry (0.2 million tons annually) are two sources of anthropogenic emissions. Before being oxidized to nitrate, NH_3 has a seven-day residence period in the atmosphere.

Inhaled, nitrous oxide (N_2O) is a colorless, non-toxic gas that causes a slight euphoria. Often referred to as "laughing gas," this gas is occasionally used as a recreational drug in addition to being used as a mild anesthetic in medicine. N_2O has a long residence time in the atmosphere of roughly four years due to its relative unreactivity. In soil and water, microbial denitrification is linked to the majority of N_2O emissions. Compared to industrial emissions, which amount to 12 million tons annually, these are roughly 18 million tons annually. Nitrate-fertilized agricultural soil can emit large amounts of N_2O, and contemporary farming methods are estimated to have contributed roughly 40% of the increase in global emissions.

While nitrogen dioxide is reddish, pungent, and irritating to respiratory and ocular membranes, NO is a colorless, odorless gas. About 430 million tons of NOx are naturally produced annually (expressed as NO; the same emissions are expressed as NO_2 and total 658 million tons annually). The three main natural processes that produce NOx emissions into the atmosphere are biomass nitrogen oxidation during fires, atmospheric nitrogen gas (N_2) fixation by lightning, and bacterial denitrification of nitrate in soil.

The primary cause of anthropogenic NOx emissions, which are roughly 83 million tons annually (represented as NO), is the burning of fossil fuels, particularly in power plants and automobiles. The majority of these emissions are NO, which the atmosphere's processes oxidize to NO_2 secondarily. The majority of atmospheric NOx gases eventually oxidize to produce nitrate (NO_3^-), an ion that is crucial for the acidification of ecosystems and precipitation.

3.4.2. Toxicity of Nitrogen Gases

Seldom do NH_3 or NOx gas concentrations get high enough to harm vegetation. The production of much more toxic gas, ozone, through photochemical reactions and acidification of precipitation and ecosystems are the main sources of environmental damage linked to NOx emissions.

Rarely are ambient NOx and NH_3 concentrations high enough to cause problems for people. In an occupational setting, guidelines for long-term exposures are 5 ppm (10 mg/m^3) for NO_2 and 25 ppm (34 mg/m^3) for NO. For short-term exposures, the occupational guidelines are 5 ppm (10 mg/m^3) and 35 ppm (47 mg/m^3), respectively. People who work in environments with high NOx exposure may experience pulmonary function impairments (Ayres and Maynard, 2006; Vallero, 2014).

3.5. ORGANIC GASES AND VAPORS

3.5.1. Emissions and Transformations

A broad class of compounds known as hydrocarbons has different configurations of hydrogen and carbon atoms in their molecules. The gas methane (CH_4) is the most basic hydrocarbon. Bigger, heavier, and more intricately structured hydrocarbons can exist as vapors, liquids, or solids. Other volatile organic compounds (VOCs) include alcohols, aldehydes, and phenols; they can also include nitrogen, oxygen, and other light elements in addition to carbon and hydrogen.

While the combined concentration of all other hydrocarbons and volatile organics is less than 1 ppb (the conversions to mg/m^3 are variable depending on the molecular weight), the background concentration of methane in the atmosphere is approximately 1.7 ppm (1 ppm = 0.65 mg/m^3). The majority of CH_4 emissions are organic and result from microbes in anaerobic wetlands fermenting organic matter. Smaller amounts are released into the atmosphere during wildfires, fossil fuel deposits, burping, and flatulent ruminant animals (like sheep and cows), and termites, which release CH_4 during the digestion of their plant-based diet. The annual natural emissions of CH_4 are approximately 1.6 billion tons worldwide, while the annual natural emissions are approximately 0.3 billion tons.

Non-methane hydrocarbons are hydrocarbons found in the atmosphere that are not CH_4. These and many other organics naturally evaporate from living plants, producing gases and vapors as well as smaller amounts that outgas from fossil fuel deposits. Usually, on hot, sunny days, forests release the most emissions. Compared to anthropogenic emissions of 186 million tons annually, natural emissions of non-methane hydrocarbons are approximately 200 million tons annually. The main man-made sources are solvent evaporation, releases from the mining and refining of fossil fuels, and unburned fuel emissions from automobiles and airplanes.

3.5.2. Toxicity of Organic Gases and Vapors

Although atmospheric concentrations of organic gases and vapors are rarely high enough to harm plants or animals, they can be toxic. These gases and vapors are significant to the environment primarily because of their part in the photochemical processes that result in toxic ozone. Furthermore, one significant greenhouse gas that contributes to global warming is CH_4. However, relatively toxic organics like formaldehyde and benzene may be significant pollutants in some workplaces.

3.6. OZONE

O_3 in the troposphere (ground-level ozone) and O_3 in the stratosphere (stratosphere) are the two distinct ozone-related environmental problems. The uppermost layer of the atmosphere, called the stratosphere, starts at a height of 8–17 km above Earth's surface and naturally contains high concentrations of ozone, depending on the season and latitude. O_3 in the stratosphere is not harmful to the atmosphere or an air pollutant. Rather, stratospheric 3 helps shield life on Earth's surface from numerous detrimental effects of exposure to this dangerous region of the electromagnetic spectrum by absorbing solar UV radiation. On the other hand, O_3 is a significant air pollutant that harms materials, plants, and people's health when it is present in the troposphere, the lower atmosphere. Through interactions with other gases, organic vapors, and terrestrial and aquatic surfaces (including vegetation), ozone is eliminated from the atmosphere.

3.6.1. Ground-Level Ozone

The most harmful of the so-called photochemical air pollutants is ground level ozone (O_3). Hydrogen peroxide (H_2O_2), peroxyacetyl nitrate (PAN), and other oxidant gases are less significant. Oxygen and other oxidant gases are abundant in oxidizing smog. Unlike primary pollutants like SO_2 and NOx, these chemicals are classified as secondary pollutants because they are not released into the atmosphere. Rather, photochemical reactions—chemical processes that need light, particularly ultraviolet wavelengths—synthesize them in the atmosphere.

If NOx and hydrocarbons are present in high concentrations—a condition that is usually caused by anthropogenic emissions—these proceed more quickly and accumulate oxidants.

In the morning, a weak atmospheric inversion frequently forms in some areas. An inversion promotes the formation of oxidizing smog in the morning and early afternoon because it is comparatively stable and prevents the in-mixing of cleaner air from above or beyond. Stronger winds usually break up the inversion later in the day, dispersing the air pollution. These inversions, along with the ozone-rich smog they produce, are a regular occurrence in the Los Angeles region, Mexico City, Vancouver, and other places.

North America's various regions have wildly differing amounts of ground-level O_3. The southwestern United States has relatively high average concentrations, approximately 100 ppb (1 ppb = 2 $\mu g/m^3$), while other regions of the United States and southern Canada typically have concentrations between 40 and 60 ppb.

O_3 air quality standards have been developed by the U.S, Canada, and other nations. These are meant to represent concentrations that would shield wild and agricultural plants from serious harm. The American O_3 standard was 120 ppb (240 $\mu g/m^3$) in 1979 after it had been set at 80 ppb (160 $\mu g/m^3$) for an average one-hour exposure for a while. The original 80 ppb standard was practically unenforceable because it was routinely exceeded over wide regions, prompting the authorities to make the change. In fact, some areas—especially in the southwest of the United States—frequently surpass the 120 ppb threshold.

There is particularly severe photochemical air pollution in the Los Angeles basin. O_3 concentrations can reach 500 parts per billion (ppb) on an hourly average, and during the summer, they usually surpass 100 parts per billion for over 15 days. In other North American cities, maximum O_3 concentrations are typically lower, averaging 150–250 parts per billion on an hourly basis. O_3 is a significant air pollutant because these concentrations are well within the range at which it can harm plants acutely. Although cities are the primary source of emissions of ozone precursors (NOx and hydrocarbons), the spread of polluted air masses to rural areas with a predominance of agriculture or natural vegetation can have a devastating ecological impact.

According to Environment Canada (1999), the lower Fraser Valley in southwest British Columbia, the Windsor-Quebec City corridor, and the southern Maritimes are the areas of Canada that are most frequently affected by ozone pollution. Bound by mountains to the east, the lower Fraser Valley is a coastal lowland that frequently experiences atmospheric inversions and receives significant emissions of ozone precursors from the Greater Vancouver area to the west. Maximum O_3 concentrations in some areas of the lower Fraser

Valley are approximately 100 parts per billion. Although the Windsor–Quebec City corridor is home to 60% of Canada's population, it also receives ozone and its precursors from air masses that blow in from densely populated areas of the United States. Local emissions from people and industries are also present in this area. O_3 concentrations in the corridor range from 110 to 160 parts per billion, with observed values as high as 190 parts per billion.

Ozone-rich air masses that are drawn in from industrialized areas of southern Ontario, Quebec, and the northeastern United States primarily affect the southern Maritime region. O_3 concentrations can reach 90–110 ppb in some areas of western Nova Scotia and southern New Brunswick.

3.6.2. Toxicity of Ozone

O_3 can irritate and damage the respiratory system and eye membranes, as well as impair lung function in humans and certain animals. In an occupational setting, the recommended long-term exposure limit for O_3 is 100 ppb (196 $\mu g/m^3$), while the short-term limit is 300 ppb (589 $\mu g/m^3$). However, at lower concentrations, O_3-related symptoms can still affect sensitive individuals. O_3 exposure can aggravate bronchitis and emphysema and cause asthmatic episodes.

Ozone leads to significant harm to a broad range of wild and cultivated plants, causing damage over extensive areas. Characteristic injuries on leaves are frequently associated with O_3 exposure, impairing the plants' photosynthetic abilities and, consequently, lowering their productivity. Severe injuries can affect most plant species after two to four hours of exposure to concentrations of 200–300 ppb O_3, while prolonged exposure to lower concentrations of 40–100 ppb may cause subtle injuries (resulting in reduced yield). However, numerous species exhibit a higher sensitivity to O_3 and experience both acute and hidden injuries at lower concentrations. Certain types of tobacco (*Nicotiana tobacco*) can suffer acute foliar injuries from as little as a two to three-hour exposure to 50–60 ppb, while spinach (*Spinacea oleracea*) may experience injury within one to two hours at 60–80 ppb. Sensitive conifer species may suffer acute injuries from 12-hour exposures to 80 ppb O_3.

In experimental chambers with ambient air or air filtered through charcoal to eliminate any O_3, researchers have grown agricultural plants. These studies have been helpful in determining the degree of harm that ambient O_3 exposure causes to agricultural crops. Crop yields were found to be lower across the board in the U.S., according to a set of field tests. The southwest saw the most damage because of its sunny weather, high NOx and hydrocarbon emissions, and particularly high O_3 concentrations. According to that study, O_3-related crop damage accounted for 2–4% of the U.S. agricultural output, meaning that annual economic losses exceeded $5 billion. As O_3 is the most significant air pollutant in agriculture, it damages vegetation over a large portion of North America. Most likely, the main air pollutant harming forests and other natural ecosystems is ozone.

3.6.3. Stratospheric Ozone

Ozone in the stratosphere shields life on Earth from the harmful effects of solar ultraviolet (UV) radiation, in contrast to ozone at ground level. This is the reason it is alarming that anthropogenic emissions of some gases are destroying stratospheric ozone.

Natural photochemical processes in the stratosphere produce ozone. They entail oxygen molecules (O_2) absorbing solar UV radiation, which produces extremely reactive oxygen atoms (O) that combine with additional O_2 molecules to form O_3. Because there is a lot of high-energy UV radiation in the stratosphere, these reactions happen rather quickly. O_3 concentrations in the stratosphere are therefore usually 200–300 ppb, or roughly 10 times higher than in the ambient troposphere.

An essential environmental service is provided by stratospheric O_3. The majority of the incoming high-energy UV radiation, which can be extremely harmful to organisms, is efficiently absorbed by it. Specifically, UV radiation can harm DNA because it is a potent absorber of this radiation. This may raise the chance of getting skin cancers, such as melanoma, a disease that frequently results in death. UV exposure also increases the risk of immune system suppression and cataract development in the eyes. Additionally, UV radiation harms plants because it degrades chlorophyll, the primary pigment involved in photosynthetic reactions. This degradation can result in lower plant productivity. UV rays also harm the cuticle of foliage, which is a waxy covering.

Numerous processes, such as reactions with reactive ions of bromine, chlorine, and fluorine, as well as with the trace gases NOx and N_2O, can destroy stratospheric O_3. There are worries regarding the stratospheric O_3 depletion because of the rise in the concentrations of some of these O_3-consuming chemicals in the stratosphere due to anthropogenic emissions. Most people agree that emissions of chlorofluorocarbons (CFCs), especially the industrial gases referred to as freons, have played a significant role in this. Due to their extreme lack of reactivity in the troposphere, CFCs eventually migrate up to the stratosphere, where they undergo photodissociation—a slow breakdown that releases free chlorine—after being exposed to UV radiation. O_3 is effectively reacted with and destroyed by the chlorine.

Extremely cold and stagnant conditions in the stratosphere, like those that occur above polar latitudes at the end of Antarctic and Arctic winters, are the most favorable conditions for the O_3-destroying reactions to proceed. During the early spring, these polar-focused O_3 depletions lead to the formation of "ozone holes.." Since the early 1980s, these occurrences have been frequently noted. The O_3 holes over Antarctica are especially large and usually result in springtime drops in O_3 concentration of 30–50%. Less significant O_3 depletions are seen above the Arctic, especially in northern Canada. Compared to Antarctica, the affected areas in the Northern Hemisphere are substantially smaller.

Although the extent of Antarctica's ozone holes varies annually, there has been a notable upward trend since 1980, when reliable data started to be gathered. Lower latitudes are impacted when the O_3-depleted air disperses during the hole breakup, even though seasonal O_3 losses only occur over polar regions. Although not nearly to the same extent as happens in the holes themselves, this momentarily lowers the stratospheric O_3 concentrations throughout the hemisphere.

3.6.3.1. The Montreal Protocol – A Success of Regulatory Action

World governments moved to address the issue shortly after it was generally acknowledged that emissions of chemicals, including CFCs, were causing the stratospheric ozone layer to deteriorate. The Montreal Protocol on Substances That Deplete the Ozone Layer is a treaty that resulted from intense negotiations that were sparked by an international meeting that the United Nations Environment Program (UNEP) arranged in Montreal in 1987. An international agreement known as the Montreal Protocol required all parties (signatory nations) to adhere to a timeline for the phase-out of CFC and other substance production and use that is known to be harmful to the ozone layer. The treaty mandated that the signatory countries cut their CFC production and consumption to half of 1986 levels by 1998, and to freeze them at 1986 levels by 1989.

At first, the governments of many nations were hesitant to ratify the protocol because they believed that tight regulations on the production and use of chemicals would be detrimental to their economies. This was especially the case for countries in the former Soviet Union, Japan, and the European Community. Nevertheless, by adamantly promoting control measures, the U.S., Canada, Sweden, and Norway were able to persuade the reluctant countries to gradually stop using ozone-depleting substances. After 40 nations ratified the Montreal Protocol, which took effect on January 1, 1989, approximately 82% of the world's CFC usage was accounted for by these nations.

In order to completely phase out the use of halons by 1994, CFCs, methyl chloroform, HBFCs (hydrobromofluorocarbons), and carbon tetrachloride by 1996, methyl bromide by 2010, and HCFCs (hydrochlorofluorocarbons) by 2030, a number of amendments were made to the Montreal Protocol. Numerous other nations, including China and India, two sizable countries that had not taken part in the original negotiations, ratified the amended protocol. As of 2009, the Montreal Protocol had been ratified by 197 nations, making it the first of its kind to receive widespread support. The Montreal Protocol Multilateral Fund was also created by the amendments to offer financial assistance to developing countries so they could quickly reduce their reliance on chemicals that deplete the ozone layer.

It has been said that the Montreal Protocol and the changes that followed it are a "success story" in terms of pollution regulation. Many developed nations—including Canada—have exceeded their initial reduction goals and accelerated their pace of change, while less developed nations have pledged to forbid the use of ozone-depleting substances in their economic systems. This success was achieved because of the following:

1. There was international recognition of a clear threat to the global environment.
2. The threat was associated with particular substances that could be easily controlled, as they were manufactured in only a few places and were used for relatively discrete purposes.
3. Economically acceptable substitutes were quickly developed to replace the use of ozone-depleting substances.

Thorough information, strong international and national institutions, a cooperative mindset, strong leadership from motivated leaders, and the accessibility of substitute technologies came together to overcome political divides in favor of pursuing a common environmental interest. For this reason, environmental regulation action has been successful with the implementation of the Montreal Protocol (Ayres and Maynard, 2006; Vallero, 2014).

3.7. AIR POLLUTION AND HEALTH

A notable instance of a natural gas leak that resulted in fatalities for people involved was the large-scale CO_2 leak from a lake in Cameroon, West Africa. Similar to bottled soda water, the deep waters of Lake Nyos, a 200-meter-deep volcanic lake, are naturally supersaturated with CO_2. A significant amount of silt appears to have slumped into the lake's steep sides one evening in 1986, churning some of the bottom water to the surface. As a dense air mass degassed from the water, it moved into the low spots in the surrounding terrain. Up to 25 kilometers from the lake, 3,500 livestock, 1,700 sleeping people, and an unknown number of wild animals were all asphyxiated by the CO_2-rich air. At concentrations above 8–10%, atmospheric CO_2 can be extremely toxic; its "normal" level is approximately 0.04%. Since plants are much less susceptible to CO_2 toxicity, this unusual and amazing natural event did not harm any vegetation. Increases in disease and mortality rates among humans have occasionally been linked to anthropogenic emissions of other gaseous pollutants. Certain individuals are more susceptible to the negative effects of air pollution than others, particularly those who have heart or chronic respiratory conditions. Exposures of people to toxic gases can occur within several contexts, including the following.

1. **The Ambient Environment:** The urban atmosphere typically contains relatively high concentrations of potentially toxic chemicals. This is true in general, but air quality is especially bad during smog events, often caused by poor dispersion during an atmospheric inversion. Consequently, city people living their normal lives are routinely exposed to higher concentrations of air pollutants than those living in cleaner, rural environments.
2. **The Working Environment:** Many people are exposed to high concentrations of pollutants as a consequence of their occupation. Of course, the specific exposures depend on the job – workers in metal smelters may be exposed to sulfur dioxide and metallic particulates, auto mechanics may be affected by exhaust fumes containing carbon monoxide and hydrocarbons, and laboratory workers may inhale various organic solvents.
3. **The Indoor Environment:** Buildings are often contaminated by gases and fumes. For example, space heaters, furnaces, and fireplaces burning wood, kerosene, or fuel oil may emit carbon monoxide into the indoor environment. All high-temperature combustions emit nitric oxide, and many synthetic materials and fabrics vent formaldehyde and other organic vapors. These chemicals can accumulate if indoor air is not exchanged frequently with cleaner, outdoor air.
4. **Tobacco Smoke:** The smoking of tobacco is a leading source of easily avoidable air pollution. Smoking is also the most important cause of preventable diseases, especially lung cancer and heart disease. People inhale a great variety of toxic gases and fumes when they smoke tobacco (and also marijuana). In addition, non-smokers are indirectly exposed to lower concentrations of those chemicals because of the lingering residues of "second-hand smoke" that may occur in indoor atmospheres.

The health effects on people of all these exposures to air pollutants are significant. The following paragraphs, however, are centered on the pollution of the surrounding urban environment.

Air pollution levels have been high for city dwellers and workers in specific factory types ever since the Industrial Revolution began in Western Europe in the middle of the 18th century. Soot, sulfur dioxide, and other emissions from burning coal and other fossil fuels have been particularly significant. The longest air inversions, which stop emissions from dispersing and produce smog heavy in SO_2 and particulates, are usually when the worst pollution exposures in urban settings take place.

Many places have long used coal to heat their homes and other buildings. Since at least 1500, the related emissions have been considered a concern in European cities and towns. Air pollution significantly increased with the start of the Industrial Revolution, which at first relied mostly on coal for energy. When an inversion occurred in Glasgow, Scotland in 1909 and a toxic haze resulted from it, approximately one thousand people may have died. This was the first clear connection established between air pollution and a significant rise in the death rate of the exposed human population.

Donora, Pennsylvania saw the most notorious "killer smog" in North America in 1948. For four days, the Donora Valley experienced an inversion and fog, but emissions from multiple factories persisted, causing the atmosphere to become extremely high in SO_2 and PM. The local population suffered from higher mortality as a result of the smog (20 deaths in a population of only 14,100). 43% of the population fell ill, with 10% becoming critically ill. The most typical symptoms were respiratory tract and eye irritation, which was occasionally accompanied by headaches, vomiting, and coughing.

When a widespread inversion and fog settled over southern England in 1952, London, England, was hit by the most infamous killer smog in history. Pollutant emissions, primarily from the burning of coal, turned London's naturally occurring "white fog" into a poisonous "black fog." The visibility was appalling; people got lost while driving or walking, and some even fell off wharves into the Thames River. Aircraft also got disoriented while attempting to taxi at the airport. Four days of smog were followed by an unusually high death toll for an additional 14 days. Approximately 3,900 deaths were linked to this toxic pollution incident overall. The majority of those afflicted were either very young or elderly, or they already had heart or respiratory conditions.

Severe instances of urban air pollution were typical in Western European and North American cities up until the early 1960s. The majority of the smogs were brought on by the widespread use of coal in furnaces and fireplaces in residences, businesses, and power utilities. In addition to harming plants, the poor quality of urban air also had an impact on animal and human health. Only specific plant species that can withstand air pollution could grow in many cities. Commonly found in urban Canada are pollution-tolerant trees such as Norway maple (*Acer platanoides),* silver maple (*A. saccharinum*), linden (*Tilia europaea*), ginkgo (*Ginkgo biloba*), and tree-of-heaven (*Ailanthus altissima*). In response to the issues posed by this type of smog, governments implemented laws requiring significant decreases in air pollution emissions, especially in urban areas. For instance, the adoption of numerous federal, provincial, and local air emission laws in Canada has significantly enhanced the quality of urban air. Since the 1960s, laws passed in the United States, Britain, and other wealthy nations have also improved the quality of the air.

The killer smogs were obviously exceptionally bad instances of air pollution. Usually, much lower concentrations of SO_2, NOx, O_3, volatile organic compounds, and particulates pollute the urban atmosphere. Numerous studies have looked into how long-term exposure to those lower levels of air pollution affects people's health. According to certain study findings, the quality of the air in modern cities has deteriorated to the point where it can harm people's health over time, particularly by raising the prevalence of lung conditions, asthma attacks, and ocular irritation. Other research hasn't discovered this to be the case, though. Effective measures have been implemented in Canada and other relatively wealthy countries, anyway. Events of air pollution similar to the ones mentioned above that were clearly dangerous and even fatal no longer happen in those countries, though they might reappear if control measures were loosened.

Unfortunately, poorly managed industrial and urban growth is leading to terrible declines in air quality in the cities of nations with rapidly expanding economies, like Brazil, China, India, Indonesia, and Mexico. These seem to be contemporary tragedies of urban air pollution, though the exact effects on human health are still poorly understood.

3.7.1. Smog in Canadian Cities

LRTAP from urban areas contributes to smog, which is a major issue in many cities and some rural areas. A toxic mixture of pollutants that appears as a brownish-yellow or greyish-white haze is commonly used to describe smog. The key components are the following:

O_3 gas, along with SO_2 and NO_x organic vapors fine particulates (< 10 μm diameter), including acidic droplets of H_2SO_4 and HNO_3, particulates of NH_4NO_3 and $(NH_4)_2SO_4$, and organics from diesel exhaust and other combustion sources.

Because it degrades building surfaces and other materials and is toxic to vegetation, smog is commonly recognized as a primary cause of environmental harm. Many people have been known to get sick and feel uncomfortable as a result of smog. Particularly at risk are young people, the elderly, and those with pre-existing lung or heart conditions (especially emphysema, bronchitis, and asthma). On days with extreme smog, however, even adults in good health may be impacted. Ozone, other gases, and the smallest particulates (<2.5 μm) are the main contributors to toxicity because they can enter the tiniest lung cavities, or alveoli, and cause irritation as well as other issues.

The data demonstrating a link between human diseases and smog, however, are epidemiological; that is, they entail establishing statistical correlations between the prevalence of particular illnesses and the concentrations of air pollutants. For instance, during periods when ozone and/or sulfate particulate concentrations are high, hospital admissions for respiratory illnesses are known to rise in southern Ontario. Even though it is rarely possible to connect a specific disease in an individual to exposure to air pollution, the Ontario Medical Association's statistical estimates suggest that smog causes approximately 9,500 premature deaths and $7.8 billion in health-related expenses in Ontario each year (OMA, 2015).

Governments in Canada have launched programs to track air pollutants and forecast their concentrations to issue "smog alerts" to the public, given the significance of smog as a stressor for Canadians living in urban areas. In cities that are prone to smog, Environment Canada regularly issues advisories in collaboration with provincial and local governments. Typically, these advisories are issued the day before a high level of ozone is predicted. In a similar vein, a few municipalities and provinces have created daily advisories through air-quality indices. The goal is to persuade individuals and organizations to take steps to mitigate air pollution and prevent needless exposure by remaining indoors and refraining from deep-breathing outdoor physical activity.

Ontario saw its worst-ever smog summer in 2005, with air so contaminated by ozone and other chemicals that it was deemed dangerous for human health for 53 days between June and September. The densely populated area's numerous automobiles and other sources of emissions, together with the unusually hot and sunny weather (which encourages the photochemical creation of ozone), and LRTAP from other areas, were the main causes of the smog. It seems that large, densely populated areas of southern Ontario and other parts of Canada are well-established sources of this kind of dangerous pollution (Ayres and Maynard, 2006; Vallero, 2014).

3.7.2. Ecological Damage Caused by Air Pollution

The first example talks about "natural" air pollution at the isolated Arctic location known as Smoking Hills. The second looks at how emissions from big smelters close to Sudbury, Ontario, affect the environment.

3.7.2.1. The Smoking Hills

The Smoking Hills are situated on the Beaufort Sea coast in the Northwest Territories. Bituminous shale seams can be found at different locations along the coast and adjacent rivers. These seams are exposed strata in steep areas where erosion is taking place. The shale has pyritic sulfur, which oxidizes to sulfate when exposed to airborne oxygen and causes the cliffs to erode. Because the reaction is exothermic, the oxidation generates heat, which, in an insulating environment, can raise the temperature to the point where the bituminous materials ignite spontaneously. These catch fire and spew out SO_2, smoldering the tundra close by. The Smoking Hills were first observed in writing in 1826 by explorer John Richardson. Still, the burns are probably thousands of years old and have been known to the local Inuit for a long time.

The SO_2-laden plumes, or air masses, are frequently blown inland at ground level by winds at the Smoking Hills, fumigating the tundra. The point where the plumes start to move inland is where the pollution is at its worst: the cliff's edge. As much as 2 ppm of SO_2 can be found at the cliff edge, and as one moves inland, the concentration drops off almost exponentially quickly. As one gets farther away from the sources of emission, the gases in the ambient atmosphere get increasingly diluted, resulting in this gradient of air pollution.

The soil has become extremely acidic due to SO_2 pollution. Freshwater ponds with acidic conditions can have a pH of 2 or lower, while areas outside of the fumigation area have a pH of 8 or higher. Metals dissolve from minerals in the soil and aquatic sediment due to the extreme acidification. Both aquatic and terrestrial organisms are toxic when exposed to high concentrations of solubilized metals. Furthermore, soil, and water both contain high concentrations of sulfate. This is mostly caused by SO_2 from the atmosphere decomposing on dry surfaces and then oxidizing to SO_1^{2-} inside the ecosystem.

The majority of plants, animals, and microorganisms find the fumigated habitats to be extremely toxic due to the high concentration of SO_2 in the air, as well as the acidity and soluble metals in the soil and water. In fumigated areas, where pollution is at its worst, vegetation never grows near the sea cliff; instead, there is complete ecological degradation. A few plants that can withstand pollution can grow farther inland where the pollution is less severe. Among these, arctic wormwood (*Artemisia tilesii*), polar grass (*Arctagrostis latifolia*), a moss (*Pohlia nutans*), and a lichen (*Cladonia bellidiflora*) are the most notable. Many species of unpolluted tundra, such as arctic willow (*Salix arctica*), mountain avens (*Dryas integrifolia*), and more than 70 other species, have been replaced by these few tolerant plants (Freedman et al., 1990).

Acidic pools at the Smoking Hills are also home to species that can withstand pollution. With pH levels as low as 1.8, even the most acidic ponds can sustain a minimum of six different types of algae. They are not found in non-acidic waterbodies and are incredibly tolerant of acidity and dissolved metals. Rich algal communities with over 90 species are supported in the unpolluted ponds, which have a pH greater than 8 and are alkaline compared to the acidic ponds.

A few other acid-tolerant invertebrates, such as the crustacean *Brachionus urceolaris* and the insect midge *Chronomus riparius*, can also be found in acidic ponds (but only at pH values higher than 2.8). Non-acidic ponds have a far more diverse and productive invertebrate fauna (Havas and Hutchinson, 1983).

The most crucial thing to take away from the Smoking Hills is that ecological harm from "natural" pollution can be just as severe as that caused by emissions from human activity. It is obvious that SO_2 pollution, regardless of its source, can harm ecosystems. Furthermore, ecosystems have been under stress from the Smoking Hills' natural pollution for a very long time—at least thousands of years—and the ecological effects are probably at a steady state now. The study of the Smoking Hills provides some understanding of the long-term effects of severe air pollution:

- It Causes a simplification of biodiversity and ecosystems to occur.
- Productivity and nutrient cycling are impaired.
- Unusual communities of pollution-tolerant species develop.

3.7.2.2. Metal Smelters Near Sudbury

When the Canadian Pacific Railroad was being built in 1883, workers with some prospecting experience found a rich body of metal-bearing ore near Sudbury, Ontario, while blasting through bedrock. Copper and nickel are the two main metals in the ore. But the mines also produce significant amounts of sulfur and selenium, as well as gold, silver, cobalt, iron, and other metals.

The rich ore bodies near Sudbury have prompted the development of one of the largest industrial complexes in the world for mining and processing. The facilities have included sulfuric acid plants, metal refineries, smelters, underground, and open-pit mines, ore-processing mills with tailings-disposal areas, and various other installations. The more than 160 thousand people who live in the Sudbury region rely heavily on the industrial activity in the area as their economic foundation.

As sulfide minerals, the metals in the Sudbury ore combine with sulfur to form compounds like iron sulfide and nickel sulfide. Because this causes the sulfides to turn into gaseous SO_2, roasting the ore at a high temperature in the presence of oxygen is a crucial step in the processing process. The residual material that is left over after roasting has a higher concentration of valuable metals that can be melted and refined into pure metals.

Massive emissions of SO_2 and metal-containing particulates have been released into the atmosphere as a result of the extensive roasting and smelting that has taken place in the Sudbury region, seriously harming the environment. Up until 1928, roasting was done in enormous open pits called "roast beds," which were filled with sulfide ore and covered in a layer of cordwood that was harvested locally. Since the oxidation is an exothermic reaction, when the wood was lit, the heat ignited the metal sulfides, releasing more thermal energy. The ore burned and smoldered in the roast bed for several months until the temperature reached a point where the reactions could no longer be stopped by adding water. The concentrates of copper and nickel were collected and sent to a refinery for additional processing after they had cooled.

The ensuing photos clearly show how this shoddy roasting method caused the terrain to be heavily fumigated at ground level with metallic particles, acidic mist, and toxic SO_2. Ecosystems close to the roast beds were severely harmed by pollution. Due to the massive soil erosion caused by the destruction of terrestrial habitats, the bedrock was exposed. The reaction of the sulfurous fumes caused the bedrock to become pitted and blackened.

The 30 roast beds that were in operation in the Sudbury area released enormous amounts of metallic particulates that were unreported along with roughly 270 thousand tons of SO_2 annually. The majority of the ecological harm in the Sudbury area was brought on by these initial ground-level emissions.

The Ontario government outlawed the use of roast beds in any capacity after 1928. After that, all roasting was done at smelters near Sudbury, in the communities of Coniston, Copper Cliff, and Falconbridge. An enormous structure with roasting chambers inside is called a smelter. Through the use of a smokestack, the majority of their waste gas and particle emissions are released high into the atmosphere, dispersing the pollutants and significantly lowering the level of ground-level pollution in the area.

At Copper Cliff, the biggest smelter was constructed in 1929. It had one smokestack at first, but in 1936 two more were added. The three stacks were replaced in 1972 by a single 381-meter "superstack," which is currently the second-tallest chimney in the world after being the tallest at the time. Coniston Smelter was shut down at the same time, and Copper Cliff became the new location for production. Another business owns the smelter at Falconbridge, which has smokestacks measuring 93 meters and 140 meters. When the superstack was put into service in 1972, pollutants could be released high enough into the atmosphere to prevent frequent occurrences of ground-level fumigations. As a result, the Sudbury region's air quality significantly improved. Elevated smokestacks aid in the dispersal and blending of pollutants into the surrounding atmosphere. This is known as the "pollution dilution solution" at times."

The superstack's emissions are evenly distributed throughout the local atmosphere due to its height. The comparatively good air quality in the area since 1972 is a reflection of the fact that little of the vented SO_2 is deposited locally. Actually, research has shown that the portion of SO_2 emissions deposited within 40 km of the superstack is only about 1%. This indicates that 99% of the SO_2 is exported over a greater distance, preventing local harm but causing precipitation over a wide area to become more acidic. Chemical detection of the superstack plume is possible at least 150 kilometers away.

Other strategies have also been used to lower the emissions of pollutants. Metal-containing dusts are recovered from smelter flue gases using devices like electrostatic precipitators, and SO_2 emissions have been decreased through the installation of wet scrubbers, the construction of sulfuric acid plants, and the construction of a facility to separate iron sulfides from the more valuable minerals of nickel and copper.

The three smelters in the Sudbury area released an average of 2.25 million tons of SO_2 annually between 1960 and 1972 when emissions reached their peak. At the time, Sudbury accounted for roughly 4% of all global releases of SO_2, making it the largest source of anthropogenic emissions in the world. Since then, the smelters' SO_2 emissions have significantly decreased, reaching roughly 0.14 × 106 t/y between 2006 and 2013. The primary cause of the decline is the high cost of investing in technologies for reducing pollution, such as flue-gas desulfurization equipment. Even with a significant reduction, SO_2 emissions are still high.

Compared to the sulfurous past, the Sudbury region's post-superstack air quality is a significant improvement. When roast beds were in use, as well as before 1972 when the smelters had shorter stacks and little pollution-abatement technology, toxic fumigations with SO_2, acidic mists, and metallic particulates were much more frequent and intense. The earlier pollutant emissions were mostly to blame for the worst of the ecological harm in the area. However, additional disruptions increased that harm, such as the clearing of forests for roast bed fuel and the lighting of wildfires by prospectors and the sparks from steam-powered railroad engines. Even though they are still significant, the Sudbury region's current emissions of SO_2 and other pollutants are widely distributed and rarely result in immediate biological harm. Because of this, there has been a significant ecological recovery since the superstack was put into service in 1972.

The roast beds and smelters in the Sudbury region seriously contaminated large tracts of land and surface water. Since the smelters are significant point sources of emissions, the extent of the ecological harm rapidly decreased as one moved farther away. Ecologists have recorded the area's vegetation damage over time. About 360 km^2 of vegetation, including a dearth of conifers in the forest, and about 100 km^2 of "severely barren" land surrounding the smelters were described as such in 1970 (Watson and Richardson, 1972). Due to its sensitivity to air pollution, the economically significant white pine (*Pinus strobus*) tree displayed diagnostic SO_2 injuries over an area of roughly 6,400 km^2.

The largest smelter is located in Copper Cliff, where pollution is particularly harmful. Nearby this facility, there are several kilometers of severely degraded habitats that barely sustain any forest. This zone's hills and slopes have lost a great deal of their vegetation, their soil has been eroded, and the exposed bedrock has turned black as a result of reactions with acidic fumigations. In this region, only a few species of plants, such as some grass species, other herbaceous plants, and some tree species with stunted, shrub-like plants, have evolved populations that can withstand pollution.

There is a gradient in space between the smelters, and the farther away from them the less severe the pollution-related stress becomes, and the less severe the damage to the vegetation. Remains of forest can be found 3–8 km from the Copper Cliff smelter, where the topography offers some protection from the pollution. Even so, this unevenly vegetated area still frequently has blackened and stripped hilltops. The trees in the remaining stands have several dead branches and other wounds, and they are stunted. Relatively pollution-tolerant species among the trees are large-toothed aspen (*Peuma cinerea*), white birch (*Betula papyrifera*), red oak (*Quercus rubra*), trembling aspen (*Populus tremuloides*), and red maple (*Acer rubium*). grandidentata). Beyond 8 km, the forest cover is practically continuous, but the stands' biodiversity and biomass are declining. As is typical of the area, mixed stands of hardwood and conifer trees can be found in the forests beyond 20 to 30 kilometers from the smelters, where pollution has less of an impact (Freedman and Hutchinson, 1980).

Pollution from the atmosphere also seriously damaged lakes near the Sudbury smelters. The region's more than 7000 waterbodies have higher than average concentrations of hazardous nickel, copper, and other metals due to SO_2 deposition, which acidified the water. There are species-poor communities of tolerant plants, algae, and zooplankton in the contaminated water bodies. Mostly because of their acidity, they don't have any fish.

Nonetheless, since the superstack was constructed in 1972, there have been significant advancements in ground-level air quality, which have led to notable ecological recoveries in the Sudbury region. Wet meadows grew where eroded soil gathered in moist basins. *Deschampsia caespitosa* or hairgrass, predominates in these and local populations of it are genetically tolerant to toxic levels of copper and nickel in the soil. The decrease in air pollution has also helped other plants, but residual soil toxicity from metals and acidification continues to hinder their recovery.

In the area, lakes are also making a comeback. Of the 44 lakes that have been observed since 1981, only 6 remained that acidic in 2004 (out of 28 lakes that were highly acidified, to pH 5.0 or less), and 14 had recovered to pH 6.0, a level that would support the majority of the aquatic biota typical of the area (Keller et al., 2007,). Smelter-related metal concentrations, like those of copper and nickel, are likewise significantly lower, despite being higher than background levels. Baby Lake had a pH of 4.0 to 4.2 in 1972, almost no algae, and neither fish nor invertebrates were present (Havas et al., 1995). But by 1985, it had recovered to pH 61.8, and by 1995, to pH 70.2, with metal concentrations also declining. The lake's water quality improved to the point where a variety of aquatic plants, invertebrates, phytoplankton, and even tiny fish were able to settle there.

Various management techniques have aided in the ecological recovery of some degraded areas since the early 1970s. The most significant of these has been the liming of soil to lessen acidity and, consequently, the metal toxicity. The addition of fertilizer, the planting of tree seedlings, and the sowing of grasses and other plants known to be tolerant of the toxic conditions have also been significant. Comparable initiatives have included liming lakes to lessen their acidity and metal toxicity, which has aided in the biota's recovery. The Sudbury region's damaged habitats have significantly improved thanks to these reclamation efforts and the natural regeneration that has happened as a result of the pollution levels having significantly dropped since the superstack was constructed.

The Sudbury case is arguably the most well-documented instance of ecological harm brought about by metals and hazardous gases released by smelters worldwide. Still, there are more instances of this kind of damage in the vicinity of other Canadian smelters that impact smaller areas, like the smelters at Flin Flon, MB, Rouyn-Noranda, QC, Trail, BC, and Yellowknife, NT.

Each of these smelters is a sizable point source of pollution emissions into the atmosphere. All of them experienced distinct spatial gradients in pollution intensity, which diminished exponentially as one got farther away from the emission source until the ambient condition was attained. These spatial gradients of toxic stress are tracked by the patterns of ecological damage.

Numerous sources, from massive power plants and smelters to individual cars and home furnaces, release gaseous air pollutants like sulfur dioxide and nitric oxide into the atmosphere. On the other hand, secondary pollutants like ozone are created in the atmosphere through photochemical reactions that combine sunlight with released hydrocarbons and nitrogen oxides. Gaseous pollutants, which are frequently combined with particulates, have the potential to seriously harm the environment and human health if their concentrations are high enough. Many governments have taken action to lower the emissions of the most significant air pollutants because these risks are now widely recognized. This is especially true for nations that are still developing, like Canada. Rich countries still emit a lot of air pollutants, but overall, these emissions are stabilizing or even declining. But in economies that are expanding quickly, like China and India, hurried, and careless industrialization is making air pollution worsen quickly (Ayres and Maynard, 2006; Vallero, 2014).

3.8. HEALTH AND ENVIRONMENTAL EFFECTS OF PRIMARY POLLUTANTS

"Primary pollutants" are those that are released straight into the atmosphere as a result of combustion, or as byproducts of the process. After describing these products, let's examine how they affect the environment and public health.

3.8.1. Carbon Dioxide (CO_2) Effects

Although carbon dioxide (CO_2) is a known greenhouse gas, it is not a pollutant in the sense that it would adversely affect human health. It has the capacity to absorb infrared radiation, which warms the atmosphere, as it escapes from the earth's surface. The connection between excessive CO_2 emissions and other greenhouse gas emissions and climate change is becoming more and more obvious.

3.8.2. Carbon Monoxide (CO) Effects

Carbon monoxide, also known as CO, is a colorless, odorless, and tasteless gas that is produced when fuel's carbon is not burned all the way through.

At much higher levels of exposure not commonly found in ambient air, CO can be poisonous, and even healthy individuals can be affected. Exposure to elevated levels of CO may result in:

1. Visual impairment.
2. Reduced work capacity.
3. Reduced manual dexterity.
4. Poor learning ability.
5. Difficulty in performing complex tasks.

The health risk posed by occasionally detectable CO levels in the ambient air is greatest for people with cardiovascular conditions like angina pectoris.

3.8.3. Sulfur Dioxide (SO_2) Effects

High concentrations of SO_2 can result in the following health problems:

1. Short-Term Exposure

i. Adults and children with asthma who are active outdoors will experience temporary breathing impairment.

ii. Individuals with asthma may experience breathing difficulties with moderate activity and may exhibit symptoms such as wheezing, chest tightness, or shortness of breath.

2. Long-Term Exposure (Along With High Levels of PM)

i. Aggravation of existing cardiovascular disease.

ii. Respiratory illness.

iii. Alterations in the lungs' defenses.

Children and the elderly are among the subgroups of the population that may be impacted by these conditions, as well as people with heart or lung diseases.

The main precursors of acidic deposition, or acid rain, which is linked to the acidification of soils, lakes, and streams as well as the accelerated corrosion of structures like buildings and monuments, are SO_2 and NOx combined. In addition, SO_2 is a major precursor to PM 2.5, a major source of visibility impairment and a health concern.

3.8.4. Nitrogen Oxides (NO_x) Effects

NOx is a representation of both nitrogen dioxide (NO_2) and nitric oxide (NO). Approximately 90% of the emissions originating from combustion devices are NO.

NO_x reacts in the air to form ground-level ozone and fine particulates, which are associated with adverse health effects.

Short-term exposures (e.g., less than 3 hours) to low levels of NO_2 may lead to changes in airway responsiveness and lung function in individuals with preexisting respiratory illnesses. These exposures may also increase respiratory illnesses in children. Long-term exposures to NO_2 may lead to increased susceptibility to respiratory infection and may cause irreversible alterations in lung structure. NO_x contributes to a wide range of environmental effects directly and when combined with other precursors in acid rain and ozone.

Increased nitrogen inputs to terrestrial and wetland systems can lead to changes in plant species composition and diversity.

Direct nitrogen inputs to aquatic ecosystems such as those found in estuarine and coastal waters (e.g., Chesapeake Bay) can lead to eutrophication (a condition that promotes excessive algae growth, which can lead to a severe depletion of dissolved oxygen and increased levels of toxins harmful to aquatic life). Nitrogen, alone or in acid rain, also can acidify soils and surface waters.

3.8.5. Particulate Matter (PM) Effects

Particles with a diameter of less than or equal to 10 µm, also known as a micrometer or millionth of a meter, have the potential to enter the lungs and cause a variety of health issues. These minuscule particles have been connected to lung and heart disease-related illnesses and deaths. Numerous health issues have been linked to long-term (e.g., exposure to these particles over a period of years. Daily peak with a shorter duration and possibly an even shorter term (e.g., Exposure to these particles for one hour or longer may also be linked to health issues.

Particles have been linked to cardiac arrhythmias, or irregular heartbeats, and heart attacks. They can exacerbate respiratory disorders like asthma and bronchitis. Children, the elderly, and those with heart or lung conditions are most vulnerable to particle exposure.

Both fine and coarse-fraction particles can be considered harmful particles, but fine particles have been more strongly associated with the most detrimental effects on health. Particles larger than 2 micro meters (µm) do not penetrate beyond the nasal cavity or trachea. Particles smaller than 0.1 µm tend to deposit in tracheobronchial tree and are removed when exhaling. Particles between 0.1 and 2.0 µm penetrate deep into the lungs and settle in respiratory bronchioles or alveolar sacs.

3.8.5.1. Health Effects

The likelihood that a particle will cause health issues is directly correlated with its size. The most dangerous particles are those with a diameter of less than 10 micrometers because they can enter your bloodstream and deep into your lungs. Exposure to such particles can affect both your lungs and your heart. Numerous scientific studies have linked particle pollution exposure to a variety of problems, including:

1. Premature death in people with heart or lung disease.
2. Nonfatal heart attacks.
3. Irregular heartbeat.
4. Aggravated asthma.
5. Decreased lung function.
6. Increased respiratory symptoms, such as irritation of the airways, coughing or difficulty breathing.

Due to increased sensitivity or potential higher exposures, individuals with heart or lung diseases, children, older adults, minority populations, and low socioeconomic status populations are most likely to be impacted by particle pollution exposure.

3.8.5.2. Environmental Effects

1. **Visibility Impairment:** Reduced visibility, or haze, is mostly caused by fine particles ($PM_{2.5}$) in many parts of the United States, including many of our beloved national parks and wilderness areas.
2. **Environmental Damage:** Particles can be carried over long distances by wind and then settle on ground or water. Depending on their chemical composition, the effects of this settling may include:

   i. Making lakes and streams acidic.

   ii. Changing the nutrient balance in coastal waters and large river basins.

   iii. Depleting the nutrients in soil.

   iv. Damaging sensitive forests and farm crops.

   v. Affecting the diversity of ecosystems.

   vi. Contributing to acid rain effects.
3. **Materials Damage:** Stone and other materials, including culturally significant items like monuments and statues, can be stained or damaged by PM. The effects of acid rain on materials are connected to a few of these effects.

3.8.6. Lead Effects

Lead exposure mostly happens when a person breathes it in or consumes lead-contaminated food, water, soil, or dust. It can have a negative impact on the kidneys, liver, neurological system, and other organs. It builds up in the blood, bones, and soft tissues.

Excessive exposure to lead may cause neurological impairments such as seizures, mental retardation, and behavioral disorders.

Even at low doses, lead exposure is associated with damage to the nervous systems of fetuses and young children, resulting in learning deficits and lowered IQ. Recent studies indicated that lead may be a factor in high blood pressure and subsequent heart disease. Lead can also be deposited on the leaves of plants, presenting a hazard to grazing animals and to humans through ingestion (Brimblecombe, 2005).

REFERENCES

1. Anderson, S. O., & Sarma, K. M. (2005). *Protecting the Ozone Layer: The United Nations History.* Earthscan Publications.
2. Ayres, J., & Maynard, R. L. (2006). *Air Pollution and Health.* World Scientific Publishing Company.
3. Barker, J. R., & Tingey, D. T. (1992). *Air Pollution Effects on Biodiversity.* Van Nostrand Reinhold.
4. Brimblecombe, P. (1996). *Air Composition and Chemistry* (2nd ed.). Cambridge University Press.
5. Brimblecombe, P. (2005). *The Effects of Air Pollution on the Built Environment.* Imperial College Press.
6. Canadian Council of Ministers of the Environment. (2014). *Canadian Environmental Quality Guidelines.* CCME. Retrieved from: http://www.ccme.ca/en/resources/canadian_environmental_quality_guidelines/index.html (accessed on 14 December 2024).
7. Environment Canada. (1999). *Ground-Level Ozone: Occurrence and Transport in Eastern North America.*
8. Environment Canada. (2013). *Canadian Ambient Air Quality Standards.* Retrieved from: http://www.ec.gc.ca/default.asp?lang=En&n=56D4043B-1&news=A4B2C28A-2DFB-4BF4-8777-ADF29B4360BD.
9. Environment Canada. (2014). *National Pollutant Release Inventory and Air Pollutant Emission Summaries and Trends.* Retrieved from: http://www.ec.gc.ca/inrp-npri/default.asp?lang=en&n=0EC58C98-1#Emission_Summaries.
10. Environmental Protection Agency (U.S.). (2014). *Air Trends.* Retrieved from: www.epa.gov/airtrends/ (accessed on 14 December 2024).
11. Freedman, B. (1995). *Environmental Ecology* (2nd ed.). Academic Press.
12. Freedman, B., & Hutchinson, T. C. (1980). Long-term effects of smelter pollution at Sudbury
13. Freedman, B., Zobens, V., & Hutchinson, T. C. (1990). Intense natural pollution affects arctic tundra vegetation at the Smoking Hills, Canada. *Ecology, 71*, 492–503.
14. Gizyn, W. (1980). *The Chemistry and Environmental Impact of the Bituminous Shale Fires at the Smoking Hills, NWT* (Master's thesis). University of Toronto, Toronto, ON.
15. Gunn, J. M. (1995). *Restoration and Recovery of an Industrial Region: Progress in Restoring the Smelter-Damaged Landscape Near Sudbury, Canada.* Springer.

16. Guxens M., Lubczyńska M.J., Muetzel R.L., Dalmau-Bueno A., Jaddoe V.W., Hoek G., van der Lugt A., Verhulst F.C., White T., Brunekreef B., et al. Air Pollution Exposure during Fetal Life, Brain Morphology, and Cognitive Function in School-Age Children. Biol. Psychiatry. 2018;84:295–303. doi: 10.1016/j.biopsych.2018.01.016.
17. Havas, M., & Hutchinson, T. C. (1983). The Smoking Hills: Natural acidification of an aquatic ecosystem. *Nature, 301*, 23–27.
18. Havas, M., Woodfine, D. G., Lutz, P., Yung, K., MacIsaac, H. H., & Hutchinson, T. C. (1995). Biological recovery of two previously acidified, metal-contaminated lakes near Sudbury, Ontario, Canada. *Water, Air, and Soil Pollution, 85*, 791–796.
19. Hemond, H. F., & Fechner, E. J. (2014). *Chemical Fate and Transport in the Environment* (3rd ed.). Academic Press.
20. Hester, R. E., & Harrison, R. M. (1998). *Air Pollution and Health.* Royal Society of Chemistry.
21. Holgate, S. T., Samet, J. M., Maynard, R. L., & Koren, H. S. (1999). *Air Pollution and Health.* Academic Press.
22. Keller, W., Yan, N. D., Gunn, J. M., & Heneberry, J. (2007). Recovery of acidified lakes: Lessons from Sudbury, Ontario, Canada. *Water, Air, and Soil Pollution: Focus, 7*, 317–322.
23. Maynard, R. L., & Brimblecombe, P. (2001). *The Urban Atmosphere and its Effects.* World Scientific Publishing Company.
24. National Weather Service (U.S.). (2015). *Meteorological Conditions and Ozone in the Polar Stratosphere.* NWS. Retrieved from: www.cpc.ncep.noaa.gov/products/stratosphere/polar/polar.shtml (accessed on 14 December 2024).
25. Netherlands Environmental Assessment Agency (NEAA). (2015). *Emission Database for Global Environmental Research (EDGAR).* NEAA. Retrieved from: http://themasites.pbl.nl/tridion/en/themasites/edgar/emission_data/edgar_32ft2000/index-2.html (accessed on 14 December 2024).
26. Ontario Medical Association (OMA). (2015). *Smog/Air Pollution: Resources.* OMA. Retrieved from: https://web.archive.org/web/20160914204453/https://www.oma.org/HealthPromotion/Smog/Pages/default.aspx (accessed on 14 December 2024).
27. Ontario Ministry of the Environment and Climate Change. (2014). *Summary of Smog Advisories.* OMECC. Retrieved from: http://www.airqualityontario.com/press/smog_advisories.php (accessed on 14 December 2024).

28. Ontario, on forest community composition. *Canadian Journal of Botany, 58*, 2123–2140.
29. Parson, E. (2003). *Protecting the Ozone Layer: Science and Strategy.* Oxford University Press.
30. Pepper, I. L., Gerba, C. P., & Brusseau, M. L. (2006). *Pollution Science* (2nd ed.). Academic Press.
31. Roberts, T. M. (1984). Effects of air pollutants in agriculture and forestry. *Atmospheric Environment, 18*, 629–652.
32. Seinfeld, J. H., & Pandis, S. H. (2006). *Atmospheric Chemistry and Physics: From Air Pollution to Climate Change* (2nd ed.). Wiley-Interscience.
33. Shriner, D. S. (1990). Responses of vegetation to atmospheric deposition and air pollution. In *Acidic Deposition: State of Science and Technology* (Vol. III: Terrestrial, Materials, Health, and Visibility Effects). Superintendent of Documents, U.S. Government Printing Office.
34. Vallero, D. (2014). *Fundamentals of Air Pollution* (5th ed.). Academic Press.
35. Wark, K. Jr., Warner, C. F., & Davis, W. T. (1998). *Air Pollution: Its Origin and Control* (3rd ed.). Addison-Wesley.
36. Watson, W. Y., & Richardson, D. H. (1972). Appreciating the potential of a devastated land. *Forestry Chronicle, 48*, 312–315.
37. Winterhalder, E. K. (1978). A historical perspective of mining and reclamation in Sudbury. *Proceedings of the 3rd Annual Meeting, Canadian Land Reclamation Association.*
38. Wise, W. (2001). *Killer Smog: The World's Worst Air Pollution Disaster.* Author's Guild.

CHAPTER 4

NATURAL TRACE COMPONENT OF THE ATMOSPHERE

Contents

INTRODUCTION

Despite their low concentration, natural trace components in the atmosphere have a significant impact on the atmosphere's chemistry. The study of the variable trace components of the atmosphere has come to be known as atmospheric chemistry. The main naturally occurring components of air are carbon dioxide, nitrogen, and oxygen. Additionally, the atmosphere contains a few naturally occurring trace gases.

The atmosphere contains almost all of the elements listed in the periodic table; however, it is most practical to use a limited number of major groupings, such as compounds containing sulfur, nitrogen, carbon, or halogen, when categorizing atmospheric species based on their chemical composition.

These classifications are obviously not mutually exclusive; for instance, carbon atoms can be found in many compounds containing sulfur. Furthermore, the backbone of almost all atmospheric halogens is carbon. The species with the general formula H_xO_y are not included in the previous list because, aside from water and hydrogen peroxide (H_2O_2), they are all radical species (e.g., hydroxyl, OH), which are important to atmospheric chemistry but do not always need their own category. All substances released into the atmosphere are eventually taken out, establishing a cycle of elements within that substance. This is referred to as the element's biogeochemical cycle. An element or compound's movement through the atmospheric, oceanic, biospheric, and land compartments, as well as the amount and rate of exchange between various reservoirs, are all included in its biogeochemical cycle. The global or regional cycles of the life elements—C, O, N, S, and P—with reservoirs including the atmosphere, the ocean, sediments, and living things are referred to as biogeochemical cycles.

When substances resulting from human activity are present at concentrations high enough above their normal ambient levels to have a detectable impact on people, animals, vegetation, or materials, it is said to be an instance of air pollution. This definition could apply to any substance, harmful or not, but it implies that the results will not be pleasant. It was once believed that large cities and industrialized areas were the only places where air pollution occurred. It is now evident that the Earth's dense urban areas are merely hotspots within a continuum of trace species concentrations. Air pollution can be broadly defined as both urban smog and the depletion of ozone in the stratosphere caused by chlorofluorocarbons.

4.1. ATMOSPHERIC LIFETIME

Consider a scenario in which we could track every single molecule that a substance released into the atmosphere. By coming into contact with airborne droplets or the Earth's surface, some may be eliminated near their point of emission. Some may be propelled high into the atmosphere and moved a considerable distance before being eliminated. An average substance's lifetime or average residence time can be calculated by averaging the life histories of all of its molecules. This residence time indicates the length of time, on average, that a representative molecule of the substance will remain in the atmosphere before being eliminated. Precipitation and the Earth's surface are the two final exits from the atmosphere. When released into the atmosphere, species must eventually depart via one of these two paths.

Dry deposition and wet deposition are the two easy categories into which atmospheric species removal processes can be divided. Dry deposition occurs when species, both gaseous and particulate, are directly transferred to the Earth's surface without the assistance of precipitation. As opposed to this, wet deposition refers to all of the following processes that bring airborne species into contact with the Earth's surface in an aqueous form (i.e., rain, snow, or fog): (1) dissolution of atmospheric gases in airborne droplets, such as cloud droplets, rain, or fog; (2) removal of atmospheric particles when they act as nuclei for the condensation of atmospheric water to form a cloud or fog droplet and are then incorporated in the droplet; and (3) removal of atmospheric particles when the particle collides with a droplet both within and below clouds.

All substances, with the exception of pure water, that are microscopic or submicroscopic in size but larger than molecules and that are present in the atmosphere as a liquid or solid under typical circumstances are referred to as PM. PM is the most complex component of the atmosphere. In addition to direct particle emissions, certain gas emissions can also result in airborne PM. These gases can either directly condense as particles or change chemically into a species that condenses as a particle. Not only must the concentration of atmospheric particles be specified, but also their size, chemical makeup, phase (i.e., liquid or solid), and morphology.

When particles enter the atmosphere, a variety of processes alter their size, quantity, and chemical makeup until eventually natural processes remove them. There are physical and chemical processes that are more effective in one particle size range than another that influence the "aging" of atmospheric particles. The typical residence time of particles in the lower atmosphere is not more than a few weeks, despite the specific processes that affect particulate aging. The primary mechanisms for removing particles very close to the ground are settling and dry deposition on surfaces; precipitation scavenging is the main mechanism for removing particles above about 100 meters above the ground.

Cloud droplets are created on condensation nuclei, which are typically soluble aerosol particles (e.g., microscopic particles of different salts) that exist in the atmosphere at concentrations of 100–3000 cm^{-3}, as air rises through a cloud and becomes slightly supersaturated with water vapor (i.e., as its relative humidity exceeds 100%).

Cloud droplets then grow through condensation of water vapor. The droplets expand and collide to form raindrops, which quickly grow as they descend and gather more cloud droplets. Conservation of mass is the fundamental physical principle that dictates how a species behaves in the atmosphere. Any hypothetical volume of air must allow for the following balance to hold:

$$\begin{bmatrix}\text{Rate of}\\ \text{accumulation}\\ \text{of species in}\\ \text{imaginary volume}\end{bmatrix} = \begin{bmatrix}\text{Rate of the}\\ \text{species}\\ \text{flowing in}\end{bmatrix} - \begin{bmatrix}\text{Rate of}\\ \text{species}\\ \text{flowing out}\end{bmatrix} + \begin{bmatrix}\text{Rate of}\\ \text{introduction of}\\ \text{species}\end{bmatrix} - \begin{bmatrix}\text{Rate of}\\ \text{removal of}\\ \text{species}\end{bmatrix}$$

This balance must hold from the smallest volume of air all the way up to the entire atmosphere.

If we let Q denote the total mass of the substance in the volume of air; F_{in} and F_{out} the mass flow rates of the substance in and out of the air volume, respectively; P the rate of introduction of the species from sources; and R the rate of removal of the species, then conservation of mass can be expressed mathematically as:

$$\frac{dQ}{dt} = (F_{in} - F_{out}) + (P - R) \qquad (4.1)$$

If the amount Q of the substance in the volume or reservoir is not changing with time, then Q is a constant and dQ/dt= 0. In order for Q to be unchanging, all the sources of the substance to the reservoir must be precisely balanced by the sinks of the substance. This means that

$$F_{in} + P = F_{out} + R \qquad (4.2)$$

In such a case steady-state conditions are said to hold.

If the volume we are referring to is the entire atmosphere, then $F_{in} = 0$ and $F_{out} = 0$, and for a substance at steady-state conditions, its rate of injection from sources must equal its rate of removal, P= R. The average residence time or lifetime τ, in terms of the quantities introduced earlier, is

$$\tau = \frac{Q}{R + F_{out}} \qquad (4.3)$$

Since at steady-state conditions $R + F_{out} = P + F_{in}$, the lifetime is also given by

$$\tau = \frac{Q}{P + F_{in}} \qquad (4.4)$$

If the entire atmosphere is taken as the reservoir, then under steady-state conditions

$$\tau = \frac{Q}{R} = \frac{Q}{P} \tag{4.5}$$

As an illustration of the concept of mean lifetime, consider all sulfur-containing compounds in the troposphere. If the average mixing ratio of these compounds is 1 part per billion by mass (ppbm) and a steady state is assumed to exist, then with the mass of the troposphere about 4×10^{21} g, the total mass of sulfur-containing compounds in the troposphere is $Q = 4 \times 10^{12}$ g. If natural and anthropogenic sources of sulfur contribute to give a total P of about 200×10^{12} g yr^{-1}, the mean lifetime of sulfur compounds in the troposphere is estimated to be:

$$\tau = \frac{4 \times 10^{12}\ \text{g}}{200 \times 10^{12}\ \text{g yr}^{-1}} = 1\ \text{week}$$

Calculations of lifetimes can be useful in estimating how far from its source a species is likely to remain airborne before it is removed from the atmosphere.

Often, a species' rate of extinction is directly correlated with its concentration (first-order loss); the more of a species there is, the faster it is likely to disappear from the atmosphere. Both dry deposition at the Earth's surface and cloud droplet scavenging typically follow this pattern. Consider a species for which there are steady-state conditions and a proportionality constant λ that determines the rate at which the species is eliminated. One important removal process for such a species, ^{85}Kr, is radioactive decay. For ^{85}Kr, then:

$$\tau = \frac{Q}{\lambda Q} = \frac{1}{\lambda} \tag{4.6}$$

Thus, knowing the radioactive decay constant alone is sufficient to estimate the lifetime of ^{85}Kr, without even needing to know its atmospheric abundance Q. As a result, estimating the lifetime of ^{85}Kr does not depend on whether it is evenly distributed throughout the atmosphere.

Even for poorly mixed species, a straightforward and precise lifetime estimate can be obtained in a first-order removal process—as long as the removal rate constant for the species can be determined with sufficient accuracy.

Let us now consider a species of mass Q in the atmosphere that is removed at two different rates from the atmosphere: k_1Q for the first process and k_2Q for the second, where k_1 and k_2 are the first-order removal coefficients. Its overall lifetime is given by:

$$\tau = \frac{Q}{(k_1 + k_2)Q} = \frac{1}{k_1 + k_2} \tag{4.7}$$

or

$$\frac{1}{\tau} = k_1 + k_2 \tag{4.8}$$

Process 1, for example, could be dry deposition at Earth's surface and process 2 cloud scavenging. We can actually associate time constants with the two individual removal processes,

$$\tau_1 = \frac{1}{k_1}, \quad \tau_2 = \frac{1}{k_2} \tag{4.9}$$

where τ_1 can be regarded as the lifetime of the species against removal process 1, and τ_2 the lifetime against process 2. From (4.8) and (4.9) we can express the overall lifetime τ in terms of the two individual removal times τ_1 and τ_2 by

$$\frac{1}{\tau} = \frac{1}{\tau_1} + \frac{1}{\tau_2} \tag{4.10}$$

Equation (4.10) shows that separate removal paths add together to give a total lifetime, like electrical resistances in parallel add to give a total resistance that is even smaller than the smallest resistance. From (4.10)

$$\tau = \frac{\tau_1 \tau_2}{\tau_1 + \tau_2} \tag{4.11}$$

If $\tau_1 \gg \tau_2$, the lifetime associated with removal by process 1 is much longer than that associated with process 2, process 2 is the more effective removal mechanism, and $\tau \cong \tau_2$. Thus, when there are several competing removal paths, in order to estimate the overall lifetime of a species, focus should always be on improving estimates for the fastest removal rate.

Removal in the troposphere of those compounds that react with the hydroxyl radical occurs according to the chemical reaction:

$$OH + A \xrightarrow{k} \text{products}$$

The parameter k is the rate constant for the reaction, such that the rate of the reaction is k[OH][A], where the brackets denote the concentration of the species contained therein.

From the analysis of atmospheric residence times, if Q is the total quantity of species A in the troposphere and its rate of removal is R =k[OH]Q, then the compound's mean lifetime is, from (4.6)

$$\tau = \frac{Q}{k[\mathrm{OH}]Q} = \frac{1}{k[\mathrm{OH}]} \qquad (4.12)$$

where [OH] is an appropriate globally averaged tropospheric concentration of OH radicals.

Let us develop the equations governing the total moles of a species i in the troposphere Q_i. The dynamic material balance can be written as

$$\frac{dQ_i}{dt} = P_i - R_i \qquad (4.13)$$

where P_i and R_i represent the source and loss rates. The terms P_i and R_i consist of the following contributions:

$$P_i \begin{cases} P_i^n & \text{natural emissions} \\ P_i^a & \text{anthropogenic emissions} \\ P_i^c & \text{chemical reactions} \end{cases}$$

$$R_i \begin{cases} R_i^d & \text{dry deposition} \\ R_i^w & \text{wet deposition} \\ R_i^c & \text{chemical reactions} \\ R_i^t & \text{transport to the stratosphere} \end{cases}$$

The loss processes are usually represented as first order; for example, $R_i^d = k_i^d Q_i$, where the first- i order rate constants, which we will denote by k's, must be specified. Thus (4.13) becomes

$$\frac{dQ_i}{dt} = P_i^n + P_i^a + P_i^c - (k_i^d + k_i^w + k_i^c + k_i^t)Q_i \qquad (4.14)$$

If the concentration of the species is not changing, then a steady state may be presumed in which

$$P_i^n + P_i^a + P_i^c - (k_i^d + k_i^w + k_i^c + k_i^t)Q_i = 0 \qquad (4.15)$$

The lifetime of species i can be calculated by either

$$\tau_i = \frac{1}{k_i^d + k_i^w + k_i^c + k_i^t} \qquad (4.16)$$

or

$$\tau_i = \frac{Q_i}{P_i^n + P_i^a + P_i^c} \qquad (4.17)$$

To use (4.16) the individual first-order rate constants for removal must be estimated, whereas in (4.17), estimates for the total mass or moles in the troposphere, which can be derived from a concentration measurement, and for the source strength terms are needed. If the k_i values are difficult to specify, mean residence times are often estimated from (4.17).

4.1.1. Characteristic Mixing Time in Atmospheric Compartments

The amount of time required to completely mix a chemical in a given volume of air can be defined as the characteristic mixing time for that volume if we focus on a specific area of the atmosphere, such as the air over a city, the air in the Northern Hemisphere, or the entire stratosphere. Designate τ_M as the characteristic mixing time. If the characteristic mixing time τ_M of a species is not small in relation to its residence time τ, then the reservoir is not well mixed for that species. Keep in mind that, depending on each species' residence period, a given reservoir may be well mixed for some species and poorly mixed for others. Furthermore, as we've seen, there are variations in the atmospheric mixing times for various directions. For instance, the typical vertical mixing time in the troposphere, or the amount of time needed to evenly mix a species from the surface to the tropopause, is approximately one week; in contrast, the characteristic horizontal mixing time in the troposphere, or the amount of time needed to thoroughly mix a constituent worldwide in the troposphere, is approximately one year. Therefore, the troposphere can be said to be well mixed for ^{85}Kr, which has a residence time of 10 years; however, the troposphere is not even well mixed vertically for sulfur compounds, which are thought to have a residence time of roughly one week.

4.2. SULFUR-CONTAINING COMPOUNDS

Less than 500 parts per million by mass (<500 ppmm) of sulfur are found in the Earth's crust, and less than 1 ppm of sulfur is present in the atmosphere overall. Nonetheless, substances containing sulfur have a significant impact on both the climate and the chemistry of the atmosphere.

The atmospheric sulfur compounds are listed in Table 4.1. The main sulfur compounds found in the atmosphere are SO_2, CS_2, H_2S, and CH_3SCH_3. There are five different oxidation states of sulfur in the atmosphere. The sulfur oxidation state of atmospheric sulfur compounds has an inverse relationship with their chemical reactivity. Reduced sulfur compounds, or those with an oxidation state of –2 or –1, have a few-day atmospheric lifetime because they are quickly oxidized by the hydroxyl radical and, to a lesser extent, by other species. Sulfur species become more soluble in water as their oxidation state increases; reduced sulfur species are more frequently found in the gas phase, while S(+6) compounds are more frequently found in particles or droplets. The removal of sulfur species by wet and dry deposition determines their residence times once they have been converted to compounds in the S(+6) state.

Table 4.1. *Microbes and Sulfur Association*

Tarting compound (reactant)	**Intermediate or end compound (products)**	**Organism**	**Average ^{34}S fractionation (product/reactant)**	**Details**	**Oxidant**	**References**
Sulfide	Sulfate	*T. neopolitanus, T. intermedius* and *T. ferrooxidans* (chemolithotrophs)	−2 to −5.5‰ (no temperature provided)	Aerobic pH 5 to 6	Carbon dioxide	Toran (1986)
	Polythionates ($S_nO_6^{2-}$) Elemental sulfur, Sulfate	*T. concretivorus* (chemolithotroph)	0.6 to 19‰ (30°C) −2.5 to 1.2‰ (30°C) −18 to −10.5‰ (30°C)	Aerobic	Carbon dioxide	Kaplan and Rittenberg (1964)
	Sulfate	*A. ferrooxidans* (chemolithotroph)	−1.5‰ −4‰ (no temperature provided)	Aerobic Anaerobic	Carbon dioxide	Thurston et al. (2010)
	Sulfate	*T. denitrificans* (chemolithotroph) *Sulfurimonas denitrificans* (chemolithotroph)	−4.3 to −1.3‰ (30°C) −2.9 to −1.6‰ (28°C)	Anaerobic, coupled to DNR, SQR pathway Anaerobic, coupled to DNR, Sox pathway	Carbon dioxide	Poser et al. (2014)
	Sulfate	*Thiomicrospira* sp. strain CVO (chemolithotroph)	1‰ (no temperature provided)	Anaerobic, coupled to DNR, no intermediates in complete oxidation of sulfide to sulfate (potentially only uses Sox pathway)	Carbon dioxide	Hubert et al. (2009)
	Elemental sulfur	*Chlorobium thiosulfatophilum* (green sulfur bacteria)	5‰ (no temperature provided)	Anaerobic	Carbon dioxide	Kelly et al. (1979)
	Thiosulfate	*Oscillatoria sp.* (Cyanobacteria) *Calothrix sp.* (Cyanobacteria)	0‰ (30°C)	Anaerobic, anoxygenic photosynthesis	Carbon dioxide	Habicht et al. (1988)

	Elemental sulfur, Sulfate	*Chromatium vinosum* (purple sulfur bacteria)	0‰ (30–35°C) 2‰ (30–35°C)	Anaerobic, anoxygenic photosynthesis		Fry et al. (1985)
	Elemental sulfur, Sulfate	*Ectothiorhodo-spira shaposh-nikovii* (purple sulfur bacteria)	±5‰ (no temperature provided)	Anaerobic, anoxygenic photosynthesis		Ivanov et al. (1976)
	Polythion-ates ($S_nO_6^{2-}$), Elemental sulfur, Sulfate	*Chromatium* sp. (purple sulfur bacteria)	4.9 to 11.2‰ (30°C) –10 to –3.6‰ (30°C) –2.9 to –0.9‰ (30°C)	Anaerobic		Kaplan and Rittenberg (1964)
Thiosulfate	Sulfate	*T. interme-dius* (chemoli-thotroph)	–4.7‰ (no temperature provided)	Aerobic		Kelly et al. (1979)
	Sulfate	*T. versutus* (chemoli-thotroph)	0‰ (28°C)	Aerobic		Fry et al. (1986)
	Elemental sulfur + Sulfate	*Chromatium vinosum* (purple sulfur bacteria)	0‰ (30–35°C)	Anaerobic		Fry et al. (1985)
	Sulfate	*Desulfovibrio sulfodismutans* (chemoli-thotroph) *D. thiozymogenes* (chemoli-thotroph; "cable bacteria")	For both bacte-ria: 0‰ (30°C; compared to the sulfonate functional group); 2 to 4‰ (30°C; compared to the sulfane functional group)	Anaerobic, dis-proportionation		Habicht et al. (1988)
Elemental sulfur	Sulfate	*Desulfocapsa thiozymogenes* (chemoli-thotroph; "cable bacteria") Enrichment culture	17.4‰ (28°C) 16.6‰ (28°C)	Anaerobic, disproportion-ation, in the presence of iron scavengers		Böttcher et al. (2001)

		Desulfocapsa sulfoexigens *Desulfocapsa thiozymogenes* (chemolithotrophs; "cable bacteria") *Desulfobulbus propionicus* (chemoorganotroph) Marine enrichments and sediments	16.4‰ (30°C) 17.4‰ (30°C) 33.9‰ (35°C) 17.1 to 20.6‰ (28°C)	Anaerobic, disproportionation		Canfield et al. (1998)
		Desulfocapsa thiozymogenes (chemolithotroph; "cable bacteria") Enrichment culture	−0.6 to 2.0‰ (28°C) −0.2 to 1.1‰ (28°C)	Anaerobic, disproportionation, attenuated isotope effect due to reoxidation by manganese oxides		Böttcher and Thamdrup (2001)
Sulfite	Sulfate	*Desulfovibrio sulfodismutans* (chemolithotroph) *D. thiozymogenes* (chemolithotroph; "cable bacteria")	9 to 12‰ (30°C) 7 to 9‰ (30°C)	Anaerobic, disproportionation		Habicht et al. (1988)

Source: https://en.wikipedia.org/wiki/Microbial_oxidation_of_sulfur.

4.2.1. Oxidation State

In covalent compounds, the oxidation states of individual atoms are determined by randomly allocating electrons to those atoms. The electrons in a covalent bond between two identical atoms are divided equally between them. When two distinct atoms are involved, the atom with the stronger electron attraction is fully allocated with the shared electrons. Since oxygen has a stronger attraction to electrons than hydrogen does in a molecule like water, when determining the oxidation states of oxygen and hydrogen in H_2O, it is assumed that the oxygen atom has all of the electrons. As a result, the oxygen has two extra electrons, putting it in an oxidation state of –2. Every hydrogen has an oxidation state of +1 and is devoid of electrons.

Rules for assigning oxidation states are

The oxidation state of an atom in an element is 0.

The oxidation state of a monatomic ion is the same as its charge.

Oxygen is assigned an oxidation state of –2 in its covalent compounds, such as CO, CO_2, SO_2, and SO_3. An exception to this rule occurs in peroxides, where each oxygen is assigned an oxidation state of 1.

In its covalent compounds with nonmetals, hydrogen is assigned an oxidation state of +1. Examples include HCl, H_2O, NH_3, and CH_4.

+In its compounds fluorine is always assigned an oxidation state of 1.

The sum of the oxidation states must be zero for an electrically neutral compound. For an ion, the sum must equal the charge of the ion. For example, the sum of oxidation states for the nitrogen and hydrogen atoms in NH_4^+ is +1, and the oxidation state of nitrogen is –3. For NO_3^-, the sum of oxidation states is 1. Since oxygen has an oxidation state of –2, nitrogen must have an oxidation state of +5. Sometimes, oxidation states are indicated with Roman numerals, for example, the sulfur atom in SO_4^{2-} is +VI. (Table 4.2).

The oxidation states of sulfur in various compounds of atmospheric importance are shown below:

$$H_2S = -2$$
$$SO_2 = +4$$
$$H_2SO_4 = +6$$
$$SO_4^{2-} = +6$$

Oxidation states of nitrogen in atmospheric species are as follows:

$$NH_3, RNH_2, R_2NH, R_3N = -3$$
$$N_2 = 0$$
$$N_2O = +1$$
$$NO = +2$$
$$HNO_2 = +3$$
$$NO_2 = +4$$
$$HNO_3, NO_3^-, N_2O_5 = +5$$
$$NO_3 = +6$$

Table 4.2. *Global Sulfur Emissions Estimates [Tg(S) yr^{-1}]*

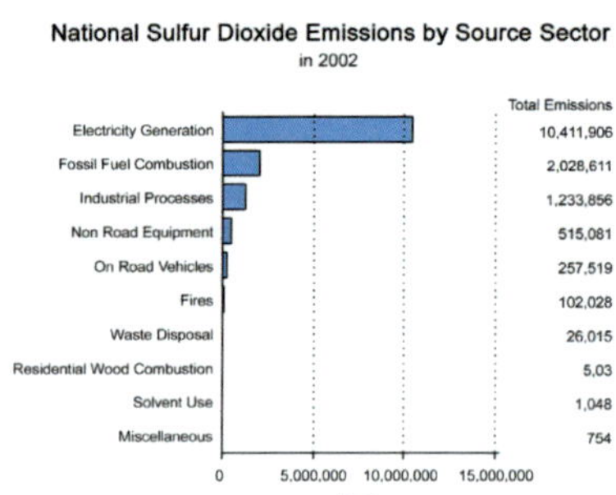

Source: https://www.marketswiki.com/wiki/images/d/dd/National_Sulfur_Dioxide_Emissions_by_Source_Sector.gif.

Estimated emissions of sulfur to the atmosphere, resulting from both natural and human sources, are shown in Table 4.2, along with an estimated division between the Northern and Southern Hemispheres. According to current estimates, total global emissions (not including sea salt) range from 98 to 120 Tg(S) yr^{-1}. Currently, human activity is responsible for approximately 75% of sulfur emissions overall, with 90% of these emissions occurring in the Northern Hemisphere. The estimate of worldwide anthropogenic sulfur emissions since 1850 is presented in Figure 4.1, and the observed mixing ratios and atmospheric lifetimes of atmospheric sulfur gases are compiled in Table 4.3.

Table 4.3. *Relative Risk, Percent Changes, and Their 95% Confidence Intervals in Daily Hospital Admissions From Hypertensive Cardiovascular Disease per 10 μg/m³ Increase in PM_{10} and SO_2*

			Total	Age groups			Gender		Season	
				18–34 years	35–64 years	≥65 years	Female	Male	Warm	Cold
PM_{10}	Model 1	Relative risk (95% CI)	1.009 (0.999,1.018)	1.04 (0.99,1.08) ab	1.011 (0.999,1.02) a	1.006 (0.994,1.017) b	**1.012 (1.002,1.02)**	1.003 (0.99,1.015)	0.990 (0.975,1.005)a	**1.014 (1.004,1.024)**a
		p-value	0.054	0.09	0.08	0.33	**0.02**	0.66	0.19	**0.003**
		Percent change (95% CI)	0.87 (−0.01,1.77)	3.73 (−0.58,8.23)	1.08 (−0.13,2.31)	0.55 (−0.56,1.68)	**1.21 (0.19,2.24)**	0.28 (−0.97,1.54)	−0.99 (−2.48,0.52)	**1.43 (0.43,2.43)**
	Model 2	Relative risk (95% CI)	**1.009 (1.001,1.017)**	1.045 (0.999,1.09) ab	1.011 (0.999,1.023) a	1.006 (0.995,1.016) b	**1.012 (1.003,1.022)**	1.002 (0.990,1.015)	0.984 (0.967,1.000)a	**1.013 (1.003,1.023)**a
		p-value	**0.03**	0.051	0.07	0.27	**0.009**	0.70	0.056	**0.008**
		Percent change (95% CI)	**0.90(0.06,1.74)**	4.49 (−0.02,9.19)	1.09 (−0.09,2.29)	0.59 (−0.45,1.64)	**1.25 (0.31,2.20)**	0.25 (−1.03,1.54)	−1.64 (−3.29,0.04)	**1.33 (0.35,2.32)**
	Model 3	Relative risk (95% CI)	**1.012 (1.003,1.020)**	**1.056 (1.011,1.104)**ab	**1.014 (1.002,1.027)**a	1.008 (0.997,1.018)b	**1.014 (1.005,1.024)**	1.007 (0.994,1.020)	0.990 (0.972,1.008)a	**1.014 (1.004,1.024)**a
		p-value	**0.006**	**0.01**	**0.02**	0.16	**0.003**	0.31	0.26	**0.007**
		Percent change (95% CI)	**1.19 (0.33,2.05)**	**5.65 (1.13,10.36)**	**1.44 (0.23,2.66)**	0.76 (−0.31,1.84)	**1.45 (0.48,2.42)**	0.68 (−0.63,2.01)	−1.02 (−2.77,0.75)	**1.36 (0.37,2.37)**
SO_2	Model 1	Relative risk (95% CI)	**1.023 (1.009,1.039)**	1.032 (0.962,1.108)	**1.031 (1.012,1.051)**a	1.016 (0.998,1.035) a	**1.019 (1.002,1.037)**	**1.030 (1.010,1.051)**	1.006 (0.961,1.053)	1.003 (0.988,1.019)
		p-value	**0.002**	0.38	**0.002**	0.09	**0.03**	**0.003**	0.80	0.69
		Percent change (95% CI)	**2.35 (0.86,3.86)**	3.22 (−3.83,10.79)	**3.14 (1.17,5.15)**	1.62 (−0.25,3.52)	**1.93 (0.21,3.67)**	**3.03 (1.001,5.10)**	0.59 (−3.94,5.34)	0.33 (−1.24,1.92)
	Model 2	Relative isk (95% CI)	**1.023 (1.009,1.038)**	1.039 (0.966,1.116)	**1.033 (1.014,1.053)**a	1.013 (0.995,1.032) a	1.014 (0.998,1.031)	**1.038 (1.017,1.059)**	1.007 (0.963,1.053)	1.011 (0.995,1.027)
		p-value	**0.001**	0.30	**0.001**	0.15	0.09	**0.001**	0.73	0.18
		Percent change (95% CI)	**2.32(0.91,3.76)**	3.85 (-3.36,11.60)	**3.33(1.36,5.34)**	[illegible]	[illegible]	**3.73(1.87,5.89)**	0.72 (-3.65,5.29)	1.07 (-0.49,2.66)
	Model 3	Relative risk (95% CI)	**1.021 (1.006,1.036)**	1.012 (0.936,1.094)	**1.031 (1.010,1.053)**a	1.012 (0.993,1.032) a	1.011 (0.993,1.029)	**1.038 (1.016,1.061)**	0.999 (0.951,1.050)	1.009 (0.992,1.026)
		p-value	**0.006**	0.77	**0.003**	0.21	0.23	**0.001**	0.99	0.30
		Percent change (95% CI)	**2.11 (0.61,3.63)**	1.18 (−6.41,9.39)	**3.14 (1.05,5.28)**	1.24 (−0.67,3.20)	1.08 (−0.67,2.86)	**3.79 (1.56,6.07)**	−0.04 (−4.87,5.05)	0.88 (−0.78,2.55)

PM10: Particulate Matter with an aerodynamic diameter <10 μm, SO_2: sulfur dioxide.
Estimates are presented per 10 μg/m3 increase in the lag 0–2 of PM_{10} and lag 0–5 of SO_2 for age, sex, and season groups. The Negative Binomial time series regression models were used to estimate effects for total, all age and season groups, and females. The poisson time series regression model was used for males.
The statistically significant estimates are highlighted in bold.
Model 1: Single pollutant model (without any adjustment).
Model 2: Adjusted for time-varying meteorological (temperature, wind speed, and dew point) and holiday variables.
Model 3: Multi-pollutant model (further adjustment for latent pollution factors by exploratory factor analysis of carbon monoxide (CO), nitrogen dioxide (NO_2), ozone (O_3), and SO_2/PM_{10}).
According to the between-group difference test, each letter (a and b) denotes a subset of groups whose effect estimates differ significantly from each other at the 0.05 level.

Source: Nouri F, Taheri M, Ziaddini M, Najafian J, Rabiei K, Pourmoghadas A, Shariful Islam SM and Sarrafzadegan N (2023) Effects of sulfur dioxide and particulate matter pollution on hospital admissions for hypertensive cardiovascular disease: A time series analysis. Front. Physiol. 14:1124967. doi: 10.3389/fphys.2023.1124967.

4.2.2. Dimethyl Sulfide (CH_3SCH_3)

The most common sulfur compound released from the world's oceans is dimethyl sulfide (DMS). Lovelock et al. (1972) conducted the first measurements of DMS in the surface ocean, demonstrating the ubiquitous presence of DMS, which is produced by marine phytoplankton, in these waters. It had long been understood that a significant flux of an ocean-to-atmosphere sulfur-containing compound was necessary to balance the world's sulfur budget. According to Lovelock's measurements, that compound was DMS. Numerous expeditions have measured DMS concentrations in ocean waters and the marine atmosphere since Lovelock's groundbreaking data. All of the oceanic DMS data that was available as of 1999, including over 15,000 individual measurements from 134 cruises, is summarized by Kettle et al. (1999). Areas with elevated DMS concentrations and the blooming of marine phytoplankton, which generates DMS, roughly correlate. The Southern Hemisphere's Antarctic waters have the highest DMS concentrations in January, while the Northern Hemisphere's oceans have the highest concentrations in July.

Dimethyl sulfoniopropionate, which is produced by marine organisms, especially phytoplankton, is thought to be the source of DMS (Andreae, 1990). Its concentration ranges from a few nanograms of S per liter to a few micrograms of S per liter in the ocean's upper layer. With an average concentration of about 100 nanograms (ng) of S per liter, the DMS surface seawater concentration is extremely nonuniform. The concentration of DMS has been found to be influenced by depth and location (Andreae and Raemdonck, 1983), as well as by diurnal (Andreae and Barnard 1984), seasonal, and Turner and Liss 1985 variations. Oceanic DMS concentrations are significantly higher than those that would be in equilibrium with atmospheric values based on DMS concentrations in the atmosphere and its Henry's law constant in seawater. A DMS flux from the ocean to the atmosphere is the outcome of this imbalance.

The amount of DMS present in the marine atmosphere has been measured numerous times. The marine boundary layer's (MBL) average DMS mixing ratio ranges from 80 to 110 ppt, but it can go as high as 1 ppb in entropic (e.g., coastal, upwelling) waters. DMS mixing ratios in the free troposphere decrease quickly with altitude to a few parts per trillion. DMS primarily reacts with the hydroxyl radical and the nitrate (NO_3) radical after being transferred across the air-sea interface and into the atmosphere. The sole source of methane sulfonic acid (MSA) in the atmosphere and the main source of SO_2 in the marine atmosphere is the oxidation of DMS.

4.2.3. Carbonyl Sulfide (OCS)

Because OCS has a long lifetime and little reactivity in the troposphere, it is the most abundant sulfur gas in the global background atmosphere. In essence, it is the sole sulfur compound that manages to make it into the upper atmosphere. The direct release of SO_2 into the stratosphere during volcanic eruptions is an exception. Table 4.4 gives the estimated global budget for OCS. Direct emissions at the surface and atmospheric conversion of CS_2 and DMS are the primary atmospheric sources of organic carbon dioxide (OCS). It is estimated that 10–20% of the total OCS source comes from biomass burning. OCS mixing ratio is comparatively constant at 500 ppt throughout the troposphere due to its long lifetime (Chin and Davis, 1995). Below the tropical tropopause, at elevations between 10 and 18 km, the point at which air enters the stratosphere, enhanced mixing ratios up to 600 ppt from emissions from burning biomass have been discovered (Notholt et al., 2003). A large portion of OCS makes its way into the stratosphere, where photolysis and oxidation produce SO_2 and ultimately sulfate aerosol. Due to their significantly shorter lifetimes than OCS, the other sulfur compounds, CS_2 and DMS, do not significantly contribute to the stratosphere's sulfur budget during periods of volcanic quiescence.

Table 4.4. *Global Budget of Carbonyl Sulfide (OCS)*

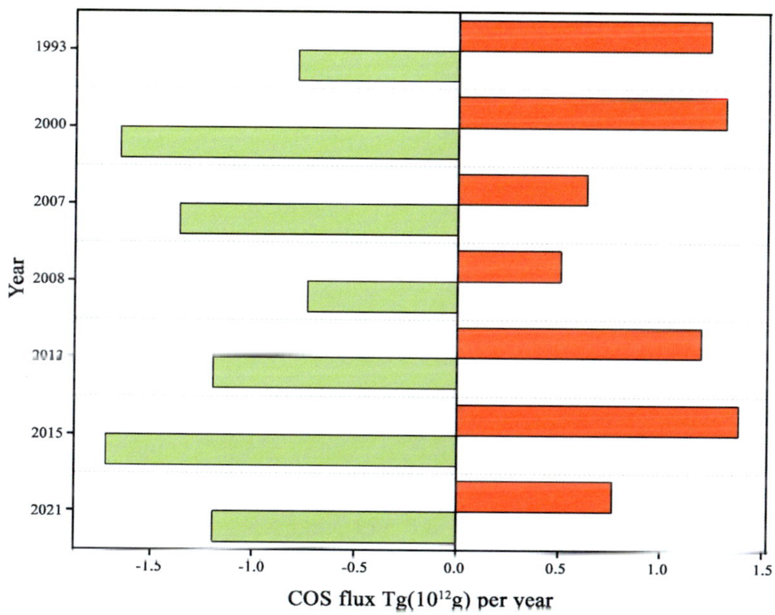

Source: Li J, Shen L, Zhang Y, Liu Y, Wu J, Wang A. Carbonyl Sulfide (COS) in Terrestrial Ecosystem: What We Know and What We Do Not. Atmosphere. *2024; 15(7):778. https://doi.org/10.3390/atmos15070778.*

Chin and Davis (1995) calculated the total amount of OCS in the atmosphere to be 5.2 Tg based on atmospheric measurement, of which 4.63 Tg is in the troposphere and 0.57 Tg is in the stratosphere. The global atmospheric lifetime of OCS is estimated to be approximately 6 years, based on the estimated global OCS source strength of 0.86 Tg yr^{-1}.

4.2.4. Sulfur Dioxide (SO_2)

The most common anthropogenic sulfur-containing air pollutant is sulfur dioxide. In the unpolluted marine boundary layer, levels of SO_2 range between 20 and 50 ppt, while mixing ratios in continental background air range from 20 ppt to >1 ppb. The ratios of urban SO_2 mixing can reach several hundred parts per billion.

4.3. NITROGEN-CONTAINING COMPOUNDS

The N≡N molecule is nearly inert due to its strong triple bond; it is also very stable chemically and does not contribute to the chemistry of the troposphere or stratosphere. Nitrous oxide (N_2O), ammonia (NH_3), nitric acid (HNO_3), nitrogen dioxide (NO_2), and nitric oxide (NO) are the main nitrogen-containing trace species in the atmosphere. The first of these, nitrous oxide (N_2O), is a colorless gas that is almost exclusively released by natural processes, primarily soil bacteria. The gas, sometimes known as "laughing gas," is used as an anesthetic. Nitric oxide (NO), which is the second, is released by both man-made and natural sources. Together with nitrogen oxide (NO), nitrogen dioxide (NO_2) is also produced in small amounts by oxidation of NO in the atmosphere and is released during combustion processes. Typically, NO_x is the total of NO and NO_2. The primary nitrogen oxide produced by high-temperature combustion is nitrate oxide, which is produced by the chemical reaction between atmospheric nitrogen and oxygen at high combustion temperatures as well as the interaction of nitrogen in the fuel with oxygen in the air. Despite being present in the atmosphere in comparatively low concentrations, other oxides of nitrogen, like NO_3 and N_2O_5, play a significant role in atmospheric chemistry. One byproduct of NO_2 oxidation in the atmosphere is nitric acid. The main sources of NH_3 emissions are natural sources. Finally, nitrate, and ammonium salts are produced by the atmospheric conversion of NO, NO_2, and NH_3, but they are not released in any appreciable amounts.

For all living things, nitrogen is a necessary nutrient. The atmosphere is this nitrogen's main source. Nevertheless, most organisms require N_2 to be "fixed," or changed into a form that they can use chemically, before they can benefit from it. (The process of chemically converting N_2 into any other nitrogen compound is known as "nitrogen fixation"). There are two different processes that lead to the "natural" fixation of N_2. One is the result of a relatively small number of microorganisms that can convert N_2 into organic nitrogen compounds, NH_3, and NH_{4+}. Ionizing phenomena, like cosmic radiation or lightning, act on N_2 to cause the other natural nitrogen fixation process in the atmosphere. Nitrogen oxides are created as a result of this process and end up as biologically beneficial nitrates on the surface of the Earth.

Human activity has resulted in the biological, industrial, and combustion-based fixations of nitrogen in addition to the fixation that occurs naturally. Legume crops, which have a symbiotic relationship with some microorganisms that can fix nitrogen, have been cultivated more by humans. In addition to being a useful food crop, legumes raise the nitrogen content of the soil. NH_3 production for use as fertilizer is the main process of industrial nitrogen fixation. Nitrogen can also fix as NO_x as a result of combustion. In the process of nitrification, ammonium is oxidized to NO_2^- and NO_3^- by microbial action. N_2O and NO are byproducts of nitrification; the result is the release of N_2O and NO to the atmosphere. Denitrification is the reduction of NO_3^- to N_2, NO_2, N_2O, or NO. The process that continuously replenishes the atmosphere with N_2 is called denitrification, and it is carried out by several bacteria. The atmospheric nitrogen cycle is shown in Figure 4.1.

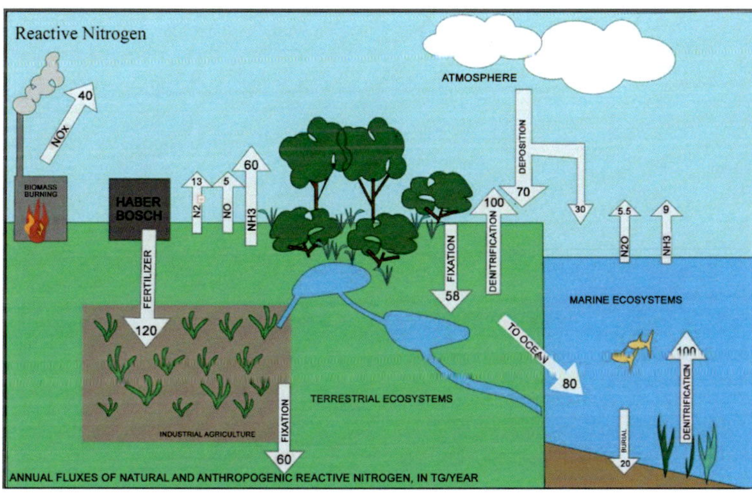

Figure 4.1. *Processes in the atmospheric cycle of nitrogen compounds. A species written over an arrow signifies reaction with the species from which the arrow originates.*

Source: By M maraviglia – Own work, CC BY-SA 4.0, https://commons.wikimedia.org/w/index.php?curid=77954029.

4.3.1. Nitrous Oxide (N_2O)

One significant atmospheric gas that is mostly released by biological sources in soils and water is nitrous oxide (N_2O). N_2O is a very strong greenhouse gas, despite having a much lower concentration than CO_2 and H_2O. This is because each molecule has a relatively large capacity to absorb energy and has a long residence time. The most significant single sources of atmospheric N_2O are tropical soils. Numerous other processes, including burning biomass, degassing irrigation water, farming, and industrial operations, also release N_2O in smaller amounts. An important source of N_2O is the oceans. About 10 Tg(N) yr^{-1} was the total preindustrial N_2O source. The IPCC (2013) estimated that the current flux of N_2O into the atmosphere resulting from anthropogenic activities is 6.9 Tg(N) yr^{-1} (Table 4.5).

Table 4.5. *Nitrogen*

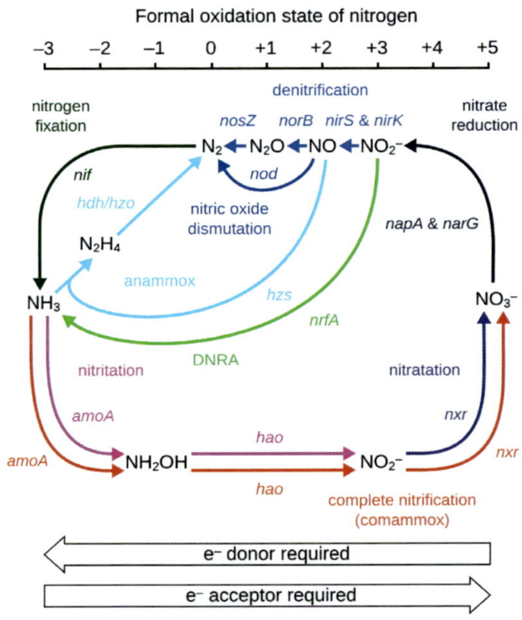

Source: By Vinicius Waldow – Own work, CC BY-SA 4.0, https://commons.wikimedia.org/w/index.php?curid=64623300.

In the troposphere, nitrous oxide is inert; however, over 90% of it is absorbed by photodissociation in the stratosphere and 10% is absorbed by reaction with excited oxygen atoms, or $O(^1D)$. The primary source of NO in the stratosphere is produced when $O(^1D)$ oxidizes N_2O. N_2O sources outweigh predicted sinks by 3.06 Tg(N) yr^{-1}.

Based on tests against observed N_2O distributions, models of stratospheric chemical transport are used to estimate the atmospheric lifetime of N_2O. N_2O has a current estimated lifetime of 131 years. Owing to its extended lifetime, N_2O concentrations are relatively uniform throughout the troposphere. Preindustrial mixing ratio of 276 ppb is indicated by N_2O ice core records. In 2011, N_2O

concentrations were 324 ppb (Figure 4.2). Assuming reasonable agreement with uncertainties, this observed atmospheric increase is consistent with an excess of 3.6 Tg(N) yr^{-1} of sources over sinks, which is consistent with the mismatch based on independent estimates of sources and sinks.

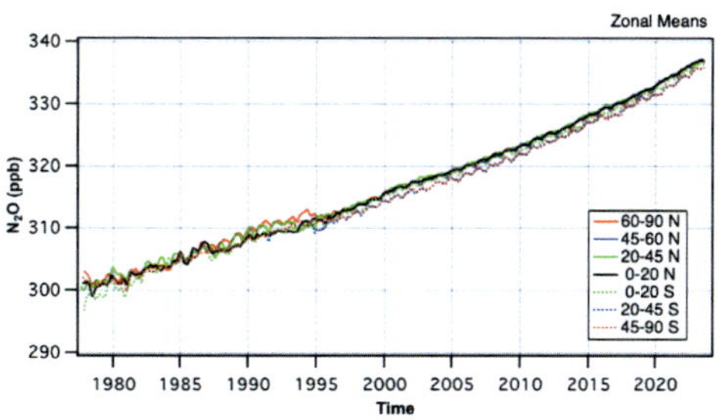

Figure 4.2. *Nitrous oxide atmospheric concentration since 1978.*

Source: By NOAA – https://www.esrl.noaa.gov/gmd/hats/combined/N2O.html, Public Domain, https://commons.wikimedia.org/w/index.php?curid=99674540.

4.3.2. Nitrogen Oxides ($NO_x = NO + NO_2$)

In atmospheric chemistry, nitrogen oxides (NO and NO_2) are two of the most significant molecules. NO_x emissions estimates in a study are provided in Table 4.6.

Table 4.6. *Study Reports*

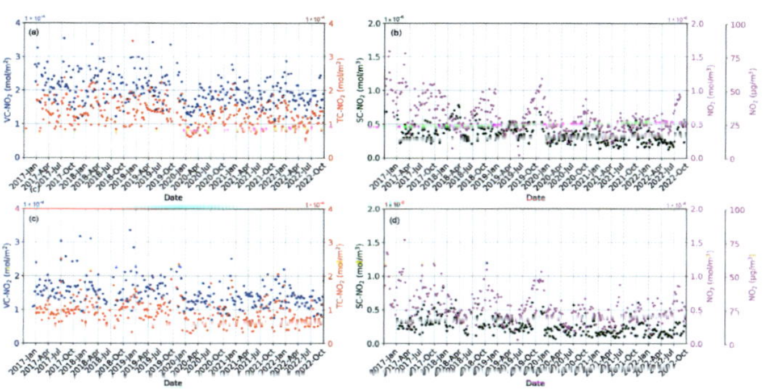

Source: Di Bernardino A, Mevi G, Iannarelli AM, Falasca S, Cede A, Tiefengraber M, Casadio S. Temporal Variation of NO_2 and O_3 in Rome (Italy) from Pandora and In Situ Measurements. Atmosphere. *2023; 14(3):594. https://doi.org/10.3390/atmos14030594.*

4.3.3. Reactive Odd Nitrogen (NO_y)

Reactive nitrogen, or NO_y, is the result of atmospheric nitrogen oxidation of NO_x and all compounds that are the product of this process. It is defined as the sum of the two oxides of nitrogen ($NO_x = NO + NO_2$). These consist of peroxynitric acid (HNO_4), peroxyacetyl nitrate (PAN) ($CH_3C(O)OONO_2$) and its homologs, alkyl nitrates ($RONO_2$), peroxyalkyl nitrates ($ROONO_2$), nitrous acid (HONO), the nitrate radical (NO_3), dinitrogen pentoxide (N_2O_5), and nitric acid (HNO_3). The primary oxidation product of NO_x in the atmosphere is nitric acid (HNO_3). HNO_3 is quickly deposited on surfaces and in water droplets due to its high-water solubility. Furthermore, HNO_3 can combine with NH_3 to form an aerosol of ammonium nitrate (NH_4NO_3). Particularly at night, the nitrate radical (NO_3) plays a significant role in the troposphere's chemistry. In the boundary layer, NO_3 can be found at mixing ratios of up to 300 ppt at night. In this context, NH_3 and N_2O are not regarded as reactive nitrogen compounds.

An essential indicator of the overall amount of oxidized nitrogen in the atmosphere is the measurement of total NO_y. The degree of interconversion between species is indicated by the concentrations of each NO_y species in relation to the total. In fact, compared to any of its constituent species, NO_y is the closest to a conserved quantity (Roberts, 1995).

Techniques that measure NO_x in nonurban locations (NO_x concentrations below 1 ppb) have only been available since the mid-1990s. Consequently, the size and dependability of the database required to define nonurban NO_x concentrations is constrained. Measurements at less-isolated rural sites typically range from a few tenths to one part per billion, and they are typically much higher than those at isolated rural sites. In remote maritime locations, NO_x measurements in the atmospheric boundary layer and lower free troposphere have typically produced mixing ratios of 0.02–0.04 ppb (20–40 ppt). When compared to remote marine locations, the somewhat higher NO_x concentrations found in remote tropical forests may be the result of biogenic NO_x emissions from soil. Mixing ratios in remote tropical forests, which are not directly impacted by biomass burning, appear to range from 0.02 to 0.08 ppb (from 20 to 80 ppt).

When one travels from urban and suburban areas to rural areas, and finally to isolated locations near the ocean and tropical forests, NO_x concentrations drop significantly. The dramatic three-order-of-magnitude or greater difference in NO_x concentrations between urban/suburban areas and remote locations provides strong evidence for the dominance of human emissions of NO_x over significant anthropogenic source regions, like North America. The median mixing ratio values range from 3 to 10 ppb, and average NO_y concentrations observed at various sites across the United States are fairly similar. These are marginally lower than the 10–1000 ppb range of NO_x mixing ratios that are commonly found in urban and suburban areas.

4.3.4. Ammonia (NH_3)

Ammonia is the most prevalent nitrogen-containing compound in the atmosphere, second only to N_2 and N_2O. It is the main basic gas. Animal waste, industrial emissions, losses of NH_3-based fertilizers from soils, and ammonification of humus followed by emission from soils are the main sources of NH_3 (Table 4.7). The ammonium (NH_4^+) ion is an important component of the continental tropospheric aerosol. Because NH_3 is readily absorbed by surfaces such as water and soil, its residence time in the lower atmosphere is estimated to be quite short, about 10 days. Wet and dry deposition of NH_3 are its main atmospheric removal mechanisms. In fact, deposition of atmospheric NH_3 and NH_4^+ may represent an important nutrient to the biosphere in some areas. Atmospheric concentrations of NH_3 are quite variable, depending on proximity to a source-rich region. NH_3 mixing ratios over continents range typically between 0.1 and 10 ppb.

Table 4.7. *Estimated Ammonia Emissions*

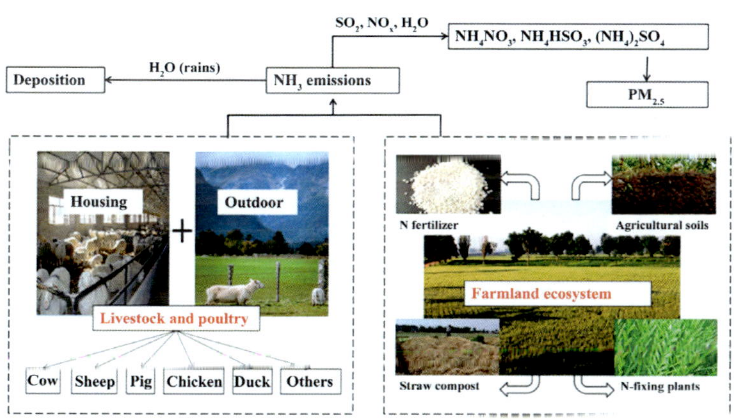

Source: Li J, Zeng W, Wan X. Management and Reduction Techniques Strategies of Ammonia Emission in Agricultural Sectors in China. Agronomy. *2023, 13(10):2555. https://doi.org/10.3390/agronomy13102555.*

4.3.5. Amines

Amines are organic compounds that contain nitrogen and are basic. The general structure of an amine is

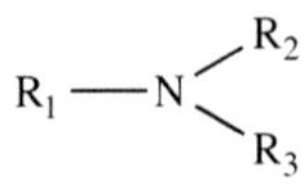

where at least one of R_1, R_2, R_3 is an organic group. Examples of amines include

CH_3NH_2	methylamine
$(CH_3)_2NH$	dimethylamine
$(CH_3)_3N$	trimethylamine

Only one of the hydrogen atoms in the NH_3 molecule is changed in primary amines, which have the general formula RNH_2. Similar to this, tertiary amines have the formula $R_1R_2R_3N$, and secondary amines have the general formula R_1R_2NH. Many different things release amines into the atmosphere (Ge et al., 2011). One to six carbon atom low molecular weight aliphatic amines are the most prevalent amines in the atmosphere. Methyl amine global flux estimates are provided in Table 4.9.

4.4. CARBON-CONTAINING COMPOUNDS

When an atom of carbon forms a bond with another element, chemical compounds known as carbon compounds are created. Compared to other elements, carbon is the most reactive and can form a wide variety of compounds. Carbon compounds are generally considered to be organic.

4.4.1. Classification of Hydrocarbons

In this brief review, we will focus on the classifications of carbon-containing compounds, especially those that are relevant to atmospheric chemistry. Since the carbon atom possesses four valence electrons, it can form bonds with one to four other atoms. The characteristics (and nomenclature) of a hydrocarbon molecule are essentially determined by the type of carbon–carbon bonding present in the molecule. The hydrocarbon molecules that share all of their carbon bonds with hydrogen atoms, with the exception of the minimum number needed for carbon-carbon bonds, can be considered the most basic in certain ways. These molecules are called alkanes or, more accurately, paraffins. C_nH_{2n+2} is the general chemical formula for alkanes. The first four alkanes having a straight-chain structure are

CH_4	methane
$CH_3–CH_3$	ethane
$CH_3–CH_2–CH_3$	propane
$CH_3–CH_2–CH_2–CH_3$	*n*-butane

A straight-chain structure is not necessary for alkanes. The name of the longest continuous chain of carbon atoms is used as the base name in the event that a side carbon chain exists. It is then modified to include the type of group. Typical examples of substituted alkanes are (the numbering system is indicated below the carbon atoms):

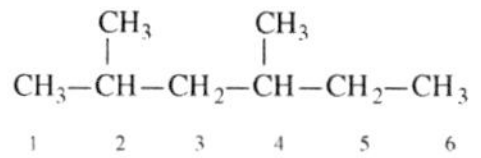

2,4-Dimethylhexane

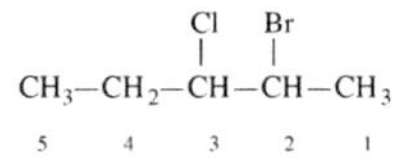

2-Bromo-3-chloropentane

A molecule is known as a cycloalkane when it has an arrangement of alkanes in a ring structure.

Alkanes typically undergo a hydrogen atom replacement reaction. When an alkane loses one of its hydrogen atoms, the carbon atom involved gains an unpaired electron and the molecule transforms into a free radical—in this example, an alkyl radical. Examples of alkyl radicals are:

$CH_3\cdot$	methyl
$CH_3—CH_2\cdot$	ethyl
$CH_3—CH_2—CH_2\cdot$	*n*-propyl
$CH_3—\dot{C}H—CH_3$	isopropyl
$CH_3—CH_2—CH_2—CH_2\cdot$	*n*-butyl

Alkyl radicals are commonly identified by the simple notation R·, where R represents the chemical formula of any alkyl group member. A free radical is a highly reactive species due to its unpaired electron. Free radicals are crucial to the chemistry of the atmosphere.

The alkenes represent the next significant class of carbon-containing compounds in atmospheric chemistry. Within this class, a pair of electrons, referred to as a double bond, is shared by two nearby carbon atoms. Alkenes can also be referred to as olefins or alkylenes. The number of the carbon atom frequently indicates the position of the carbon atom closest to the end of the molecule, which is the first of the two carbon atoms sharing the double bond. Examples of common alkenes are

$CH_2{=}CH_2$	ethene (ethylene)
$CH_3CH{=}CH_2$	propene (propylene)
$CH_3CH_2CH{=}CH_2$	1-butene
$CH_3CH{=}CHCH_3$	2-butene

Molecules with two double bonds are called dienes, an example of which is

$CH_2 = CH - CH = CH_2$
1,3-Butadiene

Molecules with a single triple bond are known as alkynes, the first in the series of which is acetylene, HC≡CH.

Benzene and its derivatives exemplify the class of organic substances called aromatic compounds. Benzene is portrayed as a cyclic triene, yet its chemistry is unique, not associated with that of the alkenes. The unique chemistry of benzene is a consequence of the fact that benzene is a hybrid of two equal resonance structures (Table 4.8).

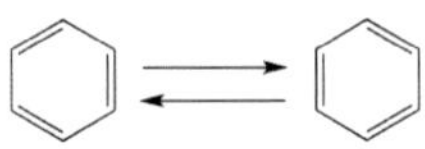

that are sometimes depicted as

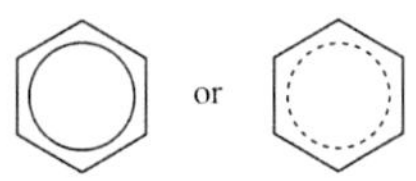

Other common aromatics are

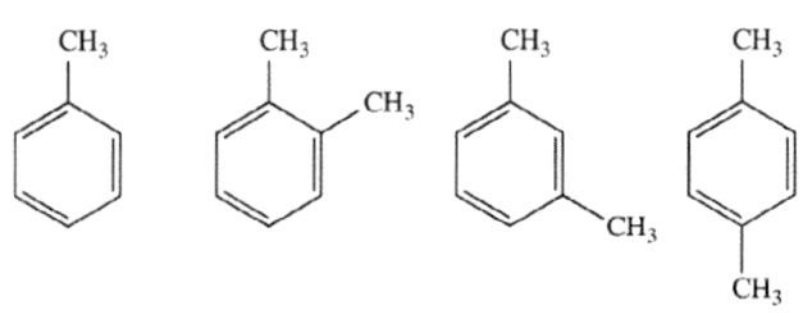

Toluene   *o*-Xylene   *m*-Xylene   *p*-Xylene

Hydrocarbons may acquire one or more oxygen atoms. Of the oxygenated hydrocarbons, two classes of carbonyls that are of considerable importance in the atmosphere are aldehydes and ketones. In each type of molecule, a carbon atom and an oxygen atom are joined by a double bond. Aldehydes have the general form

$$\mathrm{R{-}\overset{\overset{\displaystyle O}{\|}}{C}{-}H}$$

whereas ketones have the structure

$$\mathrm{R{-}\overset{\overset{\displaystyle O}{\|}}{C}{-}R}$$

Thus, the distinction lies in whether the carbon atom is bonded to one or two alkyl groups. Examples of aldehydes and ketones are

$$\mathrm{H\overset{\overset{\displaystyle O}{\|}}{C}H} \qquad \mathrm{CH_3\overset{\overset{\displaystyle O}{\|}}{C}H} \qquad \mathrm{CH_3\overset{\overset{\displaystyle O}{\|}}{C}CH_3}$$

Formaldehyde   Acetaldehyde   Acetone

Table 4.8. *Sources of Generation and Emissions of* NH_3

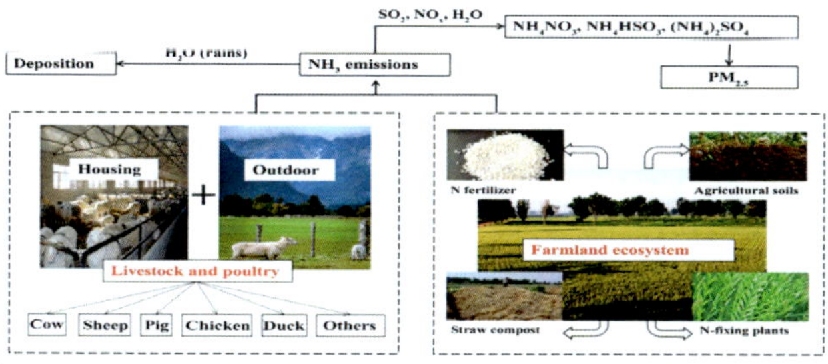

Source: Li J, Zeng W, Wan X. Management and Reduction Techniques Strategies of Ammonia Emission in Agricultural Sectors in China. Agronomy. *2023; 13(10):2555. https://doi.org/10.3390/agronomy13102555.*

4.4.2. Methane

The most prevalent hydrocarbon in the atmosphere is methane. The estimated global average CH_4 in 1750 is 722 parts per billion. The mean annual global CH_4 mixing ratio in 2011 was 1803 parts per billion. Methane is growing at a rate of about 6 parts per billion per year on average globally (Nisbet et al., 2014). The CH_4 mixing ratios over the last millennium are displayed in Figure 4.3. Except for the solid curve, which is based on atmospheric measurements made at Cape Grim, Tasmania, the data in Figure 4.4 are based on CH_4 in air bubbles trapped in ice cores in Antarctica.

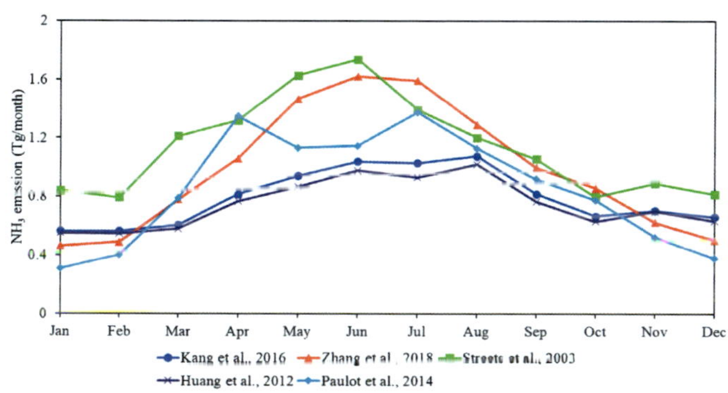

Figure 4.3. *Monthly variations in* NH_3 *emission in China.*

Source: Li J, Zeng W, Wan X. Management and Reduction Techniques Strategies of Ammonia Emission in Agricultural Sectors in China. Agronomy. *2023; 13(10):2555. https://doi.org/10.3390/agronomy13102555.*

Table 4.9. *Summary of NH_3 Emission Inventories at Different Scales*

No.	Area		Time	Source	Resolution	Method	Activity Data	Factor Differentiation	Total/Gg	Farmland/Gg	Livestock and Poultry /Gg	Human/Gg	Uncertainty Method	Uncertainty
1	National	National	2016	Anthropogenic	1 × 1 km	CHEN	Provincial	Yes	11,032.00	6475.00	3032.00	487.00	MC	(−20%, 24%)
		National	2015	Anthropogenic		CHANS	Provincial	Yes	15,600.00	5870.00	6650.00	780.00	MC	(−21%, 32%)
		National	2015	Anthropogenic	1 × 1 km	EF	Provincial	Yes	12,580.00	5580.00	2560.00	870.00	Literature-based values	±30%
2	Regional	Yangtze River Delta region	2014	Anthropogenic	1 × 1 km	EF, NARSES	Prefecture-level city	Partial	986.73	532.82	316.68	51.11	MC	(−55%, 60%)
		North China	2016	Anthropogenic		EF	Prefecture-level city	Partial	966.14	350.04	548.90	13.11	MC	(−17%, 25%)
		Sichuan and Chongqing	2004	Agricultural		IAP-N	County	No	698.76	374.94	219.60	98.59	Qualitative analysis	
3	Provincial	Fujian	2015	Anthropogenic	1 × 1 km	EF, NARSES	County	Partial	228.02	89.74	98.37	11.13	MC	±16.3%

		Jiangsu	2017	Anthropogenic		EF	Prefecture-level city	No	562.47	250.54	212.91	33.55	Qualitative analysis	
		Henan	2015	Anthropogenic	3 × 3 km	EF, NARSES	Prefecture-level city	Partial	1031.60	291.00	591.50	31.60	MC; the Au-vTool-Pro tool	(−33%, 35%)

Source: Li J, Zeng W, Wan X. Management and Reduction Techniques Strategies of Ammonia Emission in Agricultural Sectors in China. Agronomy. *2023; 13(10):2555. https://doi.org/10.3390/agronomy13102555 4.4.3 Volatile Organic Compounds*

Note: EF is the Emission Factor Method, and MC is the Monte Carlo simulation.

Table 4.11 provides an overview of the global methane budget. Global estimates are obtained from direct measurements of airborne concentrations (referred to as top-down estimates) and source emission inventories (referred to as bottom-up estimates). This discrepancy between top-down and bottom-up estimates stems from the global methane cycle's complexity. Methane emissions originate from natural sources and are caused by human activity to the tune of about two thirds. The main natural source of methane is tropical wetlands. The estimated range of total methane sources is 548 (top-down) to 678 (bottom-up) Tg CH_4 yr^{-1}. Methane is mostly sucked up from the atmosphere by reactions with the hydroxyl (OH) radical. The sum of sources exceeds the sum of sinks, according to both top-down and bottom-up budgets, which is consistent with the fact that methane is building up in the atmosphere. Methane has been responsible for approximately 20% of the total direct anthropogenic perturbation of the global radiative energy balance by long-lived greenhouse gases since preindustrial times. This increase is the result of anthropogenic emissions from fossil fuel production, agriculture, waste management, and biomass burning (Tables 4.9–4.11).

All vapor-phase atmospheric organics that are not CO and CO_2 are referred to as volatile organic compounds (VOCs). Global VOC emissions, excluding methane, are approximately 10% anthropogenic and 90% biogenic.

Table 4.10. *Classification of VOCs Pollutants*

Class	Examples of Compounds	Boiling Point Range °C
Very volatile organic compounds (VVOCs)	propane, butane, methyl-chloride	0 to 50–100
Volatile organic compounds (VOCs)	formaldehyde, toluene, acetone, isopropyl alcohol	50–100 to 240–260
Semi volatile organic compounds (SVOCs)	pesticides (chlordane, DDT), plasticizers (phthalates)	240–260 to 380–400

Source: David E, Niculescu V-C. Volatile Organic Compounds (VOCs) as Environmental Pollutants: Occurrence and Mitigation Using Nanomaterials. International Journal of Environmental Research and Public Health. 2021; 18(24):13147. https://doi.org/10.3390/ijerph182413147.

4.4.4. Biogenic Hydrocarbons

Approximately half of the ~2.4 Pg of reactive carbon compounds released into the atmosphere each year are released by terrestrial vegetation. The annual average emission rate over vegetated surfaces worldwide is approximately 7 gC m^{-2}, with some tropical regions experiencing values as high as 100 gC m^{-2}. The majority of biogenic volatile organic compounds (BVOCs) and 75% of all BVOC emissions worldwide are emitted by trees (Guenther et al., 1995; Guenther, 2002). The idea that naturally occurring VOC foliar emissions from trees and other vegetation could significantly impact the composition of Earth's atmosphere was first put forth by Went (1960). Terpenoid compounds, other volatile organic compounds (VOCs), and other reactive carbon are the three categories into which emissions of biogenic reactive carbon fall. The terpenoids are composed of five carbon atoms called hemiterpenes, 10 carbon atoms in monoterpenes, 15 carbon atoms in sesquiterpenes, and 20 carbon atoms in diterpenes. Just two hemiterpene compounds—isoprene and methyl butenol—emit significant amounts of fumes. Nearly all of the monoterpene emissions to the atmosphere are made up of fewer than 20 of the >1000 monoterpene compounds that are found in plants. α-pinene makes up about half of the total monoterpene emissions flux. The presence of an olefinic double bond in every one of the compounds in makes the molecule extremely reactive in the atmosphere. The C_{10} structure of monoterpenes has double bonds that can be found either outside (exocyclic) or inside (endocyclic) of a ring structure. Some possess both endocyclic and exocyclic double bonds, while others have structures that are open-ended. Compared to monoterpenes, sesquiterpenes (C_{15} compounds) have a more complex ring structure and are capable of having several double bonds. Certain terpenoid substances, such methyl chavicol, are oxygenated. The lifetimes of terpenes vary greatly, from a few minutes to several days, and they react quickly in the atmosphere. Deciduous trees are the primary emitters of isoprene, while coniferous trees are the primary source of terpene emissions. A significant portion of oxygenated BVOC emissions, such as methanol, acetone, ethanol, and acetaldehyde, can be attributed to coniferous forests. Additionally, some pine species release methyl butenol (MBO) in amounts comparable to isoprene (Steiner and Goldstein, 2007).

Among the biogenic hydrocarbons, isoprene (2-methyl-1,3-butadiene, C_5H_8) is special in its connection to a plant's ability to photosynthesize. When photosynthetically active radiation is present, it is released from a wide range of primarily deciduous vegetation, and its emission strongly increases with temperature. Plant species differ significantly in their emissions of isoprene and terpenoid compounds, and there also seems to be considerable variation in the biochemical and biophysical mechanisms regulating these compounds' rate of release. It has not been demonstrated that isoprene is stored in or processed by plants; rather, isoprene emissions are a species-dependent consequence of photosynthesis, photorespiration, or both. Because of this, isoprene emissions are dependent on both light and temperature; in other words, no isoprene is released in the absence of light. On the other hand, biophysical processes connected to the quantity of terpenoid material in leaf oils and resins as well as the compounds' vapor pressure appear to be the cause of terpenoid emissions. Terpene emissions therefore vary with ambient temperature but are not highly dependent on light (and they usually continue at night). Over the course of a growing season, the rate at which biogenic VOCs are produced can vary significantly due to the temperature-dependent nature of natural isoprene and terpenoid emissions. According to an analysis of emissions data by Lamb et al. (1987), a rise in ambient temperature from 25°C to 35°C can cause the rate of natural VOC emissions from deciduous trees that release isoprene to increase by a factor of 4, and from conifers that release terpene to increase by a factor of 1.5. Therefore, on hot summer days, natural VOC emissions are typically highest, all other things being equal.

There are several lists of emissions of biogenic hydrocarbons. Based on land-use data, emission rate measurements from individual plant species are scaled up to entire geographic regions using empirical algorithms that take temperature and, in the case of isoprene, light into account. To convert individual plant species emission rates, which are expressed in mg g^{-1} min^{-1}, to emission fluxes, which are expressed in mg m^{-2} min^{-1}, biomass density, which is expressed in g m^{-2} of area, is necessary. These fluxes are then multiplied by vegetation class land coverage, which is expressed in m^2, to yield total emissions. 500–750 Tg yr^{-1} (440–660 Tg C yr^{-1}) is the estimated amount of isoprene emissions worldwide (Guenther et al., 2006). According to Guenther et al. (1995), the estimated total emissions of VOCs is approximately 1150 Tg C yr^{-11}, of which 44% isoprene, 11% monoterpenes, 22.5% other reactive VOCs, and 22.5% other VOCs (Table 4.11).

Table 4.11. *Major Compounds of GOVs*

Type	Major Compound	Biological Function	Plant	Insect
GVs	Hexyl acetate	Female adult repellents		*Cydia pomonella*
	Pentadecane, Nonanal, α-farnesene	Antennal active substances	*Grapevine*	*Lobesia botrana*
	2-phenyl ethanol, Methyl salicylate, Decanal	Adult attractants	*Malus domestica*	*Argyresthia conjugella*
	4-hydroxy-4-methyl-2-pentanone, 1-hepten-3-one, 1-octen-3-ol, 3-octanone	Oviposition attractants	*Phaseolus vulgaris*	*Delia platura*
	(Z)-3-hexenol, (E)-2-hexenol, (E)-2-hexenal, Hexanal	Antennal active substances	-	*Episyrphus balteatus*
	3-octanol, 1-octen-3-ol	Antennal active substances	-	*Thanasimus formicarius*

	(E)-4,8-dimethylnona-1,3,7-triene, (E,E)-4,8,12-trimethyltrideca-1,3,7,11-tetraene, (E)-ocimene	Adult attractants	*Gossypium* spp.	*Anthonomus grandis*
	(E)-2-hexenyl acetate, (±)-Linalool, (E,E)-α-farnesene, (Z)-3-hexen-1-ol, (S)-(-)-α-pinene, Methyl salicylate	Adult repellents	Alfalfa	*Lygus hesperus*
	Methyl salicylate, Cis-hex-3-en-1-ol	Nature enemy attractants	*Solanum lycopersicum*	*Macrosiphum euphorbiae*, *Aphidius ervi*
	Nerol, α-campholenol, p-cymene, α-terpineol, Germacrene D	Adult attractants	*Corylus*	*Phytoptus avellanae*, *Myzocallis coryli*
	(+/−)-2-hexanol, 3-methyl-3-pentanol, 3,3-dimethyl-1-butanol	Female adult attractants	-	*Aleruodicus dispersus*
	β-pinene, α-pinene, 1,8-cineole	Antennal active substances	*Quercus palustris*	*Prinobius myardi*
	(Z) and (E)-β-ocimene, Undecane, Decanal, β-caryophyllene	Female adult attractants	*Salix* spp.	*Nematus oligospilus*
	Trans-4-thujanol	Adult repellents	*Picea asperata*	*Ips typographus*
GVs	(S)-linalool, 4,8-dimethyl-1,3,7-nonatriene, (E)-caryophyllene, (R/S)-(E)-nerolidol	Antennal active substances	*Oryza sativa*	*Orseolia oryzivora* Harris and Gagné
	(Z)-3-hexenol, (Z)-3-hexenyl acetate	Larvae attractants	*Populus nigra*	*Lymantria dispar*

	Dimethyl disulfide, Geraniol, Eucalyptol, Citronellol	Oviposition repellents	*Broccoli*	*Delia radicum*
	α-pinene, Trimethylindan, Cyclohexylbenzene	Adult attractants	*Baccharis salicifolia*	*Macrodactylus nigripes*
	(Z)-3-hexen-1-ol, 3-methyl-butanol, 1-pentanol. Benzaldehyde, 2-hexanone	Female adult attractants	*Prunus armeniaca*	*Capnodis tenebrionis*
	(E,E)-α-farnesene, (E)-4,8-dimethyl-1,3,7-nonatriene, (E,E)-4,8,12-trimethyl-1,3,7,11-tridecatetraene	Larvae attractants	*Coffea arabica*	*Hypothenemus hampei*
	Sulcatone	Adult attractants	*Ailanthus altissima*	*Lycorma delicatula*
	(Z)-3-hexenyl acetate, (E)-β-ocimene	Adult attractants	*Salix babylonica*	*Tuberolachnus salignus*
	Undecane, Ethyl benzaldehyde, n-hexadecanoic acid	Adult attractants	*Solanum lycopersicum*	*Tuta absoluta, Bemisia tabaci*
	Linalool, Linalyl acetate, Linalool tetrahydride	Antennal active substances	-	*Apolygus lucorum*
	p-cymene	Adult repellents	*Haplopappusmyiia gregaria*	*Haplopappus foliosus*
	Methyl salicylate, β-myrcene, (3E,7E)-4,8,12-trimethyltrideca-1,3,7,11-tetraene, α-curcumene	Adult repellents	*solanaceous plants*	*Tuta absoluta*

	(Z)-3-hexen-1-ol	Oviposition attractants	*Zea mays*	*Ostrinia furnacalis*
	Z-3-hexenyl acetate	Female adult attractants	*Triticum aestivum*	*Sitodiplosis mosellana*
GLV	α-pinene, 3-carene, γ-terpinene, Linalool	Adult attractants	*Agave tequilana*	*Scyphophorus Acupunctatus*
	(E)-2-hexenal, (Z)-3-hexen-1-ol, 1-octen-3-ol, (Z)-3-hexenyl acetate, (Z)-3-hexenyl butyrate	Female adult repellents	*Phaseolus lunatus*	*Spodoptera littoralis Boisd*
GLVs	(E)-β-caryophyllene	Larvae attractants	-	*Nilaparvata lugens*
	Myrcene, β-phellandrene, Trans-β-ocimene, Terpinolene	Female adult repellents	*Pinus sylvestris*	*Thaumetopoea pityocampa*
HIPVs	(Z)-3-hexen-1-ol, 4,8-dimethyl-1,3,7-nonatriene, 3-octanone	Cue of host location	*Phaseolus vulgaris*	*Trialeurodes vaporariorum*
	(E)-8-hydroxy 6 methyl-6-octen-3-one	Aggregation pheromone		*Oulema melanopus*
	(Germacrene D, β-caryophyllene, Methyl salicylate, E-β-ocimene, (3E)-4,8-dimethyl-1,3, 7-nonatriene	Adult attractants	*Fragaria × ananassa*	*Anthonomus rubi*
	Methyl salicylate, 2-phenylethanol	Adult attractants	*Glycine max*	*Coccinella septempunctata, Aphis glycines, Chrysoperla carnea and syrphid flies*

	Methyl salicylate	Adult attractants	*Brassica oleracea*	*Cotesia glomerata, Cotesia rubecula, Pieris caterpillars*
	Myrcene, (+)-3-carene, β-caryophyllene, (+)-α-pinene, Limonene	Adult attractants	*Pinus armandii*	*Dendroctonus armandi*
	DMNT	Female adult attractants	*Ulmus minor*	*Oomyzus gallerucae*
	Methyl jasmonate, Methyl salicylate	Nature enemy attractants	*Zea mays*	-
	1-butyl butylate, 1-hexyl acetate, 1-butyl ethanoate	Adult attractants	*Malus pumila*	*Psyttalia concolor*
	(Z)-3-hexene-1-ol	Antennal active substances	*Ricinus communis*	*Anomala corpulenta*
	Linalool, (Z)-3-hexenyl acetate, Geraniol	Male adult attractants	*Glycine max*	*Leguminivora glycinivorella*
	Linalool, DMNT	Nature enemy attractants	*Zea mays*	*Spodoptera frugiperda*
	(Z)-3-hexenyl hexanoate, (Z)-3-hexenal	Female adult attractants	*Camellia sinensis*	*Ectropis obliqua*
HIPVs	Nonanal, (Z)-3 hexenyl acetate, (E)-β-ocimene, (R)-(+)-limonene	Male adult attractants	*Gossypium hirsutum*	*Spodoptera littoralis*
	(E)-4,8-dimethyl-1,3,7-nonatriene	Nature enemy attractants	*Zea mays*	*Chilo partellus, Trichogramma bournieri, Cotesia sesamiae*

	α-pinene, β-myrcene, (Z)-3-hexenyl acetate, Limonene, (E)-ocimene, Linalool, DMNT, (E)-β-farnesene, TMTT, Indole	Cue of host location	*Zea mays*	*Polybia fastidiosuscula*
	DMNT, TMTT	Nature enemy attractants	*Anacardium occidentale*	*Diaphorina citri*
	Cis-jasmone, Cis-3-hexenyl acetate	Adult attractants	-	*Campoletis chlorideae*
	β-phellandrene, (E)-β-ocimene, (-)-β-bourbonene, β-ylangene, (E,E)-α-farnesene, Linalool, Myrtenol, (E) (-)-pinocarveol, (E)-pinocamphone, (Z)-pinocamphone, Methyl salicylate	Nature enemy attractants	*Juglans regia*	*Meliboeus ohbayashii*
	Acetic acid, 2-phenyl-ethanol	Adult attractants	*Malus pumila*	*Aphis pomi, Dysaphis plantaginea*
	β-ocimene	Female adult repellents	*Camellia sinensis*	*Ectropis obliqua*
	(Z)-3-hexenal, (E)-2-hexenal, (Z)-3-hexen-1-ol, (E)-2-hexen-1-ol, (Z)-3-hexen-1-yl acetate, 1-hexyl acetate	Female adult attractants	*Triticum aestivum*	*Aelia acuminata*
	α-pinene, Pseudocumene, Limonene	Nature enemy attractants	*Fragaria × ananassa*	*Tetranychus urticae Koch*

Source: Niu D, Xu L, Lin K. Multitrophic and Multilevel Interactions Mediated by Volatile Organic Compounds. Insects. *2024; 15(8):572. https://doi.org/10.3390/insects15080572.*

GVs: general volatiles. DMNT: (E)-4,8-dimethyl-1,3,7-nonatriene. TMTT: (E, E)-4,8,12-trimethyltrideca-1,3,7,11-tetraene. The tropics have the highest biogenic emissions due to a combination of high temperatures and high biomass densities, with isoprene being the main compound released. The maximum flux of biogenic hydrocarbon emissions in the Southeast is expected to be twice as high as in the tropics during the summer. BVOC emissions far outweigh anthropogenic VOC emissions globally.

4.4.5. Carbon Monoxide

Table 4.12 lists the attributes of CO on a global scale. A significant source of CO is the oxidation of methane (by OH), burning of biomass, industrial, and combustion processes, and the oxidation of nonmethane hydrocarbons. There are significant uncertainties in each of these estimated sources. Anthropogenic activities, such as the oxidation of CH_4 derived from human activity, are thought to be responsible for approximately two thirds of CO emissions. Reaction with OH radicals is the main sink for CO; soil uptake and stratospheric diffusion are secondary pathways.

Table 4.12. *Thermal and Physical Properties of Carbon Monoxide (CO) at Atmospheric Pressure Hide*

Temperature (°C)	Temperature (K)	Density (kg/m³)	Specific Heat (J/g°C)	Dynamic Viscosity (cg/m s)	Kinematic Viscosity (cm²/s)	Thermal Conductivity (cW/m°C)	Thermal Diffusivity (cm²/s)	Prandtl Number
−73.15	200	1.6888	1.045	1.27	0.0752	1.7	0.0963	0.781
−53.15	220	1.5341	1.044	1.37	0.0893	1.9	0.119	0.753
−33.15	240	1.4055	1.043	1.47	0.105	2.06	0.141	0.744
−13.15	260	1.2967	1.043	1.57	0.121	2.21	0.163	0.741
6.85	280	1.2038	1.042	1.66	0.138	2.36	0.188	0.733
26.85	300	1.1233	1.043	1.75	0.156	2.5	0.213	0.73
46.85	320	1.0529	1.043	1.84	0.175	2.63	0.239	0.73
66.85	340	0.9909	1.044	1.93	0.195	2.78	0.269	0.725
86.85	360	0.9357	1.045	2.02	0.216	2.91	0.298	0.725
106.85	380	0.8864	1.047	2.1	0.237	3.05	0.329	0.729
126.85	400	0.8421	1.049	2.18	0.259	3.18	0.36	0.719
176.85	450	0.7483	1.055	2.37	0.317	3.5	0.443	0.714
226.85	500	0.67352	1.065	2.54	0.377	3.81	0.531	0.71
276.85	550	0.61226	1.076	2.71	0.443	4.11	0.624	0.71
326.85	600	0.56126	1.088	2.86	0.51	4.4	0.721	0.707
376.85	650	0.51806	1.101	3.01	0.581	4.7	0.824	0.705

426.85	700	0.48102	1.114	3.15	0.655	5	0.933	0.702
476.85	750	0.44899	1.127	3.29	0.733	5.28	1.04	0.702
526.85	800	0.42095	1.14	3.43	0.815	5.55	1.16	0.705

Source: https://en.wikipedia.org/wiki/Carbon_monoxide.

The range of tropospheric CO mixing ratios is 40–200 ppb. In the troposphere, carbon monoxide has an average chemical lifetime of 30 to 90 days. The Northern Hemisphere has higher CO levels than the Southern Hemisphere, according to measurements, with the highest concentrations being found close to the surface at northern midlatitudes. In the Northern Hemisphere, the CO mixing ratio generally decreases with altitude, reaching a free tropospheric average value of roughly 120 ppb close to 45° N. CO tends to be more nearly evenly mixed vertically in the Southern Hemisphere, with a mixing ratio of roughly 60 ppb close to 45°S. Seasonality has been found to vary by approximately ±40% in the Northern Hemisphere and ±20% in the Southern Hemisphere with respect to the mean. The local spring is when the concentration is at its highest, and the late summer or early fall is when it is at its lowest.

4.4.6. Carbon Dioxide

The most significant waste product of our industrialized society is carbon dioxide. Climate is the main impact of CO_2 in the atmosphere. In August 2012, the average monthly CO_2 mixing ratio was 392 ppm, while in August 2013, it was 395 ppm. In April 2014, the average concentration of CO_2 in the Earth's atmosphere surpassed 400 parts per million for the first time in human history. There is a seasonal pattern in the CO_2 record. Every year, as carbon is absorbed by leafy plants during the spring and summer growing season in the Northern Hemisphere, atmospheric CO_2 mixing ratios fall.

4.5. HALOGEN-CONTAINING COMPOUNDS

Atmospheric halogen-containing compounds are referred to by various names:

1. **Halocarbons**—a general term referring to halogen-containing organic compounds.
2. **Chlorofluorocarbons (CFCs)**—the collective name given to a series of halocarbons containing carbon, chlorine, and fluorine atoms.
3. **Hydrochlorofluorocarbons (HCFCs)**—halocarbons containing atoms of hydrogen, in addition to carbon, chlorine, and fluorine.
4. **Hydrofluorocarbons (HFCs)**—halocarbons containing atoms of hydrogen, in addition to carbon and fluorine.

5. **Perhalocarbons**—halocarbons in which every available carbon bond contains a halogen atom (compounds saturated with halogen atoms).
6. **Halons**—bromine-containing halocarbons, especially used as fire extinguishing agents.

Sea salt's ability to produce gaseous halogens sparked the first interest in halogens in the atmosphere (Eriksson, 1959). For the past century, synthetic halocarbons have been known about. The late nineteenth century saw the first synthesis of chlorofluorocarbons, or CFCs. Halocarbons are widely used in industry as solvents, propellants, and refrigerants.

Halogen-containing compounds are a class of atmospheric chemicals that come from both natural and man-made sources. They are created by burning biomass, sea salt, biological processes in the ocean, and industrial synthesis. Depending on how they are removed, their atmospheric lifetimes can range from a few days to several centuries. Figure 4.4 displays the amounts of halogenated gases in the atmosphere according to a study.

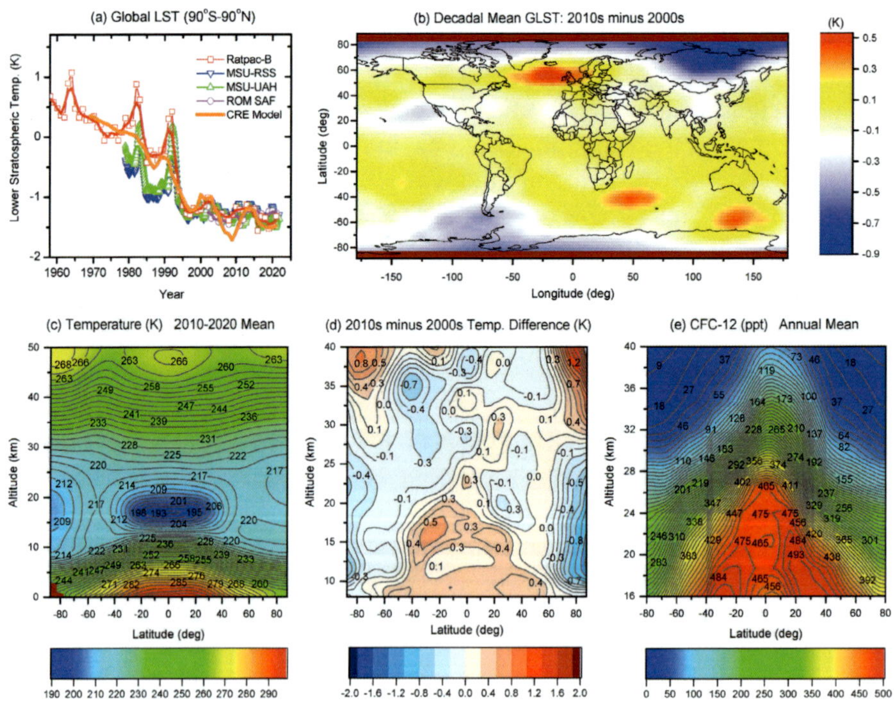

Figure 4.4. *(a): Time-series annual mean lower-stratospheric temperature anomaly datasets of the global (90° S–90° N) since 1958, obtained from multiple ground- and satellite-based data measurements (Ratpac, MSU-UAH, MSU-RSS, and ROM*

SAF), as well as temperatures calculated by the CRE equation (thick solid lines in orange). Further, also shown are the 3-year smoothing (thick solid lines in colors) to observed temperature anomalies (symbols). Note that for comparisons, the lower-stratospheric temperature anomalies are offset to the same level in the 2000s–2010s, but this offsetting has no effects on their long-term trends. (b): Map for the decadal mean global lower-stratospheric temperature (GLST) difference of the 2010s (2010–2020) minus the 2000s (2000–2010). (c): Decadal mean zonal mean latitude–altitude distribution of the temperature climatology in the troposphere and stratosphere in the 2010s (averaged from 2010 to 2020). (d): Difference in decadal mean zonal mean latitude–altitude distribution of the temperature climatology at altitudes of 8–40 km of the 2010s (2010–2020) minus the 2000s (2002–2010). (e): Representative zonal mean latitude–altitude distribution of the CF_2Cl_2 (CFC-12) concentration obtained from the NASA UARS's CLEAS dataset.

Source: Lu Q-B. Major Contribution of Halogenated Greenhouse Gases to Global Surface Temperature Change. Atmosphere. *2022; 13(9):1419. https://doi.org/10.3390/atmos13091419.*

SF_6 and $CFCl_3$ were first found in the atmosphere by Lovelock (1971) using an electron-capture detector. Molina and Rowland (1974) demonstrated in a seminal work in atmospheric chemistry for which they were awarded the 1995 Nobel Prize in Chemistry that CFCs that are resistant to removal in the troposphere could break down photolytically in the stratosphere to release Cl atoms that can catalytically destroy stratospheric ozone. Chlorofluorocarbon molecules are intrinsically valuable due to their lack of chemical reactivity, which also enables them to remain unaltered in the majority of commercial applications and eventually be released back into the atmosphere in their original gaseous state. With the CFCs, the typical tropospheric sinks of oxidation, photodissociation, and wet and dry deposition are ineffective. The mid-stratosphere (25–40 km) photodissociation is the only significant sink for $CFCl_3$ and CF_2Cl_2. In addition to being significant greenhouse gases, these same CFCs effectively absorb infrared radiation and cause stratospheric ozone depletion.

Fully halogenated halocarbons and those with one or more hydrogen atoms exhibit quite different atmospheric behaviors. Before they can enter the stratosphere, halocarbons with at least one hydrogen atom, such as CF_2HCl, $CHCl_3$, and CH_3CCl_3, are essentially broken down in the troposphere by reaction with the hydroxyl radical. These species have lifetimes in the atmosphere ranging from months to decades.

4.5.1. Chlorofluorocarbons

The collective term for a group of compounds with different ratios of carbon, hydrogen, chlorine, and fluorine atoms are HCFCs. The American Society of Heating and Refrigeration Engineers proposed the somewhat esoteric system of numbering these compounds in 1957. For the simpler HCFCs, the numbering system may be summarized as follows:

1. The first digit on the right is the number of fluorine (F) atoms in the compound.
2. The second digit from the right is one more than the number of hydrogen (H) atoms in the compound.
3. The third digit from the right, plus one, is the number of carbon (C) atoms in the compound. When this digit is zero (i.e., only one carbon atom in the compound), it is omitted from the number.

The number of chlorine (Cl) atoms in the compound is found by subtracting the sum of the fluorine and hydrogen atoms from the total number of atoms that can be connected to the carbon atoms.

CCl_2F_2	$C_2Cl_2F_4$	$CHClF_2$	CCl_3F
CFC-12	CFCl-114	CFC-22	CFC-11

4.

4.5.2. Methyl Chloride (CH_3Cl)

About 0.5 parts per billion of chlorine are supplied to the stratosphere by methyl chloride. Before 1996, it was thought that biological processes occurring in the ocean were the source of the atmospheric input of CH_3Cl.

Table 4.13. *Commonly Used Refrigerants in Residential and Automobile Applications*

Application Area	Mostly Used Refrigerant in the System	Mostly Adopting Regions	ODP	[a] GWP_{100} (IPCC5)	Comments
Residential air conditioning	R22 [**11,12,13,14,15, 16,17,18**] (retrofitting R22 are R134a (GWP-1300), R407C (GWP-1624), R407A (GWP-1923) and R404A (GWP-3942))	Developing countries	0.034	1760	Banned in January 2020

	R410A [**19**,**20**,**21**]	Developed countries	0	1923	Will be phased out soon
	R32 [**22**]	Developed countries	0	677	DAIKIN Tec. introduced
Automobile air conditioning	R134a [**23**,**24**,**25**,**26**,**27**,**28**] (retrofitting R134a are R32, R152a, R513A)	Developed and developing countries	0	1300	Replaced R12, but will be restricted very soon
	R1234yf [**30**,**31**,**32**]	Developed countries	0	<1	Low volumetric capacity

Source: Uddin K, Saha BB. An Overview of Environment-Friendly Refrigerants for Domestic Air Conditioning Applications. Energies. *2022; 15(21):8082. https://doi.org/10.3390/en15218082.*

[a] GWP_{100} (IPCC5) values based on the IPCC AR5 report (IPCC, 2014) for pure refrigerants; values for mixtures calculated based on the mixture ratio of pure refrigerants.

It is currently thought that burning biomass, tropical plants, and tropical terrestrial sources are significant. Reaction with the OH radical in the troposphere is the main sink of CH_3Cl. Table 4.13 provides an example. The best estimates indicate that the sinks are greater than the sources by 1049 Gg year^{-1}. There have been suggestions that there is a significant, as of yet unknown, source of CH_3Cl, most likely in the tropics.

4.5.3. Methyl Bromide (CH_3Br)

The most prevalent bromocarbon in the atmosphere, methyl bromide (CH_3Br), is the primary cause of bromine that reaches the stratosphere. CH_3Br is used industrially in soil and building fumigation, grain preservation in silos, and treatment of fresh fruits and vegetables due to its toxicity to a wide range of pests, including weeds, insects, and bacteria. CH_3Br is produced in the ocean, burns biomass, is released from plants and marshes, and is used as a fumigant and a leaded fuel additive. Burning biomass and fumigating soil are the two primary human-caused sources. Oceanic chemical and biological degradation, uptake by green plants, loss to soils, reaction with OH, photolysis at higher altitudes, and so on are some of the sinks for CH_3Br. With a mean interhemispheric ratio, NH/SH, of 1.3 ± 0.1, the overall mean atmospheric mixing ratio of CH_3Br is between 9 and 10 ppt (Kurylo et al., 1999). The easy release of active bromine is reflected in the rapid drop in CH_3Br mixing ratios with altitude once in the stratosphere.

Oceans are sources and sinks of CH_3Br; the net oceanic flux to the atmosphere is estimated to be negative at 21 Gg yr^{-1}. The estimated oceanic uptake of CH_3Br is currently 77 Gg yr^{-1}. This results in a total estimated annual sink of 210 Gg yr^{-1}, when combined with soil uptake (46.8 Gg yr^{-1}), atmospheric OH reaction and photolysis (86 Gg yr^{-1}). In order to sustain the net ocean sink of 21 Gg yr^{-1}, an estimated 56 Gg yr^{-1} of CH_3Br must be released into the ocean. This yields 151 Gg yr^{-1} in total estimated emissions, which causes an imbalance (missing sources) in the CH_3Br budget of 59 Gg yr^{-1}. The search is still on to find all of the natural sources of CH_3Br.

4.5.3.1. Tropospheric Lifetime of CH_3Br

The tropospheric lifetime of CH_3Br can be estimated based on its mixing ratio and its estimated total sink. With a mean tropospheric mixing ratio of 10 ppt and 1.8×10^{20} total moles of air in the atmosphere, the total number of moles of $CH_3Br \cong (10 \times 10^{-12})(1.8 \times 10^{20}) = 18 \times 10^{8}$ mol CH_3Br. With a molecular weight of 95 g mol^{-1}, this is 1.71×10^{11} g CH_3Br. Given a total sink of 210×10^{9} g yr^{-1}, the mean atmospheric residence time of CH_3Br is

$$\tau_{CH_3Br} = \frac{1.71 \times 10^{11}}{2.1 \times 10^{11}} = 0.8 \text{ yr}$$

REFERENCES

1. Amos, H. M., et al. (2012). Gas-particle partitioning of atmospheric Hg(II) and its effect on global mercury deposition. *Atmospheric Chemistry and Physics, 12,* 591–603.
2. Andreae, M. O. (1990). Ocean–atmosphere interactions in the global biogeochemical sulfur cycle. *Marine Chemistry, 30,* 1–29.
3. Andreae, M. O., & Barnard, W. R. (1984). The marine chemistry of dimethylsulfide. *Marine Chemistry, 14,* 267–279.
4. Andreae, M. O., & Crutzen, P. J. (1997). Atmospheric aerosols: Biogeochemical sources and role in atmospheric chemistry. *Science, 276,* 1052–1058.
5. Andreae, M. O., & Merlet, P. (2001). Emission of trace gases and aerosols from biomass burning. *Global Biogeochemical Cycles, 15,* 955–966.
6. Andreae, M. O., & Raemdonck, H. (1983). Dimethylsulfide in the surface ocean and the marine atmosphere: A global view. *Science, 221,* 744–747.
7. Andreae, M. O., & Rosenfeld, D. (2008). Aerosol-cloud-precipitation interactions: Part 1. The nature and sources of cloud-active aerosols. *Earth-Science Reviews, 89,* 13–41.
8. Andreae, M. O., et al. (2004). Smoking rain clouds over the Amazon. *Science, 303,* 1337–1342.
9. Artaxo, P., et al. (2002). Physical and chemical properties of aerosols in the wet and dry seasons in Rondonia, Amazonia. *Journal of Geophysical Research, 107,* 8081. https://doi.org/10.1029/2001JD000666.
10. Bergamaschi, P., et al. (2000). Inverse modeling of the global CO cycle 1. Inversion of CO mixing ratios. *Journal of Geophysical Research, 105,* 1909–1927.
11. Berresheim, H., et al. (1995). Sulfur in the atmosphere. In H. B. Singh (Ed.), *Composition, Chemistry, and Climate of the Atmosphere* (pp. 251–307). Van Nostrand Reinhold.
12. Bojkov, R. D. (1988). Ozone changes at the surface and in the free troposphere. In I. S. A. Isaksen (Ed.), *Tropospheric Ozone* (pp. 83–96). Reidel.
13. Bond, T. C., et al. (2007). Historical emissions of black and organic carbon aerosol from energy related combustion, 1850–2000. *Global Biogeochemical Cycles, 21,* GB2018. https://doi.org/10.1029/2006GB002840.
14. Bond, T. C., et al. (2013). Bounding the role of black carbon in the climate system: A scientific assessment. *Journal of Geophysical Research, 118,* 5380–5552.

15. Bouwman, A. F., et al. (1997). A global high-resolution emission inventory for ammonia. *Global Biogeochemical Cycles, 11,* 561–587.
16. Brasseur, G. P., et al. (1999). *Atmospheric Chemistry and Global Change.* Oxford University Press.
17. California Air Resources Board. (1989). *Information on Substances for Review as Toxic Air Contaminants* (Report ARB/SSD/89-01). California Air Resources Board.
18. Chang, S. G., et al. (1982). Chemical and catalytic properties of elemental carbon. In G. T. Wolff & R. L. Klimsch (Eds.), *Particulate Carbon: Atmospheric Life Cycle* (pp. 158–181). Plenum.
19. Chin, M., & Davis, D. D. (1995). A reanalysis of carbonyl sulfide as a source of stratospheric background sulfur aerosol. *Journal of Geophysical Research, 100,* 8993–9005.
20. Christian, T. J., et al. (2003). Comprehensive laboratory measurements of biomass-burning emissions: Emissions from Indonesian, African, and other fuels. *Journal of Geophysical Research, 108,* 4719. https://doi.org/10.1029/2003JD003704.
21. Chung, S. H., & Seinfeld, J. H. (2002). Global distribution and climate forcing of carbonaceous aerosols. *Journal of Geophysical Research, 107,* 4407. https://doi.org/10.1029/2001JD001397.
22. Crutzen, P. J. (1988). Tropospheric ozone: A review. In I. S. A. Isaksen (Ed.), *Tropospheric Ozone* (pp. 3–32). Reidel.
23. Dockery, D. W., et al. (1993). An association between air pollution and mortality in six U.S. cities. *New England Journal of Medicine, 329,* 1753–1759.
24. Eriksson, E. (1959). The yearly circulation of chloride and sulfur in nature: Meteorological, geochemical, and pedological implications. *Tellus, 11,* 375–404.
25. Fiore, A. M., et al. (2014). Estimating North American background ozone in U.S. surface air with two independent global models: Variability, uncertainties, and recommendations. *Atmospheric Environment, 96,* 284–300.
26. Fishman, J., et al. (1990). Distribution of tropospheric ozone determined from satellite data. *Journal of Geophysical Research, 95,* 3599–3617.
27. Fishman, J., et al. (1992). Distribution of tropospheric ozone in the tropics from satellite and ozonesonde measurements. *Journal of Atmospheric and Terrestrial Physics, 54,* 589–597.
28. Flagan, R. C., & Seinfeld, J. H. (1988). *Fundamentals of Air Pollution Engineering.* Prentice-Hall. Reprinted by Dover Publications, 2012.

29. Gantt, B., et al. (2009). A new physically-based quantification of marine isoprene and primary organic aerosol emissions. *Atmospheric Chemistry and Physics, 9,* 4915–4927.
30. Gauss, M., et al. (2006). Radiative forcing since preindustrial times due to ozone change in the troposphere and the lower stratosphere. *Atmospheric Chemistry and Physics, 6,* 575–599.
31. Ge, X., et al. (2011). Atmospheric amines–Part I. A review. *Atmospheric Environment, 45,* 524–546.
32. Gong, S. L., et al. (2002). Canadian Aerosol Module (CAM): A size-segregated simulation of atmospheric aerosol processes for climate and air quality models 2. Global sea-salt aerosol and its budgets. *Journal of Geophysical Research, 107,* 4779. https://doi.org/10.1029/2001JD002004.
33. Guenther, A. (2002). The contribution of reactive carbon emissions from vegetation to the carbon balance of terrestrial ecosystems. *Chemosphere, 49,* 837–844.
34. Guenther, A., et al. (1995). A global model of natural volatile organic compound emissions. *Journal of Geophysical Research, 100,* 8873–8892.
35. Guenther, A., et al. (2006). Estimates of global terrestrial isoprene emissions using MEGAN (Model of Emissions of Gases and Aerosols from Nature). *Atmospheric Chemistry and Physics, 6,* 3181–3210.
36. Hauglustaine, D. A., et al. (1998). MOZART, a global chemical transport model for ozone and related chemical tracers: 2. Model results and evaluation. *Journal of Geophysical Research, 103,* 28291–28335.
37. Holmes, C. D., et al. (2010). Global atmospheric model for mercury including oxidation by bromine atoms. *Atmospheric Chemistry and Physics, 10,* 12037–12057.
38. IPCC (Intergovernmental Panel on Climate Change). (1995). *Climate Change 1995: The Science of Climate Change.* Cambridge University Press.
39. IPCC. (2001). *Climate Change 2001: The Scientific Basis.* Cambridge University Press.
40. IPCC. (2013). *Climate Change 2013: The Physical Science Basis.* Cambridge University Press.
41. Janach, W. E. (1989). Surface ozone: Trend details, seasonal variations, and interpretation. *Journal of Geophysical Research, 94,* 18289–18295.
42. Jiang, Y., & Yung, Y. L. (1996). Concentrations of tropospheric ozone for 1979 to 1992 over tropical Pacific South America from TOMS data. *Science, 272,* 714–716.

43. Junker, C., & Liousse, C. (2008). A global emission inventory of carbonaceous aerosol from historic records of fossil fuel and biofuel consumption for the period 1860–1997. *Atmospheric Chemistry and Physics, 8,* 1195–1207.
44. Keeley, J. E., & Rundel, P. W. (2005). Fire and the Miocene expansion of C4 grasslands. *Ecology Letters, 8,* 683–690.
45. Kettle, A. J., et al. (1999). A global database of sea surface dimethylsulfide (DMS) measurements and a procedure to predict sea surface DMS as a function of latitude, longitude, and month. *Global Biogeochemical Cycles, 13,* 399–444.
46. Kettle, A. J., et al. (2002). Global budget of atmospheric carbonyl sulfide: Temporal and spatial variations of the dominant sources and sinks. *Journal of Geophysical Research, 107,* 4658. https://doi.org/10.1029/2002JD002187.
47. Kiehl, J. T., & Rodhe, H. (1995). Modeling geographical and seasonal forcing due to aerosols. In R. J. Charlson & J. Heintzenberg (Eds.), *Aerosol Forcing of Climate* (pp. 281–296). Wiley.
48. Kurylo, M. J., et al. (1999). Short-lived ozone-related compounds. In C. A. Ennis (Ed.), *Scientific Assessment of Ozone Depletion: 1998, Global Ozone Research and Monitoring Project, Report 44* (pp. 2.1–2.56). World Meteorological Organization.
49. Lamarque, J. F., et al. (2010). Historical (1850–2000) gridded anthropogenic and biomass burning emissions of reactive gases and aerosols: Methodology and application. *Atmospheric Chemistry and Physics, 10,* 7017–7039.
50. Lamb, B., et al. (1987). A national inventory of biogenic hydrocarbon emissions. *Atmospheric Environment, 21,* 1695–1705.
51. Lamb, B., et al. (1993). A biogenic hydrocarbon emission inventory for the U.S.A. using a simple forest canopy model. *Atmospheric Environment, 27A,* 1673–1690.
52. Lee, C., et al. (2011). SO_2 emissions and lifetimes: Estimates from inverse modeling using *in situ* and global, space-based (SCIAMACHY and OMI) observations. *Journal of Geophysical Research, 116,* D06304. https://doi.org/10.1029/2010JD014758.
53. Lelieveld, J., et al. (1997). Terrestrial sources and distribution of atmospheric sulphur. *Philosophical Transactions of the Royal Society of London B, 352,* 149–158.
54. Liao, H., et al. (2003). Interactions between tropospheric chemistry and aerosols in a unified GCM simulation. *Journal of Geophysical Research, 108,* 4001. https://doi.org/10.1029/2001JD001260.

55. Liao, H., et al. (2004). Global radiative forcing of coupled tropospheric ozone and aerosols in a unified general circulation model. *Journal of Geophysical Research, 109,* D16207. https://doi.org/10.1029/2003JD004456.
56. Lindberg, S., et al. (2007). A synthesis of progress and uncertainties in attributing the sources of mercury in deposition. *Ambio, 36,* 19–32.
57. Liousse, C., et al. (1996). A global three-dimensional model study of carbonaceous aerosols. *Journal of Geophysical Research, 101,* 19411–19432.
58. Lovelock, J. E. (1971). Atmospheric fluorine compounds as indicators of air movements. *Nature, 230,* 379.
59. Lovelock, J. E., et al. (1972). Atmospheric dimethyl sulfide and the natural sulfur cycle. *Nature, 237,* 452–453.
60. Lurmann, F. W., & Main, H. H. (1992). Analysis of the ambient VOC data collected in the Southern California Air Quality Study. Final Report, ARB Contract A832-130. California Air Resources Board, Sacramento, CA.
61. Mieville, A., et al. (2010). Emissions of gases and particles from biomass burning during the 20th century using satellite data and a historical reconstruction. *Atmospheric Environment, 44,* 1469–1477.
62. Molina, M. J., & Rowland, F. S. (1974). Stratospheric sink for chlorofluoromethanes: Chlorine atom catalyzed destruction of ozone. *Nature, 249,* 810–812.
63. Mouillot, F., & Field, C. B. (2005). Fire history and the global carbon budget: A 1 × 1 fire history reconstruction for the 20th century. *Global Change Biology, 11,* 398–420.
64. National Research Council. (1991). *Rethinking the Ozone Problem in Urban and Regional Air Pollution.* National Academy Press.
65. Nisbet, E. G., et al. (2014). Methane on the rise–again. *Science, 343,* 493–495.
66. Notholt, J., et al. (2003). Enhanced upper tropical troposphere OCS: Impact on the stratospheric aerosol layer. *Science, 300,* 307–310.
67. Olivier, J. G. J., et al. (1996). Description of EDGAR Version 2.0, A set of global emission inventories of greenhouse gases and ozone-depleting substances for all anthropogenic and most natural sources on a per country basis and on 1 × 1 degree grid. RIVM/TNO report, Bilthoven, Netherlands.
68. Oltmans, S. J., & Levy, H. II. (1994). Surface ozone measurements from a global network. *Atmospheric Environment, 28,* 9–24.
69. Palmer, H. B., & Cullis, C. F. (1965). The formation of carbon from gases. In P. L. Walker (Ed.), *Chemistry and Physics of Carbon* (Vol. 1, pp. 265–325). Marcel Dekker.

70. Pavelin, E. G., et al. (1999). Evaluation of pre-industrial surface ozone measurements using Schönbein's method. *Atmospheric Environment, 33,* 919–929.
71. Pyle, J. A., et al. (2011). Bromoform in the tropical boundary layer of the Maritime Continent during OP3. *Atmospheric Chemistry and Physics, 11,* 529–542.
72. Reid, J. S., et al. (2005a). A review of biomass burning emissions. Part II: Intensive physical properties of biomass burning particles. *Atmospheric Chemistry and Physics, 5,* 799–825.
73. Reid, J. S., et al. (2005b). A review of biomass burning emissions. Part III: Intensive optical properties of biomass burning particles. *Atmospheric Chemistry and Physics, 5,* 827–849.
74. Roberts, J. M. (1995). Reactive odd-nitrogen (NOy) in the atmosphere. In H. B. Singh (Ed.), *Composition, Chemistry, and Climate of the Atmosphere* (pp. 176–215). Van Nostrand Reinhold.
75. Schultz, M. G., et al. (2008). Global wildland fire emissions from 1960 to 2000. *Global Biogeochemical Cycles, 22,* GB2002. https://doi.org/10.1029/2007GB003031.
76. Singh, H. B., et al. (2003). *In situ* measurements of HCN and CH_3CN over the Pacific Ocean: Sources, sinks, and budgets. *Journal of Geophysical Research, 108,* 8795. https://doi.org/10.1029/2002JD003006.
77. Smith, S. J., et al. (2011). Anthropogenic sulfur dioxide emissions: 1850–2005. *Atmospheric Chemistry and Physics, 11,* 1101–1116.
78. Sokolik, I. N. (2002). Dust. In J. R. Holton (Ed.), *Encyclopedia of Atmospheric Sciences* (pp. 448–455). Elsevier.
79. Staehelin, J., & Schmid, W. (1991). Trend analysis of tropospheric ozone concentrations utilizing the 20-year data set of balloon soundings over Payerne (Switzerland). *Atmospheric Environment, 25A,* 1739–1749.
80. Steiner, A. H., & Goldstein, A. H. (2007). Biogenic VOCs in *Volatile Organic Compounds in the Atmosphere*. In R. Koppmann (Ed.), Wiley-Blackwell.
81. Streets, D. G., et al. (2003). Biomass burning in Asia: Annual and seasonal estimates and atmospheric emissions. *Global Biogeochemical Cycles, 17,* 1099. https://doi.org/10.1029/2003GB002040.
82. Turner, S. M., & Liss, P. S. (1985). Measurements of various sulfur gases in a coastal marine environment. *Journal of Atmospheric Chemistry, 2,* 223–232.

83. van der Werf, G. R., et al. (2006). Interannual variability in global biomass burning emissions from 1997 to 2004. *Atmospheric Chemistry and Physics, 6,* 3423–3441.
84. Volz, A., & Kley, D. (1988). Evaluation of the Montsouris series of ozone measurements made in the nineteenth century. *Nature, 332,* 240–242.
85. Warneck, P. (1988). *Chemistry of the Natural Atmosphere.* Academic Press.
86. Went, F. W. (1960). Blue hazes in the atmosphere. *Nature, 187,* 641–643.
87. Whitby, K. T., & Cantrell, B. (1976). Fine particles. In *Proceedings of the International Conference on Environmental Sensing and Assessment, Las Vegas, NV, 1976* (pp. 1–6). Institute of Electrical and Electronic Engineers (IEEE).
88. WMO (World Meteorological Organization). (1986). *Atmospheric ozone 1985.* Global Ozone Research and Monitoring Project, Report 16. Geneva.
89. WMO. (1990). *Report of the International Ozone Trends Panel: 1988.* Global Ozone Research and Monitoring Project, Report 18. Geneva.
90. WMO. (1998). *Scientific Assessment of Ozone Depletion: 1998.* Global Ozone Research and Monitoring Project, Report 44. World Meteorological Organization, Geneva.
91. WMO. (2002). *Scientific Assessment of Ozone Depletion: 2002.* Global Ozone Research and Monitoring Project, Report 47. World Meteorological Organization, Geneva.
92. Yamasoe, M. A., et al. (2000). Chemical composition of aerosol particles from direct emissions of vegetation fires in the Amazon Basin: Water-soluble species and trace elements. *Atmospheric Environment, 34,* 1641–1653.
93. Yokelson, R. J., et al. (2007). The Tropical Forest and Fire Emissions Experiment: Overview and airborne fire emission factor measurements. *Atmospheric Chemistry and Physics, 7,* 5175–5196.
94. Yvon-Lewis, S. (2000). Methyl bromide in the atmosphere and ocean. *IGACtivities Newsletter, International Global Atmospheric Chemistry Project,* Issue 19, 9–12.
95. Zender, C. S., et al. (2003). Mineral dust entrainment and deposition (DEAD) model: Description and 1990s dust climatology. *Journal of Geophysical Research, 107*(D24), 4416. https://doi.org/10.1029/2002JD002775.
96. Zender, C. S., et al. (2004). Quantifying mineral dust mass budgets: Terminology, constraints, and current estimates. *Eos (Transactions of the American Geophysical Union), 85*(48), 509.

CHAPTER 5

CHEMISTRY OF THE ATMOSPHERE

Contents

INTRODUCTION

The study of the chemical compositions, reactions, and processes that occur within the Earth's atmosphere is the focus of the dynamic and complex field known as atmospheric chemistry. Nitrogen (78%) and oxygen (21%), with trace amounts of other gases like argon, carbon dioxide, neon, and helium, make up the majority of the atmosphere. Another important component is water vapor, which varies substantially in concentration based on location and meteorological conditions. The study of atmospheric chemistry looks at how these gases interact with different pollutants that are brought in by both human activity and natural sources. The study of ozone, both at ground level and in the stratosphere, is one of the most important components of atmospheric chemistry. The ozone layer, which shields life on Earth from damaging UV radiation from the sun, is made up of stratospheric ozone. Tropospheric ozone, on the other hand, contributes significantly to smog and can be harmful to both the environment and human health. The greenhouse effect, which is caused by gases such as carbon dioxide, methane, and nitrous oxide trapping heat in the atmosphere and causing climate change, is another crucial topic of research.

The cycling of elements like carbon, nitrogen, and sulfur is also greatly aided by the atmosphere. Acid rain is caused by the production of acids like nitric acid and sulfuric acid that result from chemical reactions in the atmosphere. Buildings, human health, and ecosystems may all suffer as a result of this phenomenon. In addition, sunlight's interaction with atmospheric chemicals can result in the creation of reactive species like free radicals, which power numerous processes related to air pollution and the decomposition of pollutants. The development of tactics to reduce air pollution and lessen the effects of climate change depends on advancements in atmospheric chemistry. By comprehending the chemical reactions that take place in the atmosphere, scientists can enhance their ability to forecast weather patterns, identify the origins of pollution, and create technologies that lower the emissions of dangerous materials. Thus, maintaining environmental health and securing the planet's sustainability depend heavily on the study of atmospheric chemistry.

5.1. ORIGIN OF ATMOSPHERE CHEMISTRY

The formation and evolution of the atmosphere are related to Earth's beginnings. Three stages can be distinguished in the evolution of the atmosphere.

1. **Stage I:** It is estimated that a protoplanetary gas-dust cloud gave rise to Earth approximately 4.5 billion years ago. The principal gases in the cloud were hydrogen and helium. Metals and water, ammonia, methane, formaldehyde, methylacetylene, acetonitrile, and other interstellar compounds were probably present in the dust particles.

As a result of the gas dust cloud's slow gravitational contraction, radioactive decay of radionuclides, the gravitational pull of the moon that was formed at the same time, and other factors, the cloud heated up and partially melted. Earth's formation was an exothermic process. The massive volumes of highly kinetically energetic gases and vapors that were thus released were dispersed into cosmic space. Thus, hydrogen, water, ammonia, methane, noble gases, and certain sulfur compounds made up the majority of the dense atmosphere that was created approximately 3.5×10^9 years ago. The atmosphere was initially decreasing in nature because it was oxygen-free. The hydrosphere gradually began to form at the same time as a result of a massive condensation of water vapor. By dissolving carbon dioxide in liquid water that followed the formation of sedimentary rocks, this phenomenon had an impact on the evolution of the atmosphere. Due to their low atomic masses, nitrogen proliferated in the atmosphere as hydrogen and helium escaped into space.

2. **Stage II:** The second stage of evolution initiated at around 3.5×10^9 years ago lasted at about 2×10^9 years ago. Large amounts of magma saturated with dissolved gases formed during this time due to volcanic eruptions. Afterwards, magma pushed iron that was in the mantle toward the earth's surface. Because there were large amounts of reducing gases (H_2 S, H_2, and CH_4) present and no free oxygen was present, the atmosphere was still generally reducing in nature. Iron moved to the core during the later stages of stage II, preventing the ejected gases from interacting with iron and making them more oxidizing in nature. CO_2, SO_2, and H_2O vapors were among them. Still, there was no free oxygen in the atmosphere, and it remained reducing rather than oxidizing. Iron oxides eventually formed as a result of the atmosphere's oxidizing action, incorporated into sediments, and created redbeds.

3. **Stage III:** In the third stage, the present atmosphere, oxygen production began to exceed oxygen consumption via photochemical dissociation of water molecules. In the upper atmosphere, ultraviolet light was absorbed by water molecules, which then divided into hydrogen and oxygen. The primary process in the atmosphere where oxygen was accumulated was photosynthesis. During this time, multicellular marine organisms first emerged. They drifted on the water and ate phytoplankton. Heterotrophic bacteria became fewer in number as phytoplankton decreased. As a result, the atmosphere's oxygen content rose.

5.2. STRUCTURE OF THE ATMOSPHERE

The hydrosphere, which includes seas, lakes, rivers, groundwater, snow, and ice, has a mass of 1.4×10^{21} kg, while the Earth has a mass of 5.98×10^{24} kg. The total mass of the atmosphere is roughly 5.03×10^{18} kg. There is no clearly defined altitude for the upper boundary of the atmosphere. The atmospheric material gradually approaches interplanetary space as it rises in altitude, declining steadily. The envelope of gases and particles that envelops Earth is known as the atmosphere. This quantity of materials stays in our planet's orbit around the Sun and during its axis rotation.

In relation to the entire Earth's atmosphere, the relatively dense region is a very thin layer. Approximately 99% of the air is found in the lower 30 km layer of the atmosphere, and half of the atmosphere's mass is found below 5.5 km. Because of the interaction between the solar wind and Earth's natural magnetic field, the entire atmosphere is actually the same as the magnetosphere's range (Figure 5.1).

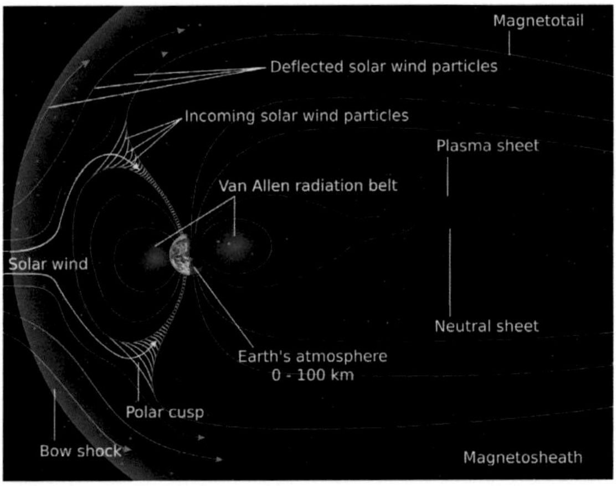

Figure 5.1. *The magnetosphere around the Earth is formed by the interaction between Earth's natural magnetic field and solar wind.*

Source: By Original: NASA Vector: Aaron Kaase, Medium69 – Structure of the magnetosphere numbered.svg, Public Domain, https://commons.wikimedia.org/w/index.php?curid=65031109.

5.2.1. Formation of the Earth Atmosphere

Hot molten rock gave rise to the Earth approximately 4.5 billion years ago. Most likely, hydrogen, helium, and a few basic hydrogen compounds like ammonia (NH_3) and methane (CH_4) made up its initial atmosphere. Similar to the atmospheres of Jupiter and Saturn today, this first Earthly atmosphere may have existed. The strong solar wind from the Sun blew away this early atmosphere because our planet didn't yet have a magnetic field to shield it. The early atmosphere of the Earth was essentially lost. A solid crust with multiple active volcanoes formed approximately 4pc billion years ago as the Earth cooled enough. Gases from our planet's hot interior have reached the surface due to the strong volcanic activity. Ammonia, carbon dioxide, and water vapor in general made up the Earth's secondary atmosphere. Similar to this early Earth's atmosphere are the current atmospheres of Mars and Venus, which are primarily composed of carbon dioxide.

As the atmosphere steadily cooled over hundreds of millions of years, the majority of the water vapor condensed and created clouds. The oceans were formed through precipitation from these clouds. The oceans absorbed the majority of the atmospheric carbon dioxide at the same time. Simultaneously, the NH_3 molecules were broken down by solar radiation, releasing the chemically inert nitrogen into the atmosphere.

Based on some theory, the oxygen was first produced in the atmosphere by photochemical dissociation of water vapor by intense ultraviolet radiation:

$$2H_2O + \text{UV radiation} \rightarrow 2H_2 + O_2.$$

Nevertheless, this was insignificant, and free oxygen was most likely created as a consequence of photosynthesis by microscopic organisms called cyanobacteria, sometimes known as blue-green algae, that lived some 2.5 million years ago. Photosynthesis uses carbon dioxide, water, and light energy releasing organic compounds (carbohydrates) and oxygen:

$$CO_2 + H_2O + \text{Sunlight} \rightarrow \text{Organic compounds} + O_2.$$

First, the surface oxidation of rocks requires a small amount of atmospheric oxygen (weathering process). After the surface rocks completely oxidized, the atmosphere's oxygen content started to rise more rapidly. As the atmospheric oxygen reached 1–2% of the present oxygen level, ozone (O_3) could form to shield Earth's surface from intense ultraviolet radiation:

$$2O_2 + \text{uv radiation} \rightarrow O_3 + O$$

and

$$O + O_2 = O_3.$$

Primitive plants began to emerge at this time, aiding in the process of photosynthesis. According to recent research, land plants began to form a few million (roughly 700 million) years ago. During that time, they would have gradually increased atmospheric oxygen content while removing carbon dioxide, which went from 0.5% to 0.04%. Over the past few million years, plate tectonics, volcanism, and climate have all caused variations in oxygen levels at different times (Holland, 2006). At last, the atmospheric oxygen content reached a stable level of approximately 21%.

5.2.2. History of the Atmospheric Chemistry

The first measurements of the composition of the atmosphere were made in the 18th century by chemists like Joseph Priestley, Antoine Lavoisier, and Henry Cavendish. In the late 19th and early 20th centuries, attention turned to trace constituents with extremely low concentrations. The discovery of ozone by Christian Friedrich Schönbein in 1840 was one particularly significant finding for atmospheric chemistry. After examining the composition of air, atmospheric science in the 20th century turned its focus to the chemical reactions that produce and destroy airborne compounds, as well as the evolution of trace gas concentrations in the atmosphere over time. The explanations of how the ozone layer is formed and maintained by Sydney Chapman and Gordon Dobson, as well as the explanation of photochemical smog by Arie Jan Haagen-Smit, are two notable examples of this. Additional research on ozone-related problems resulted in Paul Crutzen, Mario Molina, and Frank Sherwood Rowland sharing the 1995 Nobel Prize in Chemistry.

The focus is now changing once more in the twenty-first century. The study of atmospheric chemistry as a component of the Earth system is growing. The focus has shifted from studying atmospheric chemistry in isolation to studying it as a component of the atmosphere, biosphere, and geosphere as a whole. The interactions between the composition of the atmosphere and the oceans, terrestrial ecosystems, and changing climate, which impact the recovery of the ozone hole and vice versa, are significant drivers of this. One such example is the relationship between chemistry and climate (Akimoto et al., 1994).

5.3. CHEMICAL AND PHOTOCHEMICAL REACTIONS IN THE ATMOSPHERE

In order to comprehend the specific mechanisms of the chemical reactions taking place in the atmosphere, atmospheric reaction research is crucial.

The sun's radiation, mainly in the ultraviolet portion of the spectrum, and the hydroxyl radical are the two most significant factors in atmospheric chemistry. While the latter acts as the primary reactive intermediate and reservoir of daytime atmospheric chemical phenomena, the former acts as a source of high energy to a single gas molecule to initiate a series of atmospheric reactions.

Gaseous atmospheric species can be classified as follows:

1. Inorganic oxides – CO, CO_2, NO_2, SO_2

2. Oxidants – O_3
3. Reductants – CO, SO_2, H_2S
4. Organics – CH_4, some alkenes and aryl compounds
5. Photochemically active species – NO_2, formaldehyde
6. Acids (H_2SO_4), Bases (NH_3), salts (NH_4HSO_4)
7. Unstable reactive species- HO. radical, electronically excited NO_2

In addition, both solid and liquid suspended particles play a significant role as source and sink for gas-phase species, as sites for surface reactions (solid particles) and as media for aqueous phase reactions (liquid droplets).

5.3.1. Photochemistry of the Atmosphere

The photochemical reaction, which is the most important aspect of atmospheric chemistry, is a chemical reaction that is started when reactant molecules absorb light photons. The equation $E = h\nu$, where h is Planck's constant and ν is the frequency of light, gives the energy, E, of a photon of visible or ultraviolet light. Therefore, reactant molecules can be activated by the energy contained in the absorbed photons. The resulting transient excited states of the reactant molecules have different physical and chemical characteristics from the initial molecules. As a result, new molecules are created, existing molecules mix with one another or with other molecules, or electrons, protons, hydrogen atoms, or their energy are transferred to new molecules. To understand the makeup and ultimate fate of chemical species in the atmosphere, photochemical reactions are essential. Smog is primarily caused by nitrogen dioxide (NO_2), one of the most photochemically active species found in contaminated air. On absorption of light energy ($E = h\nu$) in the visible or UV region of the electromagnetic spectrum, NO_2 yields an electronically excited molecule,

$$NO_2 + h\nu \rightarrow NO_2{}^*$$

It is possible to comprehend the nature of the excited state by thinking about the addition of electrons to a molecule. An even number of electrons can be found in most molecules. With a maximum of two electrons with opposing spins per orbital, electrons have a tendency to occupy orbitals. One of these electrons may advance to a high energy vacant orbital upon the absorption of sufficiently energy-dense light. An excited singlet state can result when the promoted electron keeps its spin opposite that of its previous partner. However, in certain instances, the promoted electron undergoes a spin reversal, resulting in it having the same spin as its previous partner and forming an excited triplet state.

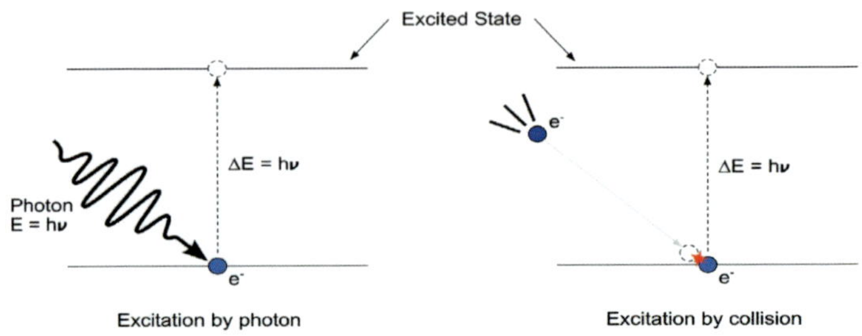

Figure 5.2. Electron excitation.

Source: By Planetphysicsguy – Own work, CC BY-SA 4.0, https://commons.wikimedia.org/w/index.php?curid=126423006.

These chemically reactive, energetically excited species actively participate in atmospheric reactions. The electronically excited molecules lose their excess energy in one of the following reactions after the light is absorbed.

Energy loss to another atom or molecule by physical quenching and dissipation of energy as heat:

$O_2* + M + h\nu \rightarrow O_2 + M$ (higher translational energy)

Dissociation of the excited molecule

$O_2* + h\nu \rightarrow O + O$

Direct reaction with another species – The energy released during de-excitation is sufficient for the reaction to take place.

$O_2* + O_3 + h\nu \rightarrow 2O_2 + O$

Luminescence – loss of energy upon emission of electromagnetic radiation

$NO_2* + h\nu \rightarrow NO_2 + h\nu$

Fluorescence is the term for the luminescence phenomenon that involves the instantaneous reemission of light. If the process is noticeably delayed, phosphorescence occurs. When an excited intermediate forms and then de-excites, releasing some of its energy as light, chemiluminescence occurs.

$O_3 + NO + h\nu \rightarrow NO_2* + O_2$ (higher energy)

Intermolecular energy transfer wherein energy is transferred from an excited species to another species, which then becomes excited.

$M* + Y + h\nu \rightarrow M + Y*$

A photosensitized reaction may take place when the second species so generated participates in chemical transformations.

Intramolecular transfer, wherein energy is transferred within the molecule itself.

$XY^* + h\nu \rightarrow XY^\dagger$ (where † stands for the excited state of the same molecule).

The process of o-nitrobenzaldehyde converting to o-nitrosobenzoic acid is an example of spontaneous isomerization, and it is used in chemical actinometers to gauge electromagnetic radiation exposure (Akimoto et al., 1994).

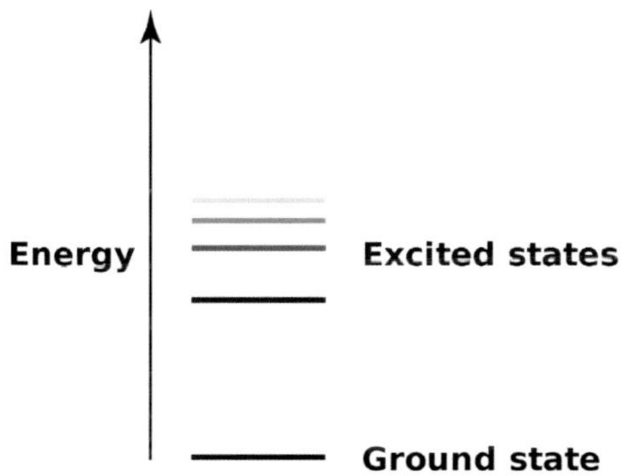

Figure 5.3. *Energy levels.*

Source: By SVG: RehuaOriginal: Rozzychan – This file was derived from: Energylevels.png, CC BY-SA 3.0, https://commons.wikimedia.org/w/index.php?curid=35220791.

Photoionization is the formation of ionic species with a loss of an electron.

$N_2{}^* \rightarrow N_2{}^+ + e^- + h\nu$

Photodissociation – radicals are produced

$N_2 + h\nu \rightarrow N\bullet + N\bullet$

The energy of the electromagnetic radiation absorbed in the infrared spectrum is insufficient to break chemical bonds. Nonetheless, they give the receptor molecules enough rotational and vibrational energy. The energy that is absorbed as infrared radiation eventually dissipates as heat, which causes the atmosphere's temperature to rise.

5.3.2. Distribution of Species in Atmosphere

The atmosphere also contains particulate matter, free radicals, and ions in addition to gaseous species.

5.3.2.1. Ions in the Atmosphere

The ionosphere is the portion of the atmosphere that is located about 50 km above sea level. The fact that the ionosphere reflects and alters radio waves used for navigation and communication makes its significance clear. The ionosphere's height, ionized particle concentration, and presence of distinct regions change over time. X-rays and UV light from the Sun have enough energy during the day to remove electrons from atoms and molecules, creating an endless supply of ions and free electrons. The principal ions present in this region are N_2^+, O_2^+, NO^+, N^+ and O^+. At the same time, some of the ions and electrons collide and re-combine to form electrically neutral atoms and molecules. Positive ions can recombine with free electrons in the dark. In lower ionosphere regions with higher species concentration, this phenomenon occurs quickly. As a result, radio waves travel farther at night because they are reflected off of lower ionosphere regions.

The Earth's magnetic field firmly confines the ions, preventing them from leaving the ionosphere. Van Allen belts are one way that these ionizing particles are affected by the Earth's magnetic field. These belts bear the name of the scientist who is credited with discovering them, Dr. James Van Allen. The earth's magnetic field traps a zone of highly energetic charged particles around the planet, forming the Van Allen Belts. If we picture them as two doughnuts, the earth's magnetic field's axis passes through the holes in the doughnuts. The outer and inner belts are separated from the main belt. Electrons form the outer belt, while protons form the inner belt due to collisions between cosmic rays in the upper atmosphere.

5.3.2.2. Free Radicals in the Atmosphere

High energy electromagnetic radiations create radicals in the atmosphere, which are atoms or groups of atoms with unpaired electrons. Because unpaired electrons have a strong pairing tendency, free radicals have a high reactivity. They take part in chain reactions in which every reaction produces a radical in its own right. The termination of a chain is marked by the reaction of a radical with another radical (chain-terminating reaction).

$$H_3C^{\cdot} + H_3C^{\cdot} \rightarrow C_2H_6$$

It is noteworthy to mention that single atoms from diatomic gases and free radicals are relatively stable and can survive for several minutes in the rarefied upper atmosphere. These species cover great distances at high altitudes before running into other reactive species. Conversely, species that are electronically excited have a limited lifespan because, rather than interacting with other species, their energy is lost through radiation. Some important radicals in the atmosphere are hydroxyl (HO•), Hydroperoxy (HOO•) and alkylperoxy (H_3 COO•) radicals. The atmospheric chemistry of the most prevalent hydroxy (HO•) and Hydroperoxy (HOO•) radicals.

1. **Hydroxyl and Hydroperoxyl Radicals in the Atmosphere:** The most significant reactive species in atmospheric chemical reactions is the hydroxyl radical, or OH•. It is typically created at higher altitudes through photolysis of water.

   $H_2O + h\nu \rightarrow HO\bullet + H$

   The wavelength of the light and the substrate's availability affect the creation of the HO• radical. As a result, summertime, and tropical regions experience higher concentrations of the HO• radical because of the high humidity levels that promote HO• formation. The diurnal plot of the variation in HO• radical concentration with the day. The plot shows that the HO• radical concentration peaks in the afternoon and progressively declines at night. Tropospheric HO• has an average daily and seasonal concentration of 2×10^5 to 1×10^6 radicals per cm^3 worldwide.

   Under laboratory conditions, HO• radical is generated by the photolysis of nitrous acid vapor and hydrogen peroxide.

   $HONO + h\nu \rightarrow HO^{\cdot} + NO$

   $HOOH + h\nu \rightarrow 2HO^{\cdot}$

   The radical is also produced by photolysis of ozone, wherein the reaction proceeds with the formation of excited oxygen atoms that combine water molecules to yield hydroxyl radical.

   $O_3 + h\nu(\lambda < 315\ nm) \rightarrow O^* + O_2$

   $O^* + H_2O \rightarrow 2HO^{\cdot}$

   Additionally, it is created as a transitional product when photochemical smog forms. An unpaired electron in a hydroxyl radical makes it extremely reactive and unstable. As a result, it interacts with a range of species found in the atmosphere in multiple ways.

2. **Sinks for Removal of Hydroxyl Radical**: A sink is a medium made up of molecules that interact with the species in question to potentially lessen their harmfulness. Methane and carbon monoxide are the two sinks that remove the OH• radical from the troposphere.

   $CH_4 + HO\bullet \rightarrow H_3C\bullet + H_2O$

   The methyl radical is highly reactive and forms methylperoxyl radical, H_3 COO'• upon reaction with O_2

   $H_3C\bullet + O_2 \rightarrow H_3COO\bullet$

Similarly, reaction with carbon monoxide produces atomic hydrogen, which then reacts with O_2 and results in the formation of hydroperoxyl radical HOO•.

$CO + HO\bullet \rightarrow CO_2 + H$

$H + O_2 \rightarrow HOO$

The hydroperoxyl radical undergoes a series of chain termination reactions, either with radicals,

$HOO\bullet + HO\bullet \rightarrow H_2O + O_2$

$HOO\bullet + HOO\bullet \rightarrow H_2O_2 + O_2$

Or with neutral species that regenerate hydroxyl radicals.

$HOO\bullet + NO \rightarrow NO_2 + HO\bullet$

$HOO\bullet + O_3 \rightarrow 2O_2 + HO\bullet$

Formaldehyde is a significant organic molecule found in the atmosphere that can react with the hydroperoxyl radical to regenerate it. Formaldehyde undergoes photolytic dissociation, producing formyl radical as the reactive intermediate. Hydroperoxyl radical is then produced by combining with atmospheric oxygen.

$HCHO + h\nu \rightarrow H + H\dot{C}O$

$H\dot{C}O + O_2 \rightarrow HOO^{\bullet} + CO$

In many different chemical reactions, the hydroperoxyl radical, HOO•, is a crucial intermediate. On the other hand, hydroperoxyl radicals, such as OH•, react more slowly with other species. Because these radicals hardly exist without hydroxyl radicals, hydroperoxyl radical kinetic studies are challenging to conduct (Akimoto et al., 1994).

5.4. REACTIONS OF ATMOSPHERIC OXYGEN

Life forms depend on oxygen in the atmosphere, particularly in the troposphere. The oxygen cycle is illustrated schematically, showing the biogeochemical reactions of oxygen in the atmosphere, lithosphere, hydrosphere, and biosphere. One of the main ingredients in the processes that take place on Earth's surface is oxygen.

It powers reactions that release energy, like the burning of fossil fuels.

CH_4 (in natural gas) $+ 2O_2 \rightarrow CO_2 + 2H_2O$

Aerobic organisms rely on atmospheric oxygen for degrading organic matter. Oxygen also plays a crucial role in chemical weathering processes.

$4FeO + O_2 \rightarrow 2Fe_2O_3$

Oxygen is returned to the atmosphere through photosynthesis by green plants and planktons.

$$CO_2 + H_2O + h\nu \rightarrow \{CH_2O\} + O_2$$

It can be demonstrated that only a small percentage of the carbon fixed by photosynthetic processes is deposited in fossil fuel beds; the majority is distributed as humic material in mineral formations. As a result, oxygen production and consumption are cyclical processes that keep oxygen levels in the atmosphere constant. Molecular oxygen exhibits peculiar behavior in that it can be excited to singlet molecular oxygen, or $^1O_2$, from its ground state, which is a triplet state with an unpaired electron and is represented as $^3O_2$. Other processes like direct photochemical excitation, energy transfer from other electronically excited molecules, ozone photolysis, and high-energy oxygen producing reactions can also produce singlet oxygen.

Under ionizing atmospheric conditions, elemental oxygen in the upper atmosphere exists in allotropic forms other than diatomic O_2 molecule. Thus, in addition to O_2, the upper atmosphere contains oxygen atoms, O; excited oxygen molecules, O_2 * and ozone, O_3. Atomic oxygen is produced via photochemical reaction, as follows.

$$O_2 + h\nu \rightarrow O + O$$

The diatomic O_2 molecule's oxygen-oxygen bond can be broken by ultraviolet radiation with enough energy in the wavelength ranges of 135–176 nm and 240–260 nm (bond energy of O_2 = 120 kcal/mol). O_2 does not exist at very high altitudes due to photodissociation, and at altitudes exceeding 400 km, less than 10% of oxygen is present in the atmosphere in its molecular form. When air is observed at sea level at altitudes above 80 km, its average molecular weight is less than 28.97g/mole, which is indicative of a high atomic oxygen concentration phenomenon. The primary source of oxygen atoms in the atmosphere is photolysis of ozone, which produces both ground state (O) and excited state (O*) oxygen atoms. Because the dissociation enthalpy needed to break an oxygen-oxygen bond in an ozone molecule is so low (26 kcal/mol), excited oxygen atoms can form more quickly.

$$O_3 + h\nu(\lambda < 308 \text{ nm}) \rightarrow O^* + O_2$$

At wavelengths of 636 nm, 630 nm, and 558 nm, excited oxygen atoms are produced as a result of the extremely energetic chemical reactions. The upper atmosphere of Earth is said to glow due in part to the light emitted in this way, a phenomenon known as airglow.

$$O + O + O \rightarrow O_2 + O^*$$

The action of ultraviolet radiation on oxygen atoms can produce the oxygen ion, or O+. In certain regions of the ionosphere, these species are prevalent. In reactions with neutral species like N_2 or O_2, oxygen ions take part and release other positive ions.

$$O + h\nu \rightarrow O^+ + e^-$$

$$O^+ + O_2 \rightarrow O_2^+ + O$$

$$O^+ + N_2 \rightarrow NO^+ + N$$

O_2^+ ions may be produced in the intermediate regions of the ionosphere upon absorption of light of wavelength 17–103 nm by O_2 molecule or from photochemical reaction of low energy X-rays.

$$O_2 + h\nu \rightarrow O_2^+ + e^-$$

$$N_2^+ + O_2 \rightarrow N_2 + O_2^+$$

5.5. REACTIONS OF ATMOSPHERIC OZONE

Because ozone is able to absorb ultraviolet radiation from the sun, which can cause damage to life forms on Earth, it creates a protective layer that envelops the Earth's surface. Molecular oxygen photodissociation reactions are responsible for the presence of ozone in upper atmospheric layers. Upon absorption of ultraviolet radiation from the sun, an oxygen molecule splits as follows:

$$O_2 + h\nu \rightarrow 2O$$

$$O + O_2 + M \rightarrow O_3 + M + \text{Heat}$$

Where M is the third body, such as a molecule of N_2, O_2 or Ar that allows formation of ozone by absorbing heat energy released during collision between atomic oxygen and oxygen. Some of the oxygen atoms react with intact ozone molecules to destroy them by concerting to O_2.

$$O + O_3 \rightarrow 2\,O_2$$

The aforementioned reaction, however, proceeds very slowly due to its 18 $kJmol^{-1}$ activation energy. In summary, ozone in the stratosphere is continually created, broken down, and reconstructed by a variety of processes during the day. The concentration of ozone at any given time of day is dependent on altitude and solar radiation intensity. In the stratosphere, the maximum ozone concentration of roughly 12 parts per million is found at a height of 25 to 30 kilometers.

Hazardous UV light is filtered out by oxygen and ozone, resulting in incoming radiation in the troposphere with wavelengths $\geq$ 290 nm. Enough energy is present in this incoming light to enable photodissociation of important tropospheric species. The heat energy released in the aforementioned reaction is what causes the temperature in the stratosphere to rise with altitude. In the stratosphere, air at a particular altitude is colder than air above it. Because cooler air is denser than hot air, it resists gravity's pull and does not rise naturally.

Because of this, the process of vertical air mixing in the stratosphere is much slower than in the troposphere.

1. Catalytic Process for Ozone Destruction

The decomposition reaction:

$$2O_3 \rightarrow 3O_2$$

is thermodynamically favorable, indicates ozone is unstable. Pollutant trace constituents such as NO, NO_2, H, HO•, HOO•, ClO•, Cl, Br, and BrO• catalyze the reaction. Furthermore, NO, hydroxyl radicals, and atomic oxygen react with stratospheric ozone.

$$O_3 + h\nu \rightarrow O_2 + O$$
$$O_3 + O \rightarrow O_2 + O_2$$
$$O_3 + HO^{\cdot} \rightarrow O_2 + HOO^{\cdot}$$

The following reaction regenerates HO• radical and further contributes to the decomposition of ozone.

$$HOO\bullet + O \rightarrow HO\bullet + O_2$$

Similarly, nitric oxide, one of the key pollutants in the troposphere is also held responsible for the destruction of ozone in the stratosphere.

$$O_3 + NO \rightarrow NO_2 + O_2$$

NO, so consumed during this reaction is recycled through the following reaction.

$$NO_2 + O \rightarrow NO + O_2$$

Atmospheric N_2O also accounts for the release of nitric oxide in the atmosphere.

$$N_2O + O \rightarrow 2NO$$

Naturally existing in the atmosphere, N_2O is a primary byproduct of the denitrification process, which releases fixed nitrogen back into the atmosphere as gas. Tropospheric ozone is regarded as an undesired air pollutant that is detrimental to the ecosystem and human health, even though it serves as a shield in the stratosphere (Akimoto et al., 1994).

5.6. REACTIONS OF NITROGEN OXIDES

Nitrogen makes up approximately 78% of the earth's atmosphere by volume, making it an endless supply of this vital element. The N_2 molecule is not reactive chemically or photochemically because of its high bond dissociation energy of 945 $kJmol^{-1}$. Therefore, upon absorption of ultraviolet radiation, nitrogen molecules do not dissociate, in contrast to oxygen, which completely dissociates into oxygen atoms in the higher regions of the thermosphere. On the other hand, photochemical reactions occur at elevations greater than or equal to 100 km to produce atomic nitrogen.

$N_2 + h\nu \rightarrow N + N$ (photodissociation)

Other reactions producing atomic nitrogen are

$N_2 + h\nu \rightarrow N_2^+ + e^-$ (photoionization)

$$N_2^+ + O \rightarrow NO^+ + N$$

$$NO^+ + e^- \rightarrow N + O$$

$$O^+ + N_2 \rightarrow NO^+ + N$$

$$NO_2 + O \rightarrow NO + O_2$$

In the aforementioned reactions, N_2^+ is created by photoionization, which occurs when an N_2 molecule absorbs a photon of light in the lower ionosphere, which is located between 50 and 85 km above Earth. The main ionic species in higher regions is NO^+, which is produced when NO is ionized. Because it is a radical, NO is an odorless, colorless gas that is also highly reactive. Three of the eight known nitrogen oxides are found in significant amounts in the atmosphere: nitrogen dioxide (NO_2), nitric oxide (NO), and nitrous oxide (N_2O). Due to their possible roles in the formation of photochemical smog, acid rain, and the stratospheric ozone layer's breakdown, NO, and NO_2 (together known as NOx) are common air pollutants.

Through a sequence of photochemical reactions, nitric oxide, or NO, is released into the atmosphere from N_2O, or NO_2, and oxygen atoms. NO formation is also influenced by other natural events such as volcanic eruptions, forest fires, and lightning strikes.

The primary human-caused sources of NO, accounting for 96% of all NOx in the atmosphere, are fossil fuel combustion in factories, furnaces, and automobiles. Nitrogen and oxygen molecules react to produce NO at extremely high temperatures found in these natural phenomena or inside internal combustion engines.

$$N_2 + O_2 \rightarrow 2NO$$

Higher temperatures are advantageous for both a high NO formation rate and an equilibrium concentration. It is hypothesized that the N_2 and O_2 molecules split apart into their individual atoms at high temperatures.

$$O_2 + M \rightarrow O + O + M$$

$$N_2 + M \rightarrow N\bullet + N\bullet + M$$

Where M stands for a third, highly energetic body that doesn't react on its own but gives off enough energy to allow the reaction to occur. After then, a series of reactions between the atoms produce NO.

$$N_2 + O \rightarrow NO + N$$
$$N + O_2 \rightarrow NO + O$$
$$N_2 + O_2 \rightarrow 2NO$$

Hydrocarbon fragments (RH) compete with N_2 for highly reactive oxygen atoms in the combustion mixture.

$$RH + O \rightarrow R\bullet + HO$$

The hydroxyl radical combines with atomic nitrogen which again leads to the formation of NO.

$$N + HO\bullet \rightarrow NO + H\bullet$$

NO plays an important role in the ozone content of stratosphere. It removes O_3 by converting itself to NO_2. NO is regenerated upon reaction of NO_2 with atomic oxygen.

$$O_3 + NO \rightarrow NO_2 + O_2$$

The color of photochemical smog is caused by nitrogen dioxide, or NO_2, a brown gas. It is created when NH_3 in the atmosphere breaks down oxidatively. Similar to NO, its radical nature is the reason for its high reactivity. In particulate matter, NO_2 is ultimately transformed into either NO_2 or NO_3. Then precipitation washes the particulates away. The production of nitric acid (HNO_3) from the dissolution of nitrate in a water droplet causes harm to the ecosystem from acid rain. Nitrogen dioxide takes part in the chemical reactions involving volatile organic compounds (VOCs) to produce part of the smog-causing PAN (peroxyacetyl nitrate). Peroxyacetic nitric anhydride is created when acetaldehyde is oxidized and the ensuing peroxyacetyl radical combines with NO_2. PAN is a highly irritating and toxic chemical for people, pets, and plants.

Dinitrogen tetraoxide N_2O_4, nitrogen dioxide's colorless dimer, and nitrogen dioxide are in balance. When combined with water, it generates nitric and nitrous acids, which oxidize a variety of metals. Because of its high reactivity, it is a crucial species in the atmosphere. It fully absorbs visible and ultraviolet light that enters the troposphere. At wavelengths below 398 nm, photodissociation takes place and generates ground state oxygen atoms as follows:

$$NO_2 + h\nu \rightarrow NO + O$$

At wavelength > 430 nm, only excited species are formed,

$$NO_2 + h\nu \rightarrow NO_2^*$$

In wavelengths ranging from 398 nm to 430 nm, both processes are possible. Since NO_2 has a much shorter half-life in direct sunlight than any other common molecular atmospheric species, it is highly likely to undergo photodissociation, which could result in a number of significant reactions, some of which are summarized below. Low levels of NO_2 have also been found in the lower stratosphere, where ozone oxidizes NO to produce the gas.

$$NO + O_3 \rightarrow NO_2 + O_2$$
$$NO_2 + O_3 \rightarrow NO_3 + O_2$$
$$O + NO_2 \rightarrow NO + O_2$$
$$O + NO_2 + M \rightarrow NO_3 + M$$
$$NO_2 + NO_3 \rightarrow N_2O_5$$
$$NO + NO_3 \rightarrow 2NO_2$$
$$O + NO + M \rightarrow NO_2 + M$$

Nitric acid, nitrates, or organic nitrogen (in atmospheres where photochemical smog forms) are the forms of nitrogen dioxide that are eliminated from the atmosphere. The anhydride of nitric acid, which is created when the reacts with water, is dinitrogen pentoxide, which is produced in the reactions mentioned above.

$$N_2O_5 + H_2O \rightarrow 2HNO_3$$

In the stratosphere, nitrogen dioxide combines with hydroxyl radicals and produces nitric acid:

$$HO\bullet + NO_2 \rightarrow HNO_3$$

Nitric acid is destroyed by hydroxyl radicals within stratosphere,

$$HO\bullet + HNO_3 \rightarrow H_2O + NO_3$$

or by a photochemical reaction,

$$HNO_3 + h\nu \rightarrow HO\bullet + NO_2$$

As a result, the aforementioned reactions demonstrate that nitric acid acts as a transient sink for NO_2 in the stratosphere. After that, nitric acid is eliminated either by precipitation or by forming particulate nitrates when it reacts with bases (ammonia, particulate lime).

In an unpolluted atmosphere, nitrous oxide, or N_2O, also known as laughing gas, is naturally occurring in trace amounts (0–3 ppm). The majority of the natural sources are released by activities in the ocean, with the microbial activity of the soil in tropical regions serving as the primary source. Both the nitrification reaction in an anaerobic atmosphere and the denitrification reaction in an aerobic atmosphere produce the gas as a byproduct. In the former case, nitrogen is reduced to N_2 from its most oxidized form, NO_3, with N_2O emerging as a byproduct. The most reduced form of nitrogen, NH_4^+, is oxidized to NO_2 – and NO_3 – during nitrification, with N_2O emerging as a byproduct. According to the photochemical reaction, N_2O is comparatively unreactive in the troposphere and its concentration in the stratosphere decreases with altitude.

$N_2O + h\nu \rightarrow N_2 + O$

and reaction with singlet atomic oxygen:

$N_2O + O \rightarrow N_2 + O_2$

$N_2O + O \rightarrow 2NO$

The ozone layer is being depleted as a result of increased nitrogen fixation and increased microbial production of N_2O. N_2O_5 and N_2O_3 are two other neutral nitrogen oxides that are found in the atmosphere. Nitrogen oxides (N_2O, NO) are easily produced through the oxidation of NH_3 and other volatile organic amines that are released during the breakdown of plant and animal materials.

5.7. PARTICLES IN THE ATMOSPHERE

The atmosphere contains a variety of particle types, including dust, smoke, soot, pollen, spores, and micron-sized aerosols. The most obvious type of air pollution is particulate matter, or simply particles. The mixture of liquid droplets and solid particles floating in the atmosphere is complex. The smaller solid particles are made up of black carbon, silver iodide, sea-salt nuclei, combustion nuclei, and black carbon formed when water evaporates from seawater droplets. The larger particles are made up of foundry dust, cement dust, wind-blown soil dust, and pulverized coal. Sulfuric acid droplets, raindrops, fog, and other mists are examples of liquid particulate matter. Additional naturally occurring airborne particles include bacteria, pollen grains, fog, and volcanic ash.

Chemically, the particulate matter comprises a variety of components including acids (nitrates and sulfates), elemental carbon, metals (Mn, Cu, Zn, Cd, Cr, Fe, Ni, K, Ca, V, Ba, As, Se, Sr), soil, fugitive dust and sea spray.

5.7.1. Characteristics of Particulates

1. **Size:** The droplets and particles have different sizes. Small particles, like aerosols, can only be found under a microscope, but large particles, like soot or smoke, can be seen. Particle sizes range from 500 μm to 2×10^{-4}. Particles with sizes between 0 and 10 μm are frequently suspended close to the source of air pollution. The possibility of a particle causing health issues is correlated with its size. The more deadly its effect, the smaller it is. $PM_{2.5}$ and PM_{10}, two size ranges, are regularly measured in the air at ambient conditions and at significant emission sources. PM_{10} and $PM_{2.5}$ particles, respectively, are defined as having aerodynamic diameters less than or equal to 10 μm and 2.5 μm, respectively. The way these two groups of particles behave in the atmosphere is different. Fine particles, or $PM_{2.5}$, have

a tendency to stay in the air longer and cover greater distances. Conversely, PM_{10}, or coarse particles, quickly settle on the surface of the earth because they are too big to stay suspended in the air. As a result, particle size serves as a helpful gauge for how long a particle hangs in the atmosphere.

2. **Concentration:** It varies from several hundred/cm^3 to 1,00,000/cm^3 and above. The concentration at any time varies with the wind source, because the source of release may be very far from site at which the particles are formed.
3. **Lifetime**: Particle's lifetime varies from few seconds to several years, that normally happen when the particles diffuse into stratosphere. Large size particulates have lower lifetime than small particles. The most common way of losing their existence is by settling down. Small particles coalesce, condense, and settle down faster than larger particles.

5.7.2. Chemical Reactions of Atmospheric Particles

Chemical reactions in the atmosphere result in the production of secondary particulate matter. Nitrogen oxides (NO_2) and sulfur oxides (SO_2) are the main species that contribute to the particulate matter's formation. These particles rise from the troposphere into the stratosphere and stay suspended between 20 and 27 kilometers above the ground.

1. **Oxidation**

i. **Chemistry of Sulfate Formation**-In the gas phase, photochemical oxidation of SO_2 during the day converts about half of it to sulfate. Oxidation rate is accelerated by NOx and hydrocarbon emissions. Droplets in the clouds are where at least half of SO_2 oxidation occurs. Soluble SO_2 in clouds is quickly converted to sulfate by water droplets, which forms tiny liquid droplets. Mechanisms: SO_2 (g) is oxidized to $H_2 SO_4$ (g) in the gas phase, and then $H_2 SO_4$ (g) is condensed. In liquid water, SO_2 (g) dissolves to produce $H_2 SO_3$ (aq), which is then converted to $H_2 SO_4$ (aq) (Figure 5.4).

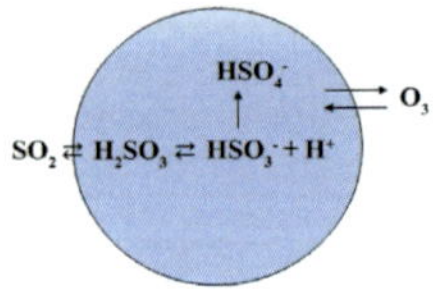

Figure 5.4. *Oxidation of SO_2 inside water droplet.*

Source: https://egyankosh.ac.in/bitstream/123456789/79934/1/Unit-5.pdf.

Sulfuric acid may also be formed through the following reactions:

$$SO_2(g) + 2OH^\bullet \rightarrow H_2SO_4(g)$$

$$SO_2(g) + O + M \rightarrow SO_3(g)$$

$$SO_3(g) + H_2O\ (aq) \rightarrow H_2SO_4$$

ii. **Chemistry of Nitrate Formation**-During the day, NO_2 can react with hydroxyl radicals to produce nitric acid. The peak daytime conversion rate of NO_2 to HNO_3 in the gas phase is approximately 10–50% per hour. The reaction of OH• with NO_2 is about 10 times faster than with SO_2 dot. Ozone and the nitrate radical participate in a sequence of reactions that convert NO_2 to HNO_3 in the gas phase during the night.

$$NO_2 + OH^\bullet \rightarrow HNO_3$$

$$NO_2 + O_3 \rightarrow NO_3 + O_2$$

$$NO_2 + NO_3 \rightarrow N_2O_5$$

$$N_2O_5 + H_2O \rightarrow HNO_3$$

2. **Acid-Base Reaction:** Lime and ammonia, which are released during biological decay processes at ground level, eventually mix with droplets of both nitric and sulfuric acids. The acid-base reaction of these acids with NH_3 affords ammonium salts as follows,

$$H_2SO_4(\text{droplet}) + 2NH_3(g) \rightarrow (NH_4)_2SO_4\ (\text{droplet})$$

$$HNO_3(\text{droplet}) + NH_3(g) \rightarrow NH_4NO_3\ (\text{droplet})$$

$$H_2SO_4(\text{droplet}) + CaO(s) \rightarrow CaSO_4\ (\text{droplet}) + H_2O$$

The evaporation of water from these droplets in low-humidity environments influences the formation of solid aerosol. Metal oxides are a significant class of inorganic particles in the atmosphere that are produced when metal-containing fuels are burned. For example, particulate iron oxide is formed during the combustion of pyrite-containing coal:

$$3FeS_2 + 8O_2 \rightarrow Fe_3O_4 + 6SO_2$$

5.7.3. Aerosols

Colloidal-sized solid or liquid particles with a diameter of less than 100 μm that are suspended in the atmosphere and in balance with the gaseous atmosphere are known as atmospheric aerosols. Aerosol particles made of silicon, sulfates, nitrates, metal oxides, ammonium salts, and carbonaceous material. The size of the particle affects the composition. Larger particles, which contain materials of mechanical origin, such as ground limestone, behave as bases, whereas fine, small particles from polluting gases, such as SO_2, are acidic. Major atmospheric phenomena like cloud formation, electrification, and fog formation are significantly influenced by aerosol particles. Aerosols in the lower atmosphere change the distribution and size of cloud droplets, which affects how well they reflect and absorb light. These particles can therefore control the earth's energy budget to a large extent. These particles significantly affect the atmosphere's visibility. Similar to molecules, particles with a diameter less than 0.1 μm scatter light through a process known as Rayleigh scattering. Their dispersal of sunlight darkens dawns and sunsets and lessens visibility (haze). Because their dimensions are similar to the wavelength of visible light, particles in the 0–1 μm^{-1} μm range exhibit interference phenomena. Aerosols also have a significant impact on the earth's climate. By reflecting light, they aid in maintaining the planet's temperature equilibrium. Condensation nuclei are minuscule particles that act as templates for atmospheric water vapor to condense, which is how raindrops are formed.

Aerosols can be hydrophobic or hydrophobic, simple molecules or complex mixtures. If the sizes of the pollutants are in the submicron range, aerosols have the ability to carry a large range of inorganic and organic pollutants over long distances. The particles serve as reaction sites for a variety of heterogeneous atmospheric chemical reactions because of their active surface area. The depletion of stratospheric ozone is caused by one of the major reactions. The aerosols multiply and create stratospheric clouds in the polar regions. The vast surface areas of the cloud offer active locations for chemical reactions to take place. The destruction of ozone in the stratosphere is facilitated by reactions like these that yield large amounts of reactive chlorine.

5.7.4. Black Carbon

Particulate matter, which includes black carbon, is released by fossil fuel-burning sources such as coal-fired power plants and diesel engines. Particles of black carbon, volatile organic compounds, carbon dioxide, organic carbon, and CO_2 are produced when fuel is not burned completely.

Black carbon particles are highly photochemically absorbent, giving the complex mixture of particulate matter released during incomplete combustion its black color.

By transforming absorbed solar radiation into heat, it adds to global warming. Black carbon is a major global environmental issue that has an impact on both the climate and human health. Black carbon inhalation can result in cancer, birth abnormalities, cardiovascular, and respiratory disorders. Black carbon is a major cause of climate change and global warming due to its absorption of solar radiation.

5.7.5. Effects of Particulate Matter on Human Health

Particles in the atmosphere that are extremely fine have the ability to enter the bloodstream, reach the respiratory system, and cause death. Exposure, both short- and long-term, raises the risk of respiratory, cardiovascular, infectious, and allergic diseases in addition to lung cancer. Deep within the body, $PM_{2.5}$ causes more harm to one's health. According to studies, there is a roughly 6% and 8% increased risk of lung cancer and cardiopulmonary disease mortality for every 10 μg/m^3 increase in fine $PM_{2.5}$ long-term exposure. Additional side effects include irritated eyes, nose, and throat; severe asthma attacks; low birth weight and preterm birth; and early death in heart and lung disease patients.

5.8. BIOGEOCHEMICAL CYCLE OF CARBON

Since carbon is the basis of all living things, it is frequently referred to as the building block of life. Carbon is an important element for the biosphere. Carbon compounds can be found in three different states: solid (like coal or diamonds), liquid (like crude oil), or gas (like carbon dioxide). There are three isotopes that occur naturally: ^{14}C is radioactive and decays with a half-life of roughly 5730 years, while ^{12}C and ^{13}C are stable. The four most important pools (or reservoirs) are where carbon is exchanged and stored. Carbon travels through the hydrosphere, atmosphere, geosphere, and biosphere (terrestrial ecosystems) as part of the carbon cycle. Carbon is stored in non-living pools as carbonate ($CaCO_3$) rocks, fossil fuels derived from dead organic matter, carbon dioxide (CO_2), carbon monoxide (CO), and carbon dioxide dissolved in water to form HCO_3^-.

These pools have significant, ongoing carbon flows. The atmosphere is taken into consideration when analyzing fluxes in atmospheric chemistry. Positive fluxes typically indicate emission to the atmosphere (the pool from which the carbon originates acts as a source), while negative fluxes indicate uptake of carbon (the pool acts as a sink). The carbon pools have the ability to release or absorb carbon into or from the atmosphere, respectively, acting as sources or sinks of carbon. In most cases, pools act as both sources and sinks at the same time. That is, carbon is taken up through one process and released into the atmosphere through another. If, over the long term, the positive and negative fluxes are equal and the pool size stays constant, the system is said to be in a state of dynamic equilibrium.

5.8.1. The Natural Carbon Cycle

Methane (CH_4) and carbon dioxide (CO_2) both contribute significantly to the carbon cycle, being the vast ocean-atmosphere and surface-atmosphere carbon fluxes that were constant at 280 parts per million during the preindustrial era until 1750. In the atmosphere, CO_2 is a long-lived greenhouse gas (LLGHG). Reconstructed prehistoric CO_2 concentrations from ice cores revealed that prior interglacial CO_2 concentrations were only slightly elevated by a few Gt C every decade. A little more than 100 parts per million was the total variation in CO_2 concentration between glacial interglacial periods. For the past 15 million years, CO_2 concentrations this high have not been attained.

The balance or imbalance between carbon sources and sinks is known as the carbon budget. The ocean, the biosphere's respiration, and geological sources are the three main natural sources of carbon. The total carbon flux on land transfers about. In natural conditions, the ocean-atmosphere flux is approximately 90 Gt C yr^{-1}. This compares to 120 Gt C yr^{-1}. Under undisturbed natural conditions, the carbon budget, or the balance between sources and sinks, is less than 0.1 Gt C yr^{-1}, which is calculated as the long-term net flux averaged over a decade or longer.

5.8.1.1. Terrestrial Processes

The biosphere absorbs carbon through autotrophs, or organisms that can synthesize their own nutrients from inorganic materials. The primary distinction between the two types of autotrophs is the energy source that each one uses for synthesis.

Light is the source of energy for photoautotrophs, which include most autotrophs like green plants, some algae, and photosynthetic bacteria. Chemoautotrophs, such as chemosynthetic bacteria, obtain their energy from chemical reactions, such as the oxidation of electron donor molecules. The majority of the CO_2 flux on Earth is absorbed by terrestrial vegetation through photosynthesis; however, the majority of that amount is respired back into the atmosphere. The soil, detritus, and terrestrial vegetation are the three most significant land carbon pools. This portion of the carbon cycle is sometimes referred to as the "fast carbon cycle" because of the relatively short time scales of the carbon fluxes between the atmosphere and these pools. Plants absorb carbon from the atmosphere through photosynthesis and return it to the atmosphere through respiration over the shortest time spans of seconds to minutes.

Gross primary production (GPP) is the amount of CO_2 that is converted to carbohydrates during photosynthesis. A portion of this carbon is respired, and the remainder is assimilated to support plant growth and function. The annual variation in autotrophic respiration (Ra) and GPP, that is Net primary production, or NPP, is the yearly growth of plants (equation 5.1). An estimate of 120 Gt C yr^{-1} is associated with terrestrial photosynthesis.

$$NPP = GPP - Ra \qquad (5.1)$$

The majority of dead biomass relocates to soil organic matter and detritus pools over longer time periods, where it can be retained for years, decades, or even centuries. It is possible to identify multiple pools of soil organic matter with various residence times. Soil microbes and fungi eventually respire this carbon back into the atmosphere, where it decomposes at a rate determined by the environmental conditions (moisture, temperature, etc.) and their chemical makeup). Carbon dioxide is released during aerobic respiration, which happens when oxygen is present. Anaerobic respiration takes place in an oxygen-limited environment, producing methane rather than CO_2.

Carbon entering the terrestrial biosphere via photosynthesis is emitted back via (i) autotrophic respiration, (ii) heterotrophic respiration (decomposers and herbivores) (iii) fires (combustion). The net ecosystem production shows the net amount of carbon removed from (or released to) the atmosphere that is gained (or lost) by the ecosystem when no other disturbances (carbon loss of the ecosystem) are considered

$$NEP = NPP - Rh \qquad (5.2)$$

Considering additional carbon losses, such as biomass transportation from agricultural areas, fires, so forth,. displays the net biome production (NBP), or carbon stored by the terrestrial biosphere. NBP ought to be 0 in an equilibrium system.

5.8.1.2. Oceanic Processes

The ocean is another major sink of carbon, under natural circumstances it removes CO_2 equivalent to ca. 70 Gt C yr^{-1} from the atmosphere. The ocean held 20 times more carbon than the terrestrial biosphere/soil compartment and roughly 60 times more than the atmosphere before the Industrial Revolution.

The removal of CO_2 involves a variety of processes that move carbon to various depths and correspond with varying residence times. This segment is sometimes referred to as the slow carbon cycle due to the accelerated nature of the participating processes. Particulate organic carbon (POC) (living and dead), dissolved organic carbon (DOC), and dissolved inorganic carbon (DIC) are the three forms of oceanic carbon. The solubility pump, which is driven by the solubility of atmospheric CO_2, the organic carbon pump, which is driven by photosynthetic uptake by marine biota and the sinking of this carbon as organic particles to deeper layers, and the $CaCO_3$ counter pump, which is driven by the release of CO_2 during the formation of $CaCO_3$ shells, are essentially the three main processes that govern carbon absorption. The solubility pump is sometimes referred to as the physical pump, whereas the carbonate pump and organic carbon pump processes are known as the "biological pumps."

5.8.1.3. The Solubility Pump

The ocean's pH is regulated by the marine carbonate buffer system, which permits the ocean to absorb CO_2 far beyond its potential capacity based only on solubility. A sequence of chemical reactions convert carbon added as CO_2 into bicarbonate (HCO_3^-) and carbonate (CO_3^{2-}), achieving this control. All three of these dissolved forms are referred to as DIC. The ocean is slightly alkaline (surface pH of 7 to 8) due to the minerals dissolved in it over geologic time. CO_2 is a weakly acidic gas. After dissolving, it combines with water to form carbonic acid, which splits into two ions: H+ and HCO_3^-. Part of the H+ then combines with CO_3^{2-} to form another HCO_3^- ion.

$$CO_2 + H_2O \rightarrow H^+ + HCO_3^- \rightarrow 2H^+ + CO_3^{2-} \tag{5.2}$$

$$CO_2 + H_2O + CO_3^{2-} \rightarrow HCO_3^- + H^+ + CO_3^{2-} \rightarrow 2HCO_3^-. \tag{5.3}$$

The air-sea gradient in pCO_2 between the atmosphere and the ocean largely determines the air-sea exchange of CO_2. The surface ocean and atmosphere take about a year to reach equilibrium. Gas exchange rates are dependent on a number of variables, including sea ice, heat flux, precipitation, and surfactants, and they rise with wind speed. At high wind speeds, the local gas exchange rates' magnitudes and uncertainties reach their maximum. On the other hand, the equilibrium CO_2 partitioning values between air and seawater, as well as the corresponding pH values of the seawater, are well-established. The solubility pump is created by the Meridional Overturning Circulation, which sinks to greater depths in colder, CO_2-rich waters near the poles during local winters (the lower the temperature, the higher the solubility).

5.8.1.4. Organic Carbon Pump

CO_2 is incorporated into organic matter during the photosynthesis process, which is known as the "organic carbon pump." A portion of the particulate organic carbon sinks to lower layers, where it is broken down and respired by bacteria. After that, it is returned to the surface layers by marine circulation as particulate organic carbon (DIC). Other POC sinks more deeply and settles on the ocean floor, where it will remain for thousands of years. Because photosynthesis produces organic matter, the effective organic pump reduces the overall carbon content in the surface layers while removing carbon from the atmosphere on a net basis.

$$CO_2 + H_2O \rightarrow CH_2O + O_2. \tag{5.4}$$

The availability of light, nutrients, and oxygen limits the organic pump's efficiency through photosynthesis. An illustration of ocean primary productivity from MODIS imagery during the year 2001.

5.8.1.5. Carbonate Pump

Carbonate ions produced by the chemical weathering of carbonate rocks and the solution of CO_2 in surface seawater play important roles in the biological pump.

The other component of the biological pump is based on the formation of carbonate and silicate shells, the ions being used by diatoms (silica-secreting organisms) and coccolithophorids (carbonate-secreting organisms).

The production of solid $CaCO_3$ (that is, "carbonate precipitation") occurs in the surface waters of the ocean, both organically – by organisms that build their shells from $CaCO_3$ – and inorganically according to the chemical equilibrium in the oceans according to the following chemical equation:

$$Ca^{2+} + 2HCO_3 \rightarrow CaCO_3 + CO_2 + H_2O \tag{5.5}$$

One CO_2 is produced in the above reaction that forms one $CaCO_3$, so even though the DIC concentration is reduced during this process, the CO_2 partial pressure will rise and affect the solubility of atmospheric CO_2 in the surface ocean. On the other hand, silica shell formation is a more efficient method of sequestering CO_2 because it doesn't produce CO_2.

5.8.1.6. Geological Processes

Over even longer periods of time, organic matter that was buried in deep sediments and shielded from deterioration gradually changed into deposits of natural gas, coal, and oil—the fossil fuels we use today. Burning these materials releases CO_2 back into the atmosphere, which has been stored for millions of years.

5.8.1.7. Silicate Weathering and Atmospheric CO_2

A small amount of carbon is absorbed in the process of weathering and carried to the ocean by surface water and stored there for a longer term. Chemical weathering of rocks is represented by reactions (5.6) and (5.7) for carbonates, and reactions (5.8) and (5.9) for silicates

$$CO_2 + H_2O + CaCO_3 \rightarrow Ca^{2+} + 2HCO_3 \tag{5.6}$$

$$2CO_2 + 2H_2O + CaMg(CO_3)_2 \rightarrow Ca^{2+} + Mg^{2+} + 4HCO_3 \tag{5.7}$$

$$2CO_2 + 3H_2O + CaSiO_3 \rightarrow Ca^{2+} + 2HCO_3 + H_4SiO_4 \tag{5.8}$$

$$2CO_2 + 3H_2O + MgSiO_3 \rightarrow Mg^{2+} + 2HCO_3 + H_4SiO_4 \tag{5.9}$$

5.8.2. Methane

Methane is an odorless, colorless gas at standard pressure and temperature. Because of its comparatively long atmospheric lifetime, it is classified as a "well mixed" greenhouse gas and one of the LLGHGs. Once emitted, CH_4 remains in the atmosphere for approximately 8.4 years before removal. Methane is a potent GHG hence being of great interest among atmospheric scientists.

5.8.2.1. Sources and Sinks of Atmospheric Methane

Geological sources of methane can be both biogenic and non-biogenic. Methane is created by anaerobic decomposition in soils and sediments on wetlands, which is the most significant biogenic source of the gas. Particularly conducive to the production of methane are warm temperatures and damp surroundings.

In general, nontidal wetlands with slow decomposition rates store carbon as peat and/or dead soil organic matter; in other words, they function as a sink for atmospheric CO_2. Due to this source's extreme temperature sensitivity, changes in the global climate that lead to warmer conditions may, depending on changes in moisture, result in a significant increase in CH_4 emissions as well as a positive feedback loop that warms the planet even more. On the other hand, a large-scale drying of the wetland areas that exist now would result in lower CH_4 emissions and probably a net release of CO_2 into the atmosphere.

Methane is released during defecation and is produced by microorganisms in the stomachs of termites and ruminant livestock that break down hard-to-digest material. Permafrost, glaciers, and ice cores are examples of non-biogenic methane sources. When permafrost melts, large volumes of methane that has been stored there are released. As global temperatures rise, frozen soils may serve as a source for the slow release of methane that has been trapped in frozen environments. Massive volumes of methane are released into the atmosphere during the mass burning of organic matter. There are other natural sources such as vegetation, termites, oceans, and CH_4 hydrates. Methane hydrate, a form of methane stored in vast quantities, is found in deep ocean deposits. Methane hydrate is a solid form of methane combined with water at moderate pressure (above 3–5 MPa) and low temperatures (less than 25°C). These conditions are found at a depth of 300–500 meters. Scientists are alerted to the warming climate when vast amounts of methane hydrate stored in deep oceans are released due to changes in temperature or pressure.

5.8.2.1.1. Sinks

The atmosphere's hydroxyl radical (×OH) is the main source of water vapor in the upper atmosphere and the largest sink for atmospheric methane.

$$CH_4 + \cdot OH \rightarrow \cdot CH_3 + H_2O. \quad (5.10)$$

Since the primary sink of CH_4 is tropospheric reactions with ×OH, the abundance of OH controls the lifetime of methane. Methane's lifetime is thereby increased when biomass burning results in lower atmospheric ×OH concentrations. Isoprene, an organic carbon compound released by plants, lowers the concentration of ×OH, prolonging the lifetime of CH_4 by competing with methane for ×OH. When CH_4 combines with ×OH, Cl, and $O(^1D)$, it is released into the stratosphere. One way that water enters the stratosphere is through methane oxidation.

$$CH_4 + O(^1D) \rightarrow \cdot CH_3 + \cdot OH. \quad (5.11)$$

Water vapor plays a significant role in the chemistry of stratospheric ozone and as a greenhouse gas. Methanotrophs—methane-consuming microorganisms—metabolize a portion of the methane found in soils. In drier soil, biological CH_4 oxidation is carried out by methanotrophic microorganisms, which use methane as part of their aerobic metabolism to produce CO_2 and water.

$$CH_4 + 2O_2 \rightarrow CO_2 + 2H_2O. \quad (5.12)$$

Under anaerobic condition methane is oxidized in the following reaction combining with nitrate/nitrites

$$CH_4 + NO_3/NO_2 + H^+ \rightarrow CO_2 + N_2 + H_2O. \quad (5.13)$$

Methane also reacts with natural chlorine gas in the atmosphere to produce chloromethane and hydrochloric acid (HCl). This process is known as free radical halogenations.

$$CH_4 + Cl_2 \rightarrow CH_3Cl + HCl \quad (5.14)$$

5.8.3. Human Disruption of Carbon Cycle

The natural GHG cycles have been changed by human activity. The natural carbon cycle is now shifting in favor of atmospheric accumulation. The majority of anthropogenic carbon compound emissions are caused by human activity, such as burning fossil fuels, changing land use (particularly deforestation and expanding agricultural lands), and producing methane and CO_2 emissions. In the modern methane cycle, anthropogenic emissions account for 60% of methane released into the atmosphere each year, with 40% of emissions coming from natural sources. Based on analysis of ice cores, the atmospheric abundance of methane prior to the industrial era varied, reaching highs of approximately 700 ppb during interglacials and lows of approximately 400 ppb during glacial periods. According to the most recent analysis from the WMO GAW program, the average global CH_4 abundance in 2011 was 1813±2 ppb. At least the last 650 years have not seen atmospheric CH_4 concentrations as high as they are now. CH_4 growth rate fell from approximately 13 parts per billion per year in the early 1980s to almost zero between 1999 and 2006. After a three-year hiatus, atmospheric CH_4 has been rising steadily since 2007.

Human activity produces a greater amount of CH_4 into the atmosphere than does natural systems. Despite a nearly zero growth rate since 1980, between 1960 and 1999, CH_4 concentrations increased on average at least six times faster than they did during any 40-year period of the two millennia prior to 1800.

The human activities that produce CH_4 include energy production from coal and natural gas, waste disposal in landfills, raising ruminant animals (e.g., cattle, and sheep), rice agriculture and biomass burning. At the same time, wetlands, and rice paddies emit about 175 Tg CH_4 annually. The warmer and moister the rice field, the more methane is produced.

Waste water treatment facilities are accountable for the release of CH_4 due to the anaerobic digestion of organic compounds in the water, which produces methane. Landfills are an important source of methane because of decomposing organic matter and anaerobic circumstances.

According to WMO GAW measurements, the atmospheric concentration of CO_2 in 2011 was 390.9±0.1 ppm, indicating a 140% increase from preindustrial levels. The natural carbon cycle has changed as a result of the extra CO_2 that humans (anthropogenic CO_2) release into the atmosphere. Since the start of the industrial era, burning fossil fuels has caused human activity to release significant amounts of carbon, mostly from long-term pools. The two main sources of anthropogenic emissions are (i) burning fossil fuels and making cement, which releases CO_2 that has been stored in the earth for hundreds of millions of years, and (ii) deforestation and the conversion of land into agricultural fields, which releases carbon that has been stored for decades to centuries. Net land-atmosphere and ocean-atmosphere fluxes have consequently diverged considerably from zero. The most erratic aspect of estimates of anthropogenic emissions is the latter part of carbon emissions linked to changes in land use.

The idea that the greenhouse effect could raise global temperatures due to CO_2 emissions from the burning and combustion of fossil fuels was first put forth by Arrhenius. Charles David Keeling began the first continuous measurement of atmospheric CO_2 concentration in 1958 at Hawaii's Mauna Loa Observatory. The first concrete proof of the atmosphere's rapidly rising CO_2 levels was provided by Keeling's measurements. Many scientists believe that Keeling's graph was the catalyst for drawing attention to the world's attention to the current increase in atmospheric carbon dioxide, which can cause a change in global temperature according to Arrhenius' theory.

Since the composition of atmospheric CO_2, specifically, its ratio of heavy to light carbon atoms, has changed due to the addition of carbon from fossil fuels, it is known that human activity is the cause of the increase in atmospheric CO_2 concentration. Furthermore, as CO_2 has increased, the atmosphere's oxygen-to-nitrogen ratio has decreased; this is to be expected since burning fossil fuels depletes oxygen. The carbon-13 isotope, a heavy form of carbon, is more prevalent in the oceans and in volcanic or geothermal emissions than it is in vegetation and, consequently, in fossil fuels that were formed from previous vegetation. The atmosphere's relative concentration of the carbon-13 isotope has been falling, indicating that vegetation and fossil fuels are the main sources of the extra carbon. A rare radioactive isotope of carbon, carbon-14, is also found in atmospheric CO_2 but not in fossil fuels. Decreases in the relative amount of carbon-14 indicated that carbon from fossil fuels was being added to the atmosphere prior to nuclear weapons testing in the atmosphere.

Atmospheric CO_2 concentration however increases with half of the rate of anthropogenic emissions. The fraction of anthropogenic emission that remains in the atmosphere is known as the airborne fraction. The airborne fraction, considering anthropogenic emissions to be relatively well constrained can be good indicators of short- and long-term changes in CO_2 sink processes.

The original theory states that roughly half of the anthropogenic emission is absorbed by the residual sink, formerly known as the "missing sink." The residual sink is the oceans' and terrestrial vegetation's increased capacity to absorb carbon as a result of environmental changes that raise the atmosphere's CO_2 burden. Stated differently, net carbon balance, that is terrestrial ecosystems have negative net biological productivity (NBP; uptake). Since inverse modeling is used to determine the residual sink, our assumptions regarding land use change emissions have a significant impact on the estimated amount of carbon absorbed by the sink. A thorough examination shows that during the past 50 years, the amount of anthropogenic CO_2 emissions that are absorbed globally by carbon sinks has doubled; however, it is still unknown how much of this is absorbed by land and how much by sea. Ocean acidification is a consequence of the ocean's net uptake of CO_2. It increases the acidity of seawater, which could have significant effects on the ocean food chain.

Aboveground biomass cannot continue to grow indefinitely, even though terrestrial vegetation is currently a net sink of CO_2. The soil has the largest potential to act as a long-term sink for atmospheric CO_2 and is ultimately the most significant location for organic carbon storage. However, modifications to soil carbon storage won't happen in a vacuum from adjustments to other trace gas exchanges, which presents significant scientific challenges for attempts to mitigate greenhouse gas emissions. It is essential to have the right understanding of how these processes react to climate variability in order to evaluate how global change affects their ability to absorb CO_2 and, in turn, reduce the amount of CO_2 in the atmosphere. The following variables affect the ability of terrestrial vegetation and oceans to absorb: (i) direct climatic effects (precipitation, temperature, changes in radiation); (ii) changes in atmospheric constituents (nucleate deposition (N, P), damage from pollutants, "CO_2 fertilization"); and (iii) changes in land use (deforestation, afforestation, agricultural management practices).

5.8.3.1. Direct Climatic Effects

The glacial-interglacial cycles best illustrate the close relationship between climate and the carbon cycle. The CO_2 concentration is lower during glacial epochs than it is during interglacials, indicating a close relationship with temperature. At higher temperatures, dissolved carbon is released from the ocean as CO_2 and dissolves at lower temperatures. In addition to short-term climate variability, such as the El Nino Southern Oscillation and Arctic Oscillation, volcanic eruptions can also cause short-term climate perturbations in the atmosphere's CO_2 level.

5.8.3.2. Changes in Atmosphere's Chemical Composition

The amount of CO_2 in the atmosphere can affect ecosystem function in a variety of ways. In-situ measurements were used in the FACE (Free-Air Carbon Dioxide Enrichment) experiments to investigate the impact of increased CO_2 concentrations on ecosystems. An increase in assimilation makes sense when there is more CO_2 available for photosynthesis. However, nutrient or environmental constraints may negate the effects of this CO_2 fertilization.

5.8.3.3. Land-use Change

Changes in land use and cover are brought about either naturally—for example, by climatic shifts—or artificially—by farming practices, the emergence of new diseases, so forth,. Particularly in the tropics, deforestation accounts for one-third of all anthropogenic CO_2 emissions at the moment. The component of the estimates of anthropogenic carbon emissions that is currently the least certain is the emission caused by changes in land use. When implemented correctly, agricultural management techniques can affect how much carbon is sequestered by agricultural soils. Depending on how big the processes are, many factors can affect the carbon fluxes between the atmosphere and the ocean. The degree of ocean circulation, sea surface temperature, salinity, ocean stratification, and ice cover all affect how effective the solubility pump is. The amount of the sinking portion of POC and DOC, which is influenced by nutrient availability and ocean circulation, essentially determines the biological pump.

The global carbon budget (GtC yr^{-1}); errors show uncertainty estimates of ±1 standard deviation rather than interannual variability, which is larger. The first line of the atmospheric increase is the result of fluxes to and from the atmosphere: numbers in parentheses denote ranges; positive fluxes are emissions or inputs into the atmosphere; negative fluxes are sinks, or losses from the atmosphere. Recall that there are strong constraints on the overall sink of anthropogenic CO_2. Consequently, there is a negative correlation between the fluxes from the ocean to the atmosphere and the land to the atmosphere; if one is higher, the other must be lower to equal the total sink, and vice versa.

5.8.4. Carbon Cycle Research in Hungary

In Hungary, research on the carbon cycle, including measurement, has a long history dating back decades, and it is currently conducted by a number of research groups. Haszpra (ed.) (2011) provides an overview of the background and outcomes of these activities for GHGs other than CO_2.

Eddy covariance (EC) technique is used at three sites in Hungary to monitor ecosystem-atmosphere CO_2 exchange over grasslands and croplands. EC measurements started at the Hegyhátsál site in 1997 to monitor CO_2 fluxes over a heterogeneous agricultural landscape. The measurements are unique in the country being the only measurements over cropland and due to the measurements height, which is 82 m providing regional scale information on the surrounding area. CO_2 mixing ratio measurements started in 1994 at the site, followed by measurements of other GHGs (CH_4, CO, H_2, N_2O, SF_6). Soil CO_2 and CH_4 emissions were measured at sites of different soil and vegetation types including agricultural land (Tóth et al., 2010), and grassland (Horváth et al., 2008).

REFERENCES

1. AFEAS Workshop. (1995). *Proceedings of the Workshop on the Atmospheric Degradation of HCFCs and HFCS* (November 1993, Boulder, CO). Sponsored by National Aeronautics and Space Administration, National Oceanic and Atmospheric Administration, and the Alternative Fluorocarbons Environmental Acceptability Study.
2. Akimoto, H., Nakane, H., & Matsumoto, Y. (1994). The chemistry of oxidant generation: Tropospheric ozone increase in Japan. In J. G. Calvert (Ed.), *Chemistry of the Atmosphere: The Impact of Global Change* (pp. 261–273). Blackwell Scientific Publications.
3. Anfossi, D., & Sandroni, S. (1993). Surface ozone at mid latitudes in the past century. *Il Nuovo Cimento, 17*(2), 199–207.
4. Baird, C., & Cann, M. (1995). *Environmental Chemistry* (5th ed.). WH Freeman and Company.
5. Beaton, S. P., Bishop, G. A., Zhang, Y., Ashbaugh, L. L., Lawson, D. R., & Stedman, D. H. (1995). On-road vehicle emissions: Regulations, costs, and benefits. *Science, 268*(5212), 991–993.
6. Calvert, J. G. (1995). Review of the activities related to environmental chemistry within the International Union of Pure and Applied Chemistry. Report available from the Secretariat of the IUPAC, Bank Court Chambers, 2–3 Pound Way, Templars Square, Cowley, Oxford, OX4 3YF, UK.
7. Cantrell, C. A., Shetter, R. E., Calvert, J. G., Parrish, D. D., Fehsenfeld, F. C., Goldan, P. D., Kuster, W., Williams, E. J., Westberg, H. H., Allwine, G., & Martin, R. (1993). Peroxyl radicals as measured in ROSE and estimated from photo stationary state deviations. *Journal of Geophysical Research, 98*(D11), 18355–18366.
8. DeMore, W. B., Sander, S. P., Golden, D. M., Hampson, R. F., Kurylo, M. J., Howard, C. J., Ravishankara, A. R., Kolb, C. E., & Molina, M. J. (1997). *Chemical Kinetics and Photochemical Data for Use in Stratospheric Modeling* (Jet Propulsion Laboratory Publication No. 97-4, 266 pp.). Pasadena, California: Jet Propulsion Laboratory.
9. Girard, J. E. (2005). *Principles of Environmental Chemistry* (2nd ed.). Jones and Bartlett Publishers.
10. Haszpra, L. (1995). Carbon dioxide concentration measurements at a rural site in Hungary. *Tellus, 47B*, 14–22.
11. Haszpra, L. (2011). *Atmospheric Greenhouse Gases: The Hungarian Perspective*. Springer.

12. Haszpra, L., Barcza, Z., Bakwin, P. S., Berger, B. W., Davis, K. J., & Weidinger, T. (2001). Measuring system for the long-term monitoring of biosphere/atmosphere exchange of carbon dioxide. *Journal of Geophysical Research, 106*(D3), 3057–3070.
13. Holland, H. D. (2006). The oxygenation of the atmosphere and oceans. *Philosophical Transactions of the Royal Society B, 361*, 903–915.
14. Horváth, L., Grosz, B., Machon, A., Balogh, J., Pintér, K., & Czóbel, S. (2008). Influence of soil type on N_2O and CH_4 soil fluxes in Hungarian grasslands. *IPCC Fourth Assessment Report: Climate Change* (Communications in Ecology, Volume 9, Supplement 1), 75–80. https://doi.org/10.1556/ComEc.9.2008.S.11.
15. Hutchinson, T. C., & Havas, M. (1994). Ecological impacts of acid deposition. In J. G. Calvert (Ed.), *The Chemistry of the Atmosphere: Its Impact on Global Change* (pp. 297–315). Blackwell Scientific Publications.
16. Kley, D., Geiss, H., & Mohnen, V. A. (1994). Concentrations and trends of tropospheric ozone and precursor emissions in the USA and Europe. In J. G. Calvert (Ed.), *The Chemistry of the Atmosphere: Its Impact on Global Change* (pp. 245–259). Blackwell Scientific Publications.
17. Kuptyug, V. (1990). *Chemistry and the Environment: The IUPAC Programme* (M. H. Freemantle, Ed.). Rocardo Press Ltd.
18. Manahan, S. E. (2010). *Environmental Chemistry* (9th ed.). CRC Press.
19. May, L. (2010). Atomism before Dalton. In *Atoms in Chemistry: From Dalton's Predecessors to Complex Atoms and Beyond* (Vol. 1044, pp. 21–33). American Chemical Society.
20. Miller, G. T. (2012). *Living in the Environment: Principles, Connections, and Solutions*. Brooks Cole.
21. Pöschl, U. (2005). Atmospheric aerosols: Composition, transformation, climate and health effects. *Angewandte Chemie International Edition, 44*(46), 7520–7540.
22. Tóth, E., Barcza, Z., Birkás, M., Gelybó, G., Zsembeli, J., Bottlik, L., Davis, K. J., Haszpra, L., Kem, A., Kljun, N., Koós, S., Kovács, G., Stingli, A., & Farkas, C. S. (2010). Measurements and estimations of biosphere-atmosphere exchange of greenhouse gases–Arable lands. In L. Haszpra (Ed.), *Atmospheric Greenhouse Gases: The Hungarian Perspective* (pp. 75–98). Springer.
23. U. S. National Research Council (NRC). (1993). *The Atmospheric Effects of Stratospheric Aircraft: Interim Assessment Report of the NASA High-Speed Research Program, Panel on Atmospheric Effects of Stratospheric Aircraft.*

24. World Meteorological Organization (WMO). (1999). *Scientific Assessment of Ozone Depletion: 1998* (Report No. 44). Global Ozone Research and Monitoring Project. Geneva, Switzerland.

CHAPTER 6

OZONE CHEMISTRY

Contents

INTRODUCTION

Ozone, having the chemical formula O_3, is an inorganic molecule. It smells very strongly and is a light blue gas. It is an oxygen allotrope that breaks down into O_2 (dioxygen) in the lower atmosphere. It is far less stable than the diatomic allotrope O_2. Within the Earth's atmosphere, ultraviolet (UV) light and electrical discharges combine to form ozone from dioxygen. It is found in very small amounts throughout the atmosphere, with the highest concentration occurring high in the stratospheric ozone layer, which is responsible for absorbing the majority of ultraviolet (UV) radiation from the Sun.

The smell of ozone is similar to that of chlorine and can be detected in the air at as low as 0.1 parts per million (ppm) by most people. The O_3 structure of ozone was discovered in 1865. It was subsequently demonstrated that the molecule was weakly diamagnetic and had a bent structure. Ozone is normally a pale blue gas that, when heated to cryogenic temperatures, condenses into a dark blue liquid and, eventually, a violet-black solid. Due to ozone's instability in relation to common dioxygen, it can decompose explosively in liquid and concentrated gas forms when exposed to high temperatures, physical shock, or rapid warming to boiling point. Therefore, it is only used in small concentrations for commercial purposes.

Ozone has numerous industrial and consumer uses involving oxidation because it is a potent oxidant—far more potent than dioxygen. However, above concentrations of roughly 0–1 ppm, ozone damages plant and animal tissues, including mucous and respiratory tissues. This is due to its high oxidizing potential. Because of this, ozone is a strong respiratory hazard and pollutant close to the ground. However, a higher ozone layer concentration (between two and eight parts per million) is advantageous because it blocks harmful UV light from reaching the Earth's surface.

6.1. HISTORY

Dutch chemist Martinus van Marum discovered an odd smell in 1785 while performing electrical sparking experiments above water. He dismissed it as the result of electrical reactions, not realizing that he had actually produced ozone.

Christian Friedrich Schönbein noticed the same strong odor 50 years later and identified it as the smell that frequently follows a lightning strike. After successfully isolating the gaseous chemical in 1839, he gave it the name ozone, derived from the Greek word *ozein*, which means to smell. Schönbein is therefore widely acknowledged as the person who discovered ozone. He also observed that the smell of ozone and phosphorus were similar, and in 1844 he established that the result of reacting white phosphorus with air is the same. He mocked a later attempt to refer to ozone as electrified oxygen by suggesting that it be called phosphorized oxygen, which is derived from white phosphorus. It wasn't until 1865 that Jacques-Louis Soret determined the formula for ozone or O_3, and Schönbein confirmed it in 1867.

For much of the second half of the 19th century and well into the 20th, ozone was considered a healthy component of the environment by naturalists and health-seekers. Beaumont, California, had as its official slogan "Beaumont: Zone of Ozone," as evidenced on postcards and Chamber of Commerce letterhead. Naturalists working outdoors often considered the higher elevations beneficial because of their ozone content. "There is quite a different atmosphere [at higher elevation] with enough ozone to sustain the necessary energy [to work]," wrote naturalist Henry Henshaw, working in Hawaii. Seaside air was considered to be healthy because of its believed ozone content. The smell giving rise to this belief is in fact that of halogenated seaweed metabolites and dimethyl sulfide.

Ozone's fresh smell evoked associations with purifying properties, which likely contributed too much of its appeal. Scientists observed negative consequences. When exposed to ozonized air, frogs became lethargic, birds gasped for air, and the oxygen content of rabbits' blood dropped, as reported by James Dewar and John Gray McKendrick in 1873. This was a detrimental effect. Schönbein himself stated that breathing in ozone caused small mammals to perish as well as chest pains, mucous membrane irritation, and breathing difficulties. The positive health effects of ozone have, through repeated exposure, become widely accepted, according to Leonard Hill and Martin Flack's 1911 statement in the Proceedings of the Royal Society B. However, precise physiological evidence supporting ozone's benefits has been virtually nonexistent up until this point. As far as the physiological effects of ozone are concerned, the only fully established information is that it can cause lung irritation and edema, as well as death if inhaled in relatively high concentrations for an extended period of time.

Ozone was tested at Queen Alexandra Military Hospital in London during World War I as a potential wound disinfectant. For a maximum of 15 minutes, the gas was directly applied to the wounds. Both human tissue and bacterial cells were harmed as a result of this. It was discovered that other sanitizing methods, like antiseptic irrigation, were superior.

Because it is difficult to apply analytical chemistry techniques to the explosive concentrated chemical, it was unknown until the 1920s whether trace amounts of oxozone, O_4. Georg-Maria Schwab was the first to accurately analyze and successfully solidify ozone in 1923, effectively disproving the oxozone hypothesis. The Riesenfeld group discovered additional, previously unmeasured physical characteristics of pure concentrated ozone in the 1920s.

6.2. PHYSICAL AND CHEMICAL PROPERTIES OF OZONE

The use of advanced oxidation processes (AOPs) for the purification of water and air has gained popularity over the past few decades. The AOPs are defined as processes that involve the generation and reactions of hydroxyl radicals. It has been observed that the AOPs work well as pre-treatment or treatment procedures for contaminated media, as well as for the purification of air and water. One could argue that the use of ozone was among the first, if not the first, of these processes to be thoroughly studied. The primary purpose of the ozonation process, which is now widely employed worldwide, is to purify drinking water.

All of the ozone's properties that are involved in the process of water ozonation must be considered to fully comprehend it. In general, the steps involved in ozonation are as follows: ozone generation, ozone introduction to water, ozone dissolution, and finally the chemical reactions between ozone and the dissolved materials. The physical characteristics of the water control the ozone's dissolution. As we'll see later, this dissolution itself causes partial ozone breakdown, which starts a series of radical reactions and serves as a source of various oxidants, including the hydroxyl radical mentioned earlier. Subsequently, the residual dissolved molecular ozone initiates diverse reactions with the constituents in the water. Its breakdown into radical species is mostly caused by inorganic compounds. Conversely, organic materials react directly with ozone as well as with radicals.

Ozone reacts with organics in a number of main ways. When an unsaturated bond is present, the substance under investigation's characteristics, particularly the makeup of the atoms next to the unsaturated bond, may allow for the reactions of electrophilic substitution, nucleophilic addition, and cycloaddition. We'll also go over reactions involving aliphatic organics, such as amines, sulfur, and organic compounds containing heteroatoms in the carbon group. The interactions of molecular ozone with saturated organic substances of various chemical structures are of interest and should be clarified, despite their lower reaction rate in comparison to unsaturated hydrocarbons.

Ozone's properties and its widespread use make it an important topic to discuss separately, along with health and safety concerns related to its use.

In conclusion, it is imperative to acknowledge that a deeper understanding of the molecular structure of ozone can provide light on its physical and chemical properties, particularly with regard to its ultimate source. We're going to start at this point.

6.2.1. Molecular Structure of Ozone

Ozone is a triatomic oxygen compound, and its molecular makeup contributes to its high reactivity. In 1952, the ozone molecule's structure was determined through a hybridization of four resonance structures, as illustrated in Figure 6.1. The true molecular structures are found in between the resonance structures, which are essentially the extreme forms of a molecule that do not exist. These structures are employed when the real chemical structure cannot be described by any conventional structural formula and the molecule's experimentally determined properties simultaneously correspond to all the resonance structures. The case of the ozone molecule structure investigation accurately reflects this state of affairs. The length of the bonds between the oxygen atoms in an ozone molecule has been calculated to be 1.278 Å, given that the length of a single bond between two oxygen atoms in hydrogen peroxide is 1.49 Å and the length of a double bond in an oxygen molecule is 1.21 Å. Because the bonds are both longer than double bonds and much shorter than single bonds should be, the hybrid of the first and second resonance structures reflects this phenomenon.

Nevertheless, the molecular structure of ozone cannot yet be precisely outlined by those two resonance structures alone. The O-O-O angle of 120° is expected to result from the electron clouds of the atoms repelling each other; however, the experimentally determined angle is marginally smaller, measuring 116.78°. The mutual attraction of positively and negatively charged oxygen atoms, as suggested by third and fourth resonance structures, can also account for the lower angle. This latter characteristic explains why, depending on the nature of the organic substrate, the ozone molecule can behave as an electrophilic or nucleophilic agent in reactions with unsaturated organics; the mechanism of these reactions will be covered later.

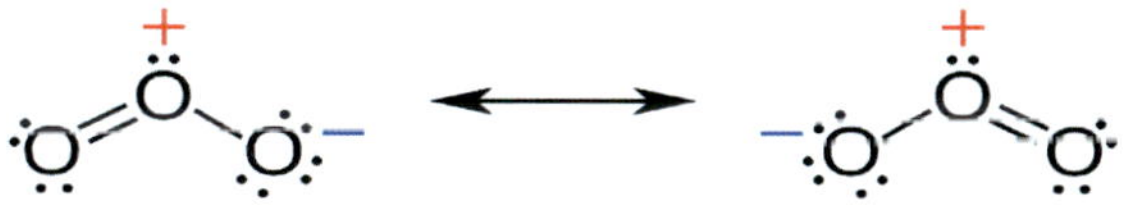

Figure 6.1. *Resonance structures of ozone molecule.*

Source: By D. Ilyin: vectorization – File:Ozone-resonance-Lewis-2D.png by Ben Mills, CC0, https://commons.wikimedia.org/w/index.php?curid=149150037.

A cyclic ozone molecule with electrically neutral oxygen atoms joined by single bonds is an alternate structure for the molecule. Despite being created in a lab setting with high-frequency lasers, this structure is uncommon in the natural world. The angle strain in triangular structures, as in the cases of cyclopropane, ethylene oxide, and aziridine, is one of the primary causes of this. For energetic reasons, the O-O-O angle should be near 120° when three atoms have the same electronegativity; however, in the case of a cyclic structure, a triatomic molecule is shaped like an equilateral triangle, where the angles can only be 60°. In this manner, the electron clouds that generate σ-bonds are sharply twisted away from the triangle's straight lines. This state of affairs is frequently denoted as a bent bond or banana bond. Because of the aforementioned strain, these bonds are far less stable, and the cycle is simple to open at relatively small energies (Schubert et al., 1956).

6.2.2. Physical Properties of Ozone

6.2.2.1. Basic Physical Data

Ozone is a colorless gas at standard conditions and a pale blue gas at concentrations found in the Earth's atmosphere. Over 1.5 times more ozone is present in the atmosphere than there is in the air. By nature, gaseous mixtures of ozone with air, nitrogen, or oxygen are quite similar to perfect combinations. Ozone-air mixtures become explosive at concentrations of 10–11% by volume or higher because ozone breaks down with significant energy emissions. In that regard, working with ozone at lower concentrations is regarded as safe.

6.2.2.2. Solubility of Ozone

Similar to oxygen, ozone does not dissolve easily in water. It is possible to characterize the solubility of ozone in water as a function of water temperature at 1 atm of pressure. The solubility values are displayed as a solubility ratio (S), which is the ratio of the airborne and waterborne ozone concentrations (measured in milliliters per liter).

According to the available experimental data, ozone dissolves according to Henry's law, which states that the concentration of dissolved ozone at a fixed temperature increases steadily as the partial pressure of ozone in the gaseous phase increases. This trend is almost linear.

6.2.3. Chemical Properties of Ozone

6.2.3.1. Ozone-initiated Radical Reactions in Water

Oxidation-reduction, or more simply redox reactions, are radical reactions that are triggered by ozone. e. reactions in which the oxidation state of the reagents changes and electrons are transferred between the reacting species. Ozone is effectively broken down in these reactions, though it may also serve as a springboard for the emergence of much more potent oxidants than the one it breaks down into, such as the hydroxyl radical, an even more potent oxidizing agent than fluorine. In general, there are two main ways that the breakdown of ozone, which occurs when it dissolves in water and starts the formation of different radical species, can happen. These mechanisms are called Gordon-Tomiyasu-Fukutomi and Hoigné-Staehlin-Bader, respectively, after the authors of the corresponding models. Although the latter pathway is far more accurate at higher pH values, the former is thought to be the most commonly accepted (Benson and, Axworthy, 1957).

The mechanism of ozone decomposition in water can be categorized into three main stages, as is typical with radical reactions: the beginning of the reaction, the actual chain reaction, and the breakdown of the radical reaction chain. More radicals are created from reaction precursors by the radicals formed in the initiation step, and these newly formed radicals act as the catalysts for further reactions. When there is enough reaction substrate present, a radical reaction can be said to be self-sustaining—at least initially. The radical chain can be inhibited and broken down in a number of ways. For example, the radical reaction can be stopped by introducing an agent known as a radical scavenger, which can adsorb radicals without effectively oxidizing them, or by continuously forming and accumulating less reactive species. Tert-butyl alcohol, humic substances, and carbonate anion are a few examples of radical scavengers. Radical scavengers introduce radicals into a substance's molecule without allowing the reaction chain to continue sequentially.

6.2.3.2. Hoigné-Staehlin-Bader Mechanism

According to the Hoigné-Staehlin-Bader model, ozone decomposition and the start of radical species formation start with an interaction between ozone and the hydroxide anion in water, which is mostly caused by the hydrolysis of water molecules. Hydroperoxide radical and superoxide radical ions are created as a result.

$$O_3 + OH^- \rightarrow HO_2^{\bullet} + {}^{\bullet}O_2^-,\ k_1 = 70\ M^{-1}\ s^{-1} \qquad (6.1)$$

The value of the reaction rate constant 1 k indicates that the ozone decomposition initiation step is the limiting stage due to its somewhat slow nature. Equation (6.2) suggests that the hydrogen peroxide radical produced during reaction (6.1) may also dissociate to form a proton and a superoxide radical ion.

$$HO_2^{\bullet} \rightarrow H^+ + {}^{\bullet}O_2^-,\ k_2 = 7.9 \times 10^5\ s^{-1} \qquad (6.2)$$

The superoxide radical ion, which is produced during these reactions, acts as the catalyst for the ongoing chain reaction by giving the ozone molecule its extra electron. Equation (6.3) illustrates how it reacts with ozone, creating an ozonide radical ion in the process:

$$O_3 + {}^{\bullet}O_2^- \rightarrow {}^{\bullet}O_3- + O_2, k_3 = 1.6 \times 10^9 \text{ M}^{-1} \text{ s}^{-1} \quad (6.3)$$

As a result, ozonide radical ions can enter reaction (6.4), which can be seen from reaction Equation (6.5) to form the hydroxyl radical, a potent and well-known oxidant:

$$^{\bullet}O3^- + H^+ = HO_3^{\bullet}, k_4 = 5.2 \times 10^{10} \text{ M}^{-1} \text{ s}^{-1} \quad (6.4)$$

$$k_{-4} = 2.3 \times 10^2 \text{ s}^{-1}$$

$$HO_3^{\bullet} \rightarrow {}^{\bullet}OH + O_2, k_5 = 1.1 \times 10^5 \text{ s}^{-1} \quad (6.5)$$

The hydroxyl radical can initiate the following reactions (6.6–6.9):

$$^{\bullet}OH + O_3 \rightarrow HO_4^{\bullet}, k_6 = 2.0 \times 10^9 \text{ M}^{-1} \text{ s}^{-1} \quad (6.6)$$

$$HO_4^{\bullet} \rightarrow HO_2^{\cdot} + O_2, k_7 = 2.8 \times 10^4 \text{ s}^{-1} \quad (6.7)$$

$$2\, HO_4^{\bullet} \rightarrow H_2O_2 + 2\, O_3, k_8 = 5.9 \times 10^9 \text{ M}^{-1} \text{ s}^{-1} \quad (6.8)$$

$$HO_4^{\bullet} + HO_3^{\bullet} \rightarrow H_2O_2 + O_3 + O_2, k_9 = 5.0 \times 10^9 \text{ M}^{-1} \text{ s}^{-1} \quad (6.9)$$

Reactions (6.8) and (6.9) can be viewed as an inhibition stage in the radical reaction chain because ozone is at least partially restored and radical species are not produced as byproducts.

6.2.3.3. Gordon-Tomiyasu-Fukutomi Mechanism

The initiation reaction in this scheme differs slightly from the preceding one because only the hydroperoxide anion radical is produced as a result of a distinct electron transfer pattern:

$$O_3 + OH^- \rightarrow HO_2^{-\bullet} + O_2, k_{10} = 40 \pm 2 \text{ M}^{-1} \text{ s}^{-1} \quad (6.10)$$

After that, the ozone molecule and the hydroperoxide anion radical react to produce the ozonide ion radical and the hydroperoxide radical:

$$O_3 + HO_2^{\bullet -} \rightarrow {}^{\bullet}O_3^- + HO_2^{\bullet}, k_{11} = 2.2 \times 10^6 \text{ M}^{-1} \text{ s}^{-1} \quad (6.11)$$

Hydroperoxide radical can also react with hydroxide anion, as the following reaction equation suggests:

$$HO_2^{\bullet} + OH^- = {}^{\bullet}O_2^- + H_2O, K_a = 10^{-4.8} \quad (6.12)$$

Superoxide radical, thus formed, reacts accordingly to Equation (6.3):

$$^{\bullet}O_3^- + H_2O \rightarrow {}^{\bullet}OH + OH^- + O_2, k_{13} = 20 \text{ to } 30 \text{ M}^{-1} \text{ s}^{-1}. \quad (6.13)$$

Hydroxyl radical can consequently step into reaction with ozonide anion radical or molecular ozone (Equations. 6.14 to 6.16):

$$^{\bullet}OH + {}^{\bullet}O_3^- \rightarrow {}^{\bullet}O_2^- + HO_2^{\bullet},\ k_{14} = 6 \times 10^9\ M^{-1}\ s^{-1} \quad (6.14)$$

$$^{\bullet}OH + {}^{\bullet}O_3^- \rightarrow O_3 + OH^-,\ k_{15} = 2.5 \times 10^9\ M^{-1}\ s^{-1} \quad (6.15)$$

$$^{\bullet}OH + O_3 \rightarrow O_2 + HO_2^{\bullet},\ k_{16} = 3 \times 10^9\ M^{-1}\ s^{-1} \quad (6.16)$$

As well-known radical scavengers, dissolved carbonate ions efficiently block the radical reaction pattern as it is above described. Equations (6.17) and (6.18) describe their action:

$$^{\bullet}OH + CO_3^{2-} \rightarrow OH^- + {}^{\bullet}CO_3^-,\ k_{17} = 4.2 \times 10^8\ M^{-1}\ s^{-1} \quad (6.17)$$

$$^{\bullet}CO_3^- + O_3 \rightarrow CO_2 + {}^{\bullet}O_2^- + O_2 \quad (6.18)$$

There have been many arguments made that reactions (6.17) and (6.18) cannot be completely regarded as preventing the breakdown of ozone because reaction (6.18) is thought to produce a superoxide ion radical, which is what is seen to start the ozone breakdown process. Therefore, by including two more reactions, the existing mechanism can be marginally enhanced.

$$^{\bullet}OH + H_2O_2 \rightarrow HO_2^{\bullet} + H_2O,\ k_{19} = 2.7 \times 10^7\ M^{-1}\ s^{-1} \quad (6.19)$$

$$^{\bullet}OH + HO_2^- \rightarrow HO_2^{\bullet} + OH^-,\ k_{20} = 7.5 \times 10^9\ M^{-1}\ s^{-1} \quad (6.20)$$

The dissociation of hydrogen peroxide results in the formation of the hydroperoxy anion in reaction (6.20).

At higher concentrations, hydrogen peroxide can inhibit ozone decomposition even though it can act as an initiator because mass transfer becomes the reaction's limiting step.

It is also known that hydroxyl radicals are more active than hydroperoxide radicals, which are created during these reactions.

6.3. OZONE IN EARTH'S ATMOSPHERE

Dobson units are commonly used to express total ozone levels in the atmosphere, which are defined as the amount of ozone in a given vertical column. Point measurements are expressed as concentrations in μg/m^3 or as mole fractions in nmol/mol (parts per billion, ppb). In the 1920s, research on the amount of ozone in the atmosphere began.

6.3.1. Ozone Layer

A portion of the Earth's stratosphere known as the ozone layer or ozone shield is responsible for absorbing the majority of the sun's UV rays. Compared to other regions of the atmosphere, it has a higher concentration of ozone (O_3), but it is still relatively small in comparison to other gases in the stratosphere. Less than 10 parts per million of ozone are found in the ozone layer, compared to an average of 0.3 parts per million in the Earth's atmosphere overall. The ozone layer is mostly located in the lower stratosphere, between 15 and 35 kilometers (9 and 22 miles) above Earth, though its thickness varies with the seasons and the location.

The ozone layer was first identified in 1913 by Henri Buisson and Charles Fabry, two French physicists. With the exception of the absence of radiation at the ultraviolet end of the spectrum, which has a wavelength of approximately 310 nm, measurements of the sun's radiation are generally consistent with the spectrum of a black body with a temperature in the range of 5,500–6,000 K (5,230–5,730°C). It was concluded that something in the atmosphere was absorbing the missing radiation.

In the end, the spectrum of the absent radiation was identified as ozone, the only known chemical. The British meteorologist G thoroughly investigated its characteristics. M. B. Dobson, who created the Dobsonmeter, a basic spectrophotometer that allowed for ground-based stratospheric ozone measurements. Dobson set up a global network of ozone monitoring stations between 1928 and 1958, and these stations are still in operation today. The "Dobson Unit" (DU), a convenient measure of the amount of ozone overhead, is named in his honor.

The Sun's medium-frequency ultraviolet light, which has a wavelength range of about 200 nm to 315 nm, is absorbed by the ozone layer to the extent that it would otherwise potentially harm exposed life forms close to the surface.

Research on the atmosphere in 1985 showed that chemicals released by industry, particularly chlorofluorocarbons (CFCs), were causing the ozone layer to thin.

Bans on the chemicals resulted from worries that rising UV radiation from ozone depletion threatened life on Earth, including an increase in skin cancer cases among humans and other ecological issues. The most recent data suggests that ozone depletion has slowed or stopped. September 16 has been declared the International Day for the Preservation of the Ozone Layer by the United Nations General Assembly.

A thin layer of ozone exists on Venus as well, situated 100 kilometers above the planet's surface.

6.3.1.1. Sources

Sydney Chapman, a British physicist, made the discovery of the photochemical processes that create the ozone layer in 1930. When ultraviolet light strikes ordinary oxygen molecules with two oxygen atoms (O_2), it splits the molecules into individual oxygen atoms (atomic oxygen); the atomic oxygen then combines with unbroken O_2 to form ozone, or O_3. This process creates ozone in the Earth's stratosphere. The ozone-oxygen cycle is the ongoing process that results from the unstable (but long-lived in the stratosphere) ozone molecule splitting into an individual oxygen atom and an O_2 molecule when exposed to ultraviolet light.

Chemically, this can be described as:

$$O_2 + h\nu_{uv} -> 2O$$

$$O + O_2 <-> O_3$$

The stratosphere contains about 90% of the ozone in the atmosphere. The highest ozone concentrations, ranging from 2 to 8 parts per million, are found between 20 and 40 kilometers (66,000 and 131,000 feet). Squeezed to the pressure of sea level air, all of the ozone would be just 3 millimeters (1/8 inch) thick.

6.3.1.2. Ultraviolet Light

Ozone, despite having a very low concentration in the ozone layer, is essential to life as it filters out ultraviolet (UV) radiation from the Sun that is harmful to living things. Nitrogen screens out very short or vacuum UV (10–100 nm). Based on wavelength, UV radiation that can penetrate nitrogen is classified into three groups: UV-A (400–315 nm), UV-B (315–280 nm), and UV-C (280–100 nm).

All living things are severely harmed by UV-C, which is completely blocked by a mixture of dioxygen (. < 200 nm) and ozone (> roughly 200 nm) at a height of about 35 km (115,000 ft). Sunburn is primarily caused by UV-B radiation, which can be damaging to the skin. Prolonged exposure can also lead to immune system suppression, cataract development, and genetic damage, all of which can result in conditions like skin cancer. UV-B radiation is effectively blocked by the ozone layer, which has a maximum absorption at 250 nm and a wavelength range of 200–310 nm. At 290 nm, the intensity of radiation is 350 million times stronger at the top of the atmosphere than it is at the surface of the Earth. However, some UV-B, especially at its longest wavelengths, reaches the skin's surface and plays a crucial role in mammals' skin's ability to produce vitamin D.

Since most UV-A is transparent to ozone, the majority of this longer-wavelength UV radiation that reaches Earth's surface does so. While this kind of UV radiation is much less damaging to DNA, it can still result in physical harm, early skin aging, indirect genetic damage, and skin cancer.

6.3.1.3. Distribution in the Stratosphere

Globally, the ozone layer's thickness varies, with the layer typically being thinner near the equator and thicker near the poles. The amount of ozone in a column over a specific area is referred to as thickness, and it varies depending on the season. These variations can be attributed to variations in solar intensity and patterns of atmospheric circulation.

Most ozone is created over the tropics and is carried towards the poles by patterns of stratospheric winds. These patterns, referred to as the Brewer-Dobson circulation, cause the ozone layer to thin in the fall and thicken in the spring in the northern hemisphere. In the tropics, solar UV radiation produces ozone by means of circulation, which pushes ozone-deficient air from the troposphere into the stratosphere, where photolysis of oxygen molecules by the sun produces ozone. After that, the air rich in ozone is transported to higher latitudes where it descends into the lower atmospheric layers.

According to research, October is when ozone levels are lowest in the United States and highest in the spring months of April and May. As ozone moves from the tropics to higher latitudes, its total amount increases. However, its concentrations are higher in high northern latitudes than in high southern latitudes. Springtime ozone columns in high northern latitudes can occasionally exceed 600 DU and average 450 DU, while the Antarctic's usual maximum before anthropogenic ozone depletion was 400 DU. This difference happened naturally as a result of the northern hemisphere's larger mountain ranges, higher contrasts between land and ocean temperatures, and weaker polar vortex and stronger Brewer-Dobson circulation. The ozone hole phenomenon has caused a rise in the disparity between high northern and southern latitudes since the 1970s. The Antarctic has the lowest levels of ozone in the summer months of September and October, while the Arctic has the highest levels in the spring months of March and April.

6.3.1.4. Depletion

Free radical catalysts such as nitric oxide (NO), nitrous oxide (N_2O), hydroxyl (OH), atomic chlorine (Cl), and atomic bromine (Br) can cause the ozone layer to thin. All of these species have natural sources, but in recent decades, man-made organohalogen compounds, particularly CFCs and bromofluorocarbons, have been released in large quantities, which has led to a notable increase in the concentrations of chlorine and bromine. These incredibly stable substances can withstand the ascent to the stratosphere, where ultraviolet light releases Cl and Br radicals. After that, every radical is free to start and catalyze a chain reaction that can destroy more than 100,000 ozone molecules. The most significant ozone-depleting substance (ODS) released by human activity as of 2009 was nitrous oxide.

Reduced absorption of UV radiation is the result of ozone breakdown in the stratosphere. As a result, hazardous UV radiation that has not been absorbed can now reach the Earth's surface at a higher intensity. Since the late 1970s, there has been an average global decrease of approximately 4% in ozone levels. Much larger seasonal declines have been observed for about 5% of the Earth's surface, around the north and south poles; these areas are known as ozone holes. In actuality, ozone holes are areas in the ozone layer where the ozone is thinner. At the polar points of Earth's axis, the ozone layer is at its thinnest. Joe Farman, Brian Gardiner, and Jonathan Shanklin first reported the annual ozone depletion above the Antarctic in a paper published in Nature on May 16, 1985.

Attempts at regulation have included, but have not been limited to, the United States Environmental Protection Agency's implementation of the Clean Air Act. Ozone pollution is one of the six criteria pollutants for the National Ambient Air Quality Standards (NAAQS), which were mandated by the Clean Air Act. Since counties, cities, and tribal regions are required to follow these standards and the EPA also offers support to each region in order to regulate contaminants, this regulation has shown to be effective. Educating the public about the existence and control of ozone depletion and pollutants has also shown to require effective information presentation. The ozone case was explained to laypeople using clear, bridging metaphors from popular culture that connected it to current dangers that were relevant to everyday life. The discussion's use of specific metaphors, such as "ozone shield" and "ozone hole," proved to be very helpful. The ozone situation was also perceived as a pressing concern, as opposed to global climate change. Concerns about skin cancer and the ozone layer's thinning were shared by laypeople.

Aluminum oxide ($A_{12}O_3$) nanoparticles are created when satellites burn up during re-entry into Earth's atmosphere, and they stay in the atmosphere for decades. An estimated 17 metric tons (about 30 kg of nanoparticles per 250 kg satellite) were expected for 2022 alone. Significant ozone depletion may eventually result from growing satellite constellation populations.

Bad ozone has been shown to exacerbate respiratory conditions like asthma, COPD, and emphysema and to pose serious health risks. It can also cause respiratory effects, such as difficulty breathing. For this reason, a lot of nations have implemented laws to enhance beneficial ozone and stop the growth of harmful ozone in cities or residential areas. The European Union has tight regulations on what products can be purchased, distributed, or used in certain locations in order to protect the ozone layer (keeping it healthy). Over time, it is anticipated that the ozone will recover with proper regulation.

The European Community turned down a proposal to outlaw aerosol sprays that contained CFCs and destroyed the ozone layer, but the U.S., Canada, and Norway passed laws prohibiting them in 1978. CFCs were still in use in the U.S. in 1985 when the Antarctic ozone hole was discovered, as well as in other uses like industrial cleaning and refrigeration. The production of CFCs was limited to 1986 levels with commitments to long-term reductions following the negotiation of an international treaty known as the Montreal Protocol. Developing countries were granted a 10-year phase-in period as a result. Since then, the treaty has been changed to forbid the production of CFCs in developed nations starting in 1995 and later in developing nations. The 197 nations that make up the world have all ratified the treaty. In developed nations like the U.S., only stockpiled or recycled CFCs were permitted for use as of January 1, 1996. Efforts to guarantee that there would be alternative chemicals and technologies for all ODS uses made the production phase out possible.

On August 2, 2003, scientists declared that the international regulation of substances that deplete the ozone layer may be slowing down the process of global ozone layer depletion. Three satellites and three ground stations confirmed, in a study sponsored by the American Geophysical Union that the upper-atmosphere ozone-depletion rate has slowed significantly over the past 10 years. Because of the gases already in the stratosphere and the use of ODSs by countries that have not outlawed them, some breakdown was predicted to continue. The atmospheric lifetimes of certain ODSs, such as CFCs, can reach over 100 years. The ozone layer is predicted to replenish to 1980 levels by the middle of the twenty-first century. A gradual trend toward "healing" was reported in 2016.

In some applications, compounds with C–H bonds—such as hydrochlorofluorocarbons, or HCFCs—have been developed to replace CFCs. These substitute substances are less likely to endure long enough in the atmosphere to enter the stratosphere, where they might have an impact on the ozone layer, and they are also more reactive. HCFCs are also being phased out even though they are less harmful than CFCs and may harm the ozone layer. Hydrofluorocarbons (HFCs) and other substances that do not obliterate stratospheric ozone are gradually replacing these.

There is a concentration gradient between the atmosphere and the ocean as a result of the lingering effects of CFCs building up in the atmosphere. This organohalogen compound can function as a time-dependent tracer and dissolve into the surface waters of the ocean. This tracer follows chemical, physical, and biological pathways to aid in the study of ocean circulation by scientists.

6.3.2. Low Level Ozone

One type of air pollution is low level ozone, also known as tropospheric ozone. It is created by sunlight reacting with air that contains hydrocarbons and nitrogen oxides, either right at the source of the pollution or several kilometers downwind. It is not directly released by automobile engines or industrial processes.

Certain hydrocarbons, like aldehydes, react with ozone directly, removing them from the atmosphere, but the byproducts are also important parts of smog. In addition to helping to remove hydrocarbons from the air, ozone photolysis by UV light is the initial step in the formation of smog-causing substances like peroxyacyl nitrates, which can be extremely irritating to the eyes. The atmospheric lifetime of tropospheric ozone is about 22 days; its main removal mechanisms are being deposited to the ground, the above-mentioned reaction giving HO•, and by reactions with OH and the peroxy radical HO_2•.

There is proof that elevated ground-level ozone and pollution, which obstruct photosynthesis and limit the overall growth of certain plant species, have significantly reduced agricultural yields. In addition to the primary rule intended to protect public health, the U.S. Environmental Protection Agency (EPA) has proposed a secondary regulation to lessen crop damage.

6.3.2.1. Low Level Ozone in Urban Areas

A few instances of cities with high ozone levels are Mexico City, Mexico; Denver, Colorado; and Houston, Texas. With a reading of about 125 nmol/mol, Mexico City is far more dangerous than Houston, which has a reading of about 41 nmol/mol.

The most dangerous form of ozone pollution in cities is low level ozone, also known as tropospheric ozone, and it is generally rising. Denser populations are more affected by ozone pollution in urban areas, which is made worse by a large number of vehicles that release NO_2 and VOCs, the main pollutants that contribute to dangerously high ozone levels. Because urban ozone pollution increases heat-related mortality during heat waves and temperature rises, it is especially concerning. Ground level ozone pollution can be 20% higher than usual during heat waves in urban areas. The summer and fall months in urban areas see higher than average levels of ozone pollution, which could be attributed to traffic and weather patterns. Despite the fact that those living in poverty are less likely to be the ones contributing to pollution levels, they are nonetheless more negatively impacted by it overall.

Denver, Colorado, is one of the many U.S. cities with high levels of ozone, as was previously mentioned. As per the American Lung Association, the Denver–Aurora region ranks 14th in the United States with respect to ozone pollution. S. In this region, the issue of elevated ozone levels is not new. In accordance with the 1997 8-hour ozone standard, the EPA designated the Denver Metro and North Front Range as non-attainment areas in 2004, but this designation was later postponed until 2007.

The non-attainment standard denotes a region's noncompliance with EPA air quality guidelines. In response, the Colorado Ozone Action Plan was developed, and many of its recommendations were put into practice. The first significant adjustment was the expansion of vehicle emission testing throughout the state to include more counties, such as parts of Larimer and Weld County, that had not previously required emissions testing. Additionally, adjustments have been made to reduce emissions of Nitrogen Oxides (NOx) and Volatile Organic Compounds (VOCs), which should aid in lowering ozone levels.

The oil and gas industry, located in the Denver-Julesburg Basin (DJB), which is home to most of Colorado's major metropolitan areas, is a major cause of the region's high ozone levels. In addition to being produced naturally in the Earth's stratosphere, man-made processes also produce ozone in the troposphere. As was briefly mentioned above, photochemistry is the process by which NOx and VOCs combine with sunlight to produce ozone. One hour elevated ozone events (<75 ppb) "occur during June–August indicating that elevated ozone levels are driven by regional photochemistry." According to an article from the University of Colorado-Boulder, "Oil and natural gas VOC emission have a major role in ozone production and bear the potential to contribute to elevated O_3 levels in the Northern Colorado Front Range (NCFR)."

Plans to assess emission controls for major industrial NOx sources and statewide control requirements for new oil and gas condensate tanks and pneumatic valves are included in the 2008 Colorado Ozone Action Plan. A more detailed strategy to aid in reducing NOx emissions was included in the 2011 release of the Regional Haze Plan. These initiatives take many years to bear fruit and become more challenging to execute. Of course, there are other explanations for the persistence of high ozone levels. These include mountains along the NCFR that have the potential to trap emissions and an increasing population that means more vehicle emissions.

6.3.2.2. Ozone Cracking

Any polymer with olefinic or double bonds in its chain structure, like nitrile rubber, styrene-butadiene rubber, and natural rubber, is vulnerable to attack by ozone gas. These polymers are particularly prone to attack, which leads to longer and deeper cracks over time. The rate at which cracks grow is determined by the load that the rubber component is carrying as well as the amount of ozone present in the atmosphere. Antiozonants, like waxes, can be added to such materials to provide long-term protection or to form a protective film by bonding to the surface and blending in. For instance, ozone cracking in automobile tires used to be a major concern, but it doesn't exist in tires made today. On the other hand, ozone generated in compressed air systems can harm a number of vital products, including O-rings and gaskets. Reinforced rubber fuel lines are also vulnerable to attack, particularly in the engine compartment where electrical components produce some ozone. Rubber goods can cause ozone cracking if they are stored next to a DC electric motor. Ozone is created by the motor's commutator, which produces sparks.

6.3.3. Ozone as a Greenhouse Gas

Ozone was present at ground level prior to the Industrial Revolution, but even background concentrations far from sources of pollution are significantly higher today, with peak concentrations being far higher than pre-industrial levels. As a greenhouse gas, ozone absorbs part of the infrared radiation that the earth emits. It is challenging to measure the potency of ozone as a greenhouse gas because its global concentrations are not constant. Nonetheless, the most widely recognized scientific evaluations of climate change indicate that about 25% of carbon dioxide's radiative forcing is attributed to tropospheric ozone.

Tropospheric ozone has an annual global warming potential of 918–1022 tons carbon dioxide equivalent/tons of ozone. This indicates that the radiative forcing effect of ozone in the troposphere is about 1,000 times stronger than that of carbon dioxide per molecule. Tropospheric ozone, on the other hand, degrades in the atmosphere far faster than carbon dioxide due to its short half-life. This indicates that tropospheric ozone has a much lower global warming potential over a 20-year period—roughly 62 to 69 tons carbon dioxide equivalent per ton of tropospheric ozone.

Tropospheric ozone has strong radiative forcing effects on regional scales but weak global effects due to its short lifetime. Indeed, in certain parts of the globe, the radiative forcing of tropospheric ozone can reach up to 150% that of carbon dioxide. For instance, it has been demonstrated that between 1955 and 2000, the upper Southern Ocean's interior warming was approximately 30% caused by an increase in ozone in the troposphere.

6.4. OZONE POLLUTION

When solar radiation is present, reactions between nitrogen oxides (NO_x) and volatile organic compounds (VOCs) released into the atmosphere can occur, resulting in the formation of ozone in the troposphere. Historically, ozone concentrations were high enough to cause acute effects like irritation of the nose and eyes, respiratory disease emergencies, and impairment of lung function before clean air regulations were put in place to control NO_x and VOCs. Daily fluctuations in ozone concentrations have been positively correlated with the incidence of asthma and the daily non-accidental mortality rate at or above current regulatory standards. New research indicates that exposure to ozone, both short- and long-term, at levels below current regulatory limits, is linked to a higher risk of dying from cardiovascular and respiratory disorders. The pathophysiology underpinning the epidemiologic correlations between ozone exposure and mortality and morbidity focuses on ozone's potent oxidative properties, which can cause oxidative damage to cells and the fluid lining the airways, as well as immune-inflammatory reactions both inside and outside the lung. These new discoveries significantly increase the difficulties currently facing ozone pollution control.

For instance, in 2016, ozone was responsible for 90% of the United States' non-compliance with national ambient air quality standards, while particulate matter and other regulated pollutants accounted for only 10% of the problem. Ozone concentrations have increased and will continue to rise in many regions of the world due to climate change's creation of atmospheric conditions that favor ozone formation. Ozone causes tens of millions of ER visits due to asthma each year, as well as several hundred thousand premature deaths worldwide. Globally, more drastic cuts in the use of fossil fuels are required to reduce greenhouse gas emissions, NOx, and VOC emissions, and ozone pollution. In the meanwhile, therapeutic, and preventive measures are required to lessen the harmful effects of ozone, particularly in those who are more vulnerable. To assess the effectiveness of antioxidants and ozone-scavenging substances that have demonstrated encouraging outcomes in animal research, interventional trials in humans are required.

6.4.1. Impact of Climate Change on Ground-Level Ozone

One of the greenhouse gases in the atmosphere is ozone itself. Thus, rising ground-level ozone levels are a factor in global warming. However, there are two primary physicochemical mechanisms through which a warming climate promotes the formation and accumulation of ozone in the atmosphere. First, changes in humidity and wind patterns brought on by global warming cause surface cyclone frequency to decline in some regions of the world. As a result, the atmosphere becomes more stagnant, which slows down the release of NOx and VOCs and lengthens the time it takes for ozone to be produced. Second, rising atmospheric temperatures generally promote ozone-forming reactions. The number of ozone-standard exceedance days in the eastern United States may rise by 68% by the year 2050 due to warming alone, according to predictions based on these climate-induced changes in temperature and atmospheric stability (air stagnation). According to a different study, the majority of the United States would see an increase in ground-level ozone due to regional climate changes and globally enhanced ozone. More specifically, it is expected that by 2050, the daily 8-hour maximum ozone concentration will reach 87 ppb, up from 79 ppb in 2012. Similarly, another predictive analysis estimated that the average ozone concentration during the 2040s compared to 2000 would increase by 0.43 ppb (95% CI: 0.14–0.75) due to climate change alone. This analysis was conducted by integrating data from climate model outputs and historical meteorology and ozone observations across 19 urban communities in the Southeast United States. (Chang et al., 2010).

An extended ozone season is another effect of climate change. For instance, the summer months in the U.S. are typically when high ozone concentrations occur. But in several Octobers in the 2000s and in 2010 over the Southeast, ozone levels in the fall reached summer levels. This was ascribed to increased emissions from water-stressed plants in a drying and warming environment of biogenic isoprene, a VOC precursor of ozone. This finding implies that in the regions with significant biogenic VOC emissions, an extension of the ozone season from summer to fall may result from future occurrences of a drying and warming fall.

6.4.2. Ambient Concentrations in Reference of Air Quality Standards

In normal atmospheric conditions (temperature and pressure), ozone is in the gas phase and is commonly quantified as the mixing ratio, or parts per million (ppm) or parts per billion (ppb). 1 ppb of ozone is equivalent to 1.97 μg/m^3 at standard conditions of temperature (25°C) and pressure (1 atmosphere). Ozone is one of the regulated air pollutants in many countries and has a recommended limit by the World Health Organization (WHO) due to its widely acknowledged health effects, which include producing breathing difficulties, inciting asthma attacks, decreasing lung function, and increasing the incidence of respiratory diseases. For ambient (outdoor) ozone, the current WHO Air Quality Guidelines are 100 μg/m^3 (~50 ppb), measured as the maximum moving average over eight hours in a day.

The National Ambient Air Quality Standards (NAAQS) for ozone in the United States are currently set at 120 parts per billion (ppb) for a 1-hour standard and an 80 parts per day maximum for a 1-hour standard. The following justifications explain why there are two standards with various averaging times. (McDonald BC, de Gouw JA et al., 2018).

In the past, the NAAQS only included a 1-hour standard because, in large cities like Los Angeles, California, a sharp peak in ozone concentration usually persisted for 1 hour or slightly longer during the afternoon and evening hours. This peak concentration was high enough to have immediate effects, including respiratory tract and eye irritation, decreased lung function, dyspnea, and an increase in ER visits.

Unfortunately, because of the regional transport of ozone precursors, this feature is not present in the majority of the United States, as the hours of elevated ozone concentrations are prolonged. Ozone concentrations averaged over a longer period of time (e.g., 8 hours instead of 1 hour) within a day are a more health-relevant indicator of ozone exposure, according to epidemiological studies. In fact, as of June 15, 2005, areas designated under the 8-h ozone standard—which encompasses the majority of the United States, with the exception of 10 states—are no longer subject to the 1-h ozone standard.

Abruptly, since the U.S. Clean Air Act Amendment was passed in 1970, tremendous efforts have been made to limit the emissions of the two precursors of ozone (including other criteria pollutants). The overall amount of NOX and VOC emissions in the country were decreased by 61 and 54%, respectively, between 1980 and 2017. As a result, the national average of daily maximum 8-hour ozone averages, which were measured at 200 monitoring sites across the United States, decreased by 32%. In 2017, 100.6 million people, or nearly one in every three people, lived in U.S. counties where ozone levels exceeded the NAAQS standard of 70 parts per billion (ppb). In contrast, much fewer people—58 million in total—lived in counties where $PM_{2.5}$, PM_{10}, SO_2, or lead exceeded the NAAQS. This decrease was observed both nationally and in some ozone "hot spots," such as Los Angeles and Atlanta, Georgia. There would be even more people living in areas where ozone levels are higher than the health-based limit if the WHO's recommended 8-hour limit of 50 ppb were followed.

The average amount of ozone in the atmosphere appears to be rising globally, despite the fact that ambient ozone concentrations have been declining in the U.S., Western Europe, and Japan over the past few decades. For instance, measurements of ozone taken between 1994 and 2013 at the Tibetan Plateau's Mt. Waliguan Observatory, a global "background" site, revealed an increasing trend at 0–3 parts per billion annually during the spring and fall. An increase in stratosphere-to-troposphere transport was responsible for about 60% of the springtime ozone increase, while increasing anthropogenic emissions of ozone precursors from Asia were the main cause of the autumnal ozone increase at this site. Because ozone is typically thought to be too reactive to be transported long distances, this finding is concerning. But this also shows that emissions of VOCs and NO_x from more populated areas can travel great distances and impact the level of "background" ozone. More significantly, this data points to an increasing ozone trend along the air trajectories from precursor sources to the background site as well as in the areas where NO_x and VOCs were first released.

While particulate matter, particularly $PM_{2.5}$, has received most of the attention in recent years when it comes to air quality in rapidly developing economies like China and India, "invisible" ozone pollution is becoming more widely acknowledged as a serious health risk. Ground-level ozone concentrations measured in certain Chinese monitoring sites showed an increasing trend in the past few years, while annual $PM_{2.5}$ average concentrations showed a decreasing trend in many Chinese cities. For instance, from 2004 to 2015, observations at a rural site (Dadianzi) 100 km northeast of Beijing revealed a consistent rise in the annual averages of the 8-hour daily maximum ozone concentrations. A recent analysis of ground-level ozone concentrations recorded at almost 1,000 locations throughout China consistently revealed an increase in summertime ozone in northeastern China between 2013 and 2017. This rise was linked, as previously mentioned, to both decreases in $PM_{2.5}$ concentrations and adjustments in human emissions of ozone precursors. Anthropogenic NO_x emissions have declined significantly in the majority of China's urban areas (a 21% nationwide reduction) between 2013 and 2017 as a result of the country's vigorous efforts to primarily control $PM_{2.5}$.

On the other hand, VOC emissions have not changed all that much. In VOC-limited environments, which have been present in many Chinese cities, reducing NO_x would normally result in an increase in ozone. All of these explain why, in many Chinese cities, the daily public air quality reports during the summer months have frequently shown more days when the ozone standard was exceeded than when $PM_{2.5}$ or other regulated pollutants exceeded the standards. This trend in China seems to be similar to that seen in the U.S., where violations of the ozone standard have been documented more frequently than violations of $PM_{2.5}$ or other pollution standards in more locations. (Li et al., 2019)

6.4.3. Human Exposure and Dosimetry

The basic tenet of toxicology is that "dosage makes poison." In epidemiologic studies of air pollution, ambient concentrations are frequently used as a stand-in for exposure or, more precisely, dose. This method ignores variations in breathing rates between and within individuals as well as variations in concentration between indoor and outdoor settings. Ozone is one of the common or U.S. EPA-defined criteria pollutants with special properties that can cause significant errors when using ambient ozone concentration as a proxy for dose.

First of all, building surfaces have a greater ability to scavenge ozone due to its chemical reactivity. Indoor ozone levels in sealed buildings with closed doors and windows are usually less than 20% of outdoor levels. On the other hand, indoor concentrations can exceed 70% of outdoor concentrations in structures with leaks or frequently open windows. Since most people spend most of their time indoors, using outdoor concentration as a stand-in for ozone exposure would cause those who live or work in more airtight buildings to be overexposed compared to those who live in less airtight buildings. The U.S. population residing in buildings with varying indoor-outdoor air exchange rates had ozone effect estimates that varied, which was explained by this systematic exposure assessment error1. The affordability and feasibility of personal monitoring or indoor monitoring has increased due to recent advancements in small and inexpensive ozone sensors. Future research on ozone epidemiology can benefit from more precise ozone monitoring, which can also support data-driven individual preventive measures.

Second, in the majority of the locations, outdoor ozone concentrations show a significant seasonal variation. This makes it more difficult to evaluate in epidemiological studies the long-term health effects of exposure. Warm-season averages have been widely utilized in place of annual averages for other pollutants, such as $PM_{2.5}$, on the grounds that the health risks associated with lower-level ozone during the winter months are minimal. As a result, some ozone control measures have only been put into effect during the months of photochemical smog. For instance, in order to lower VOC emissions that contribute to the formation of ozone, gasoline in the United States is formulated with a higher oxygen content (usually with an increased fraction of ethanol) during the warmer months. Nevertheless, a growing body of research indicates that there might not be a cutoff ozone concentration below which the risk is "zero." Thus, it might be necessary to reevaluate the ozone control strategy that ignores the cold months entirely.

Third, there is usually a clear diurnal pattern to outdoor ozone, with afternoon and early evening high concentrations. Ambient monitoring stations typically report hourly concentrations of ozone.

These hourly data are computed as moving averages for regulatory purposes in order to determine the maximum concentrations (based on moving averages) of 1 and 8 hours within a day. Different toxicological assumptions apply when using concentrations with different averaging times in epidemiological studies. While an 8-hour daily max concentration assumes that lower concentrations during the remaining 16 hours do not contribute to an adverse effect, a 1-hour max concentration is used to evaluate the acute effect of peak exposure.

Another possibility is that the 1- or 8-hour max concentrations were chosen merely because those values were available from earlier epidemiologic studies on the short-term effects of ozone. However, the 24-hour average may be another useful indicator of daily exposure when discussing exposure to low-level concentrations (such as concentrations below the current air quality standard), at least for specific results. (Sun et al., 2018)

6.5. HEALTH EFFECTS OF OZONE ON RESPIRATORY DISEASES

According to the global burden of disease linked to ozone exposure, there were 254,000 excess deaths worldwide in 2015.

This is 10 times lower than the 4.2 million deaths in 2015 that were linked to particulate matter 2.5 ($PM_{2.5}$). Compared to the number of deaths linked to $PM_{2.5}$, mortality from ozone exposure may seem insignificant, but it is more widely distributed globally. In the Asia-Pacific area, air pollution is becoming a more significant issue due to the aging population. In many nations, ongoing efforts have reduced air pollution, which has improved the quality of the air.

Nevertheless, annual ozone levels are still rising despite a decline in the average annual concentrations of nitrogen dioxide and fine particulate matter since 2000. Individually, it is advised that those who are vulnerable stay indoors on days when the ozone level is elevated. While quitting smoking is primarily the result of individual efforts, national policy and regulations also play a role in reducing ambient air pollution. There is currently no standard in Korea for long-term standards like mean annual concentrations, but the 8-hour average ozone standard is set at 60 parts per billion (ppb).

6.5.1. Short-Term Exposure to Ozone and Respiratory Diseases

The link between brief ozone exposure and respiratory health outcomes, such as death and exacerbations, is well-supported by data. Time-series or case-crossover study designs that seek to determine whether there is an increase in mortality, hospital admission, or emergency department (ED) visits soon after ozone level increases are commonly used to assess the short-term effects of ozone exposure.

6.5.1.1. Mortality

Numerous studies have shown that short-term exposure to ozone increased worldwide mortality rates for cardiovascular, respiratory, and overall conditions. In the U.S. Medicare population data, short-term warm-season ozone exposure was linked to an increased risk of death. This study also showed that the risk existed at concentrations lower than the national air quality standards in place at the time. There have also been reports from East Asia linking brief exposure to mortality. Additionally, this study demonstrated that in Korea, Japan, and Taiwan, ambient ozone was linked to a higher risk of death over a comparatively small range of levels.

When weather conditions were taken into account, the risk of dying from lung cancer rose during the warm season, particularly for men and older patients. Given that lung cancer primarily affects the elderly, particular management should be taken into consideration along with close observation. For example, elderly patients should make every effort to stay indoors during the day during the warm/summer season to minimize their exposure to ozone. Despite research demonstrating the advantages of physical activity even in contaminated areas, such a recommendation may cause people to engage in less physical activity. As a result, the future guidelines should carefully consider both the advantages and disadvantages.

6.5.1.2. Respiratory Outcomes

Short-term ozone exposure was linked to a higher risk of hospitalizations for chronic obstructive pulmonary disease (COPD), according to a recent systemic review and meta-analysis. Additionally, a recent study found a correlation between the incidence of COPD and the ozone level. A greater number of ED visits and hospitalizations is indicative of a worse prognosis for COPD patients and has a significant financial impact on both the patients and the country's medical expenses. Another chronic respiratory condition that warrants attention is interstitial pulmonary fibrosis (IPF), which is the most prevalent and severe kind of idiopathic interstitial pneumonia. Although many studies have shown that air pollution from traffic is linked to unfavorable outcomes in IPF patients, such as hospital admission and mortality, the effects of ozone should still be thoroughly studied. Exposure to ozone has been shown to exacerbate asthma attacks. A recent study found that children in the Pittsburgh area saw a rise in ED visits for asthma at lag day 1 (odds ratio, 1.12; 95% confidence interval [CI], 1.03–1.22; $p<0.05$) after short-term exposure to ambient ozone. It's interesting to note that when age-specific effects of ozone and $PM_{2.5}$ on ED visits for respiratory diseases like COPD, asthma, and acute respiratory infections were assessed, it was discovered that older people were more susceptible to ozone and that children were more susceptible to $PM_{2.5}$. According to this study, ozone exposure may increase in older subjects because of changes in their bodies' structures and functions.

6.5.1.3. Mechanism of Injury

Exposure to ozone causes inflammation and hyper-responsiveness in the bronchi, while oxidative stress in the respiratory tract causes allergic sensitization, changes in the structure of the tract, and compromised host defense. Lung function is subsequently compromised and airway obstruction results from the increased airway responsiveness. Additionally, the activation of local reflex responses and sensory nerves in the respiratory tract reduces inspiratory capacity and causes pain during inspiration.

6.5.2. Association with Long-Term Exposure to Ozone in Epidemiological Studies

Compared to the association of short-term exposure, the epidemiological evidence of a long-term exposure to ozone is remarkably weak. There have been relatively few epidemiological studies that have demonstrated the association, especially for incidence, even though numerous toxicological studies and controlled exposure studies have consistently suggested the association. This scant evidence differs from that of other pollutants as well. Large-scale cohort studies conducted in North America, Europe, and Asia have consistently shown a link between long-term exposure to nitrogen dioxide and particulate matter and lung cancer as well as cardiovascular and respiratory disorders, but the results for ozone have been less clear.

The U.S Environmental Protection Agency's (U.S EPA) comprehensive review report on ozone came to the inconsistent conclusion that long-term exposure to ozone has only suggestive causal relationships. Compared to other significant endpoints like mortality, cardiovascular diseases, and lung cancer, the association between long-term exposure to ozone and a few respiratory outcomes, such as the incidence of asthma, was comparatively stronger. A few recent studies have discovered associations with subclinical cardiovascular and respiratory endpoints, as well as broadened the outcomes of interest. To estimate ozone exposure at the individual level, these studies used sophisticated exposure prediction techniques. To address the lack of individual exposure measurements, one of the most recent developments in epidemiological studies evaluating the effects of long-term air pollution is the development of exposure prediction approaches. To estimate long-term concentrations at the individual level in people's homes and places of employment, cohort studies of air pollution have produced basic to sophisticated exposure prediction models. Ozone prediction models were recently adopted and have contributed to the discovery of new evidence of the association, as there have been comparatively fewer cohort studies for ozone than for other pollutants. Most studies of ozone used warm-season averages of 1 or 8-hour maximum to represent large production of ozone derived by chemical reaction with sunlight and high temperature, whereas long-term exposure to air pollution was generally assessed as average concentrations of 24-hour means over one to multiple years.

6.5.2.1. Mortality

Cohort studies conducted in the United States, Europe, and Asia found no correlation between long-term exposure to ozone and non-accidental mortality, despite some studies in these regions finding a modest association. Using a previously validated exposure prediction model, a large cohort study involving 60 million Medicare beneficiaries over 65 in the continental United States from 2000 to 2012 estimated warm-season individual-level ozone exposure based on zipcode-level addresses.

They discovered that a 1% (hazard ratio [HR], 1.01; 95% confidence interval [CI], 1.01–1.01) increase in non-accidental mortality was linked to a 10 ppb increase in ozone. An exposure prediction model was used to track the estimated warm-season postal-code residential concentrations for over a decade in studies involving 2 to 3 million participants in the Canadian Census Health and Environment Cohort (CanCHEC).

The results showed increases in non-accidental mortality of 8% (HR, 1.08; 95% CI, 1.02–1.24) and 7% (HR, 1.07; 95% CI, 1.06–1.08). The American Cancer Society (ACS) cohort in the United States had an estimated risk for respiratory mortality that was generally higher than other cohort studies in Canada, England, and France that found no such association. In the first ACS study, average city-wide regulatory monitoring data for the years 1997–2000 were assigned to 450 thousand residents of U.S. metropolitan areas, who were followed up on for 18 years. They discovered that a 10 ppb increase in ozone was linked to a 4% increase in respiratory mortality (HR, 1.04; 95% CI, 1.01–1.07), but that ozone was not associated with overall mortality. Increases in pneumonia and COPD mortality were found to be 10% (HR, 1.10; 95% CI, 1.03–1.18) and 14% (HR, 1.14; 95% CI, 1.08–1.21) in a follow-up study that focused on specific respiratory mortality and enlarged the population. The results of studies on the correlation with cardiovascular death were generally contradictory.

6.5.2.2. Respiratory and Cardiovascular Outcomes

The incidence of asthma in children provided the most evidence of the long-term effects of ozone on respiratory health. Children living in communities with decreased ozone concentration over almost 20 years from 1-year concentration at baseline showed decreased asthma incidence compared to those living in the communities with increased ozone, according to the Children's Health Study (CHS), which has been monitoring about 4,000 children in 12 communities in California, USA, since 1993. Children in communities with high ozone concentrations had higher incidence when they exercised more than those who exercised less, according to early CHS data with a 5-year follow-up. The association with the onset of asthma was also demonstrated by a Canadian study based on the Quebec Integrated Chronic Disease Surveillance System cohort with a 2-year follow-up. In a study involving 1 million children, the summertime ozone concentrations were estimated at birth by zip-code addresses. An increase of 10 parts per billion in ozone was linked to a 38% rise in the onset of asthma (HR, 1.38; 95% confidence interval, 1.34–1.42). The results of other studies that looked into the relationships between respiratory infections, allergic sensitization, lung development, and lung function were largely contradictory.

A spatio-temporal prediction model was recently developed by the Multi-Ethnic Study of Atherosclerosis and Air Pollution (MEAS Air) to estimate the long-term ozone concentrations at the individual level of approximately 6,000 healthy adults over 45 who reside in six major U.S. cities. This study found a greater increase in percent emphysema at baseline (0.43 for 10 ppb increase in ozone; 95% CI, 0.10–0.80) and during follow-up (0.60; 95% CI, 0.27–0.93) using% emphysema obtained by cardiac computed tomography scan. Long-term exposure to ozone was linked to respiratory symptoms, exacerbations, emphysema severity, and lung function in a cohort of heavy smokers with or without COPD.

There have been few findings regarding cardiovascular outcomes. A few recent cohort studies examined subclinical atherosclerosis in healthy adults from MESA Air and college students participating in the Testing Responses on Youth (TROY) study, as measured by carotid intima media thickness (CIMT) from ultrasound imaging.

Based on residential addresses and regulatory monitoring data, the TROY calculated exposures for various periods and discovered associations using exposures from elementary school (10.1 μm; 95% CI, 1.8–18.5), but not for exposures during early childhood or lifetime. Baseline exposure in the MESA Air was linked to a higher increase in CIMT over a 10-year period (18.7 μm; 95% CI, 4.7–32.3).

6.5.3. Immune-Inflammatory Responses and Emerging Effects

Because ambient ozone is a strong oxidizer, it is well known that it damages cells and mucous lining the airways, which triggers immunological-inflammatory reactions in the lungs. According to recent research, innate immunity plays a role in ozone-induced airway inflammation. In mice, this includes the involvement of innate lymphoid cells (ILCs). Exposure to ozone leads to elevated levels of fibrinogenic proteins, IL-6, IL-8, and tumor necrosis factor-α (TNF-α) in human airway epithelial cells, as well as increased expression of these mRNAs in human alveolar macrophages and interleukin-1β (IL-lβ), IL-6, and IL-8. A landmark study by Koren and colleagues showed that a two-hour acute exposure to ozone (0.4 ppm) increased the number of polymorphonuclear leukocytes (PMN) in bronchial alveolar lavage (BAL) fluid by eight times and raised the level of inflammatory mediators in human lower airways. Krishna along with others further demonstrated that IL-8 acted as a partial mediating factor in the neutrophil influx caused by ozone in human peripheral airways. Significant neutrophilic inflammation was caused by ozone exposure, as evidenced by elevated myeloperoxidase (MPO) levels in the supernatant of induced sputum samples from healthy participants.

The circulatory system may experience "spill over" immune-inflammatory reactions to ozone, which could help to explain new findings regarding ozone's effects on the heart and brain. Following the release of the 2013 EPA assessment, several North American studies have further confirmed strong positive associations between short-term exposure to ozone and one or more of the following mortality classifications: cardiovascular, dysrhythmia, cardiometabolic, and ischemic heart disease.

These associations are robust when controlling for co-pollutants. A weak but significant correlation between ozone and hospital admission and stroke mortality was found in a meta-analysis of 53 studies. Ozone was found to have significant correlations with the incidence of ischemic stroke in Seoul and non-myocardial infarction out-of-hospital cardiac arrests in Helsinki. Although Jerrett et al in the original analysis of an American Cancer Society cohort found that ozone exposure was associated with respiratory but not cardiovascular mortality, in the follow-up study using the same cohort, Turner et al. found a significant association of long-term exposure to ozone with cardiovascular mortality. A recent cohort study by Lim et al. further confirmed this association between long-term exposure to ozone and increased cardiovascular mortality. However, there have also been epidemiological studies reporting null findings between long-term ozone exposure and cardiovascular mortality in Europe (Jerrett et al., 2009).

6.6. EFFECTS OF TROPOSPHERIC OZONE AND PARTICULATES ON PLANTS

Plants are susceptible to damage from photochemical air pollution, which may have negative consequences for agriculture and other natural resources. Ground-level ozone poses a special risk because it has been shown in multiple studies to cause substantial harm. Peroxyacetyl nitrate ($CH_3C(O)O_2NO_2$) and other air pollutants that are co-produced with O_3 are also phytotoxic, though it can be challenging to distinguish their unique effects from O_3's. The effects of particulates on plants are evaluated below, with an emphasis on the mechanisms and mitigation of these effects that have improved with time.

6.6.1. Effects of Tropospheric Ozone on Health and Yields of Plants

We observed that substantial losses in crop quality and yield, such as 10–36% for wheat and 7–24% for rice, may be caused by tropospheric O_3. A meta-analysis comprising 48 studies carried out between 1980 and 2019 regarding soybean exposure to tropospheric O_3 revealed elevated levels of chlorophyll degradation and foliar damage. There was a reduction of 21% in leaf area, 14% in leaf biomass, 23% in shoot biomass, and 17% in root biomass. Sustained exposure to O_3 at a dose of approximately 150 μg m^{-3} resulted in a 28% reduction in seed yield. In an Argentine study, sensitive soybeans were exposed to O_3 at a concentration of 274 μg m^{-3} for 7 days. This led to a 25% decrease in below-ground biomass, a 30% reduction in nodule biomass, and a 21% decrease in biological nitrogen fixation. In tests conducted on low-fertility soils, where seed and seed protein production were reduced by 10% and 12%, respectively, the effects were more pronounced. These effects in soybeans exposed to O_3 at 160 μg m^{-3} for 7 days were associated with reductions in carbon metabolism and detoxification capacity in the soybean roots. According to a study on historical losses in U.S. maize and soybean crops due to air pollution, yields have increased by an average of 20% as a result of better control over O_3, SO_2, PM, and NO_2. It seemed that PM and NO_2 were more harmful than O_3 and SO_2 among these pollutants. All told, the yield improvement was worth about $5 billion USD (Wilson et al., 2019).

In the Chinese province of Henan, observations made between 2015 and 2018 revealed that the annual yield losses of wheat exposed to O_3 at concentrations above 80 μg m^{-3} were 12.8, 8.9, 10.8, and 14.1%. These amounted to $2,141, $1,32, $1,68, and $2,16 billion in losses annually, respectively. To extrapolate these losses to other Chinese crops, a model was created. Estimated losses in wheat were 50 million tons annually, primarily in winter wheat (48 million tons), 21 million tons in rice, 18 million tons in maize, and 1.6 million tons in soybeans, based on a 4-year average of the tropospheric concentration of O_3. According to a different modeling study, current O_3 concentrations decreased yield by 6.9% for rice and 10% for wheat. It is obvious that tropospheric O_3 has a major negative impact on food security in some nations, and this could get worse if stratospheric ozone were to super recover.

According to a modeling study on the effects of measured O_3 concentrations on grapes in Portugal's Demarcated Region of Douro, grape productivity decreased by 27% and sugar content by 32% during two years of high O_3 concentrations. Some grape-growing regions of the world experienced similar effects (Ascenso et al., 2021). Plants other than crops can experience lower yields as a result of air pollution exposure. Tropospheric air pollutants like O_3

can have an impact on forests, which are significant providers of wood and fiber. Sacchelli et al conducted an analysis of the effects of O_3 on forest production in Italy and determined that, in 2005, the average cost of possible O_3 damage to Italy's forests ranged from 31.6–27.1 million € (i.e., 10–17 € ha^{-1} $year^{-1}$). The profitable forest areas saw a 1% decrease as a result of this damage. The yearly national production of roundwood, firewood, timber for poles, and wood for pulp and paper was estimated to have decreased by 7.5, 7.4, 5.0, and 4.8, respectively. Two species of oak (*Quercus pubescens* and *Qilex*) and pine (*Pinus nigra*) are inhibited in growth by current levels of O_3, according to a study on the effects of O_3 on trees in Mediterranean forests in Istria and Dalmatia. According to a climatological modeling study conducted in European forests, the growing season has been extended by approximately 7 days per decade due to climate change. This has led to an increase in the total phytotoxic dose of O_3 that trees absorb during the season, outweighing the advantages of an 11.6% reduction in tropospheric O_3 concentration brought about by pollution control measures between 2000 and 2014.

Plants have been the subject of more in-depth research on the possible effects of O_3 in the environment than animals because of their sensitivity. Nonetheless, the effects of O_3 on amphibians were the subject of a recent study. Tadpoles of the midwife toad (*Alytes obstetricans*) were subjected to airborne O_3 at concentrations as high as 180–220 $\mu g\ m^{-3}$ for eight hours every day, starting from an early developmental stage (limbs not yet formed) and ending with metamorphosis. This is comparable to the highest concentrations recorded over a 10-year period in the Sierra de Guadarrama Mountains. The tadpoles' successful development and infection with the aquatic fungus *Bactrachochytrium dendrobatidis* (Bd), which causes chytridiomycosis, were the measured responses. The actual dose could not be determined because O_3 airborne concentrations were measured in the exposure chambers but not in the water that the tadpoles were in. The findings indicated that the tadpole's development was delayed and its susceptibility to Bd was increased at the highest level of airborne exposure. Since this study is preliminary, more research is required to clarify any possible effects.

To summarize, variations in UV-B radiation in the future will have an impact on ground-level O_3 and other pollutants. Outside of major cities, the return of stratospheric O_3 to 1980 levels is predicted to contribute 1–2 $\mu g\ m^{-3}$ to ground-level O_3, but super-recovery under certain future climate scenarios may result in larger increases. The actual O_3 concentrations will depend on local and regional air quality control measures, the effects of climate change and the Montreal Protocol on stratospheric O_3, and potential offsets from further reductions in NO and VOC emissions. These effects may have an impact on forests and food security.

6.6.2. Toxicological Mechanisms

The production of free radicals in plant tissues is the mechanism through which ozone affects plants. The plant's stomata allow ozone to enter the leaf, where it condenses into ROS such as $O_2^{\bullet-}$, H_2O_2, OH, $^1O_2$, as well as reactive carbonyl species like methylglyoxal and malondialdehyde. These reaction products not only cause damage to cell components but also activate defense mechanisms by stimulating signaling systems like isoprene release. These defenses take the form of chemical buffers like ascorbic acid, glutathione, phenolic chemicals, flavonoids, proline, and other amino acids, carotenoids, tocopherols, polyamines, and sugars, as well as physical actions like closing the stomata and biochemical reactions like the release of superoxide dismutase, catalase, and peroxidases to destroy the ROS. There are certain indirect negative effects of tropospheric air pollutants on plants. For instance, visual, and chemical signals that mediate interactions between plants and organisms that depend on plants or are essential to the sustainability of plant communities can be impacted by air pollution. Tropospheric O_3, for instance, has the ability to alter or destroy biogenic volatile chemicals, which can prevent pests or pollinators from being drawn to plants. O_3 may also disrupt sensory organs, pollinator ability to detect nectar sources, and biological control organisms' ability to detect their intended hosts. Additionally, plant pigment ratios, the phenology (seasonal development) of individual flowers and entire plants, or the timing of flowering may all change as a result of physiological reactions to O_3 damage, which could have an impact on pollinators and host recognition. By directly harming airborne pollen, stimulating repair mechanisms, and rerouting resources toward cell repair rather than reproduction, pollutants can also have an impact on plant reproduction. Furthermore, pollen can become more allergenic, which could have negative effects on people's health.

6.6.3. Effects of Particulates on Plants

Although the effects of PM on plants were once thought to be minimal and insignificant, recent research indicates they might be significant, especially in areas that are heavily polluted. In a study of water-extractable chemicals gathered from $PM_{2.5}$ samplers along Hungarian roadsides, the OECD-227 guideline test was used to determine the sensitivity of 13 common roadside plants. The study's endpoints included shoot weight, shoot height, growth rate, photosynthetic pigments, visible signs of damage on the plants, and peroxidase enzyme activity. Roadside plant communities experienced significant additional stress due to particulate pollution caused by traffic. Since the study was carried out in the middle of winter close to sources, the majority of PM was likely primary rather than secondary and therefore unaffected by variations in UV radiation. It's unclear if UV-dependent secondary aerosols would have resulted in comparable plant damage. Throughout the 4-month study, the total ambient O_3 concentration above the threshold of 80 μg m^{-3} h^{-1} was 453 μg m^{-3} h^{-1}. The levels of ambient PM2.5 and PM10 varied from 45–412 μg m^{-3} and 103–580 μg m^{-3}, in that order. The controls were treated with ethylene diurea, an ozone-stress mitigator, after being cleared of particulates and shielded from O_3. Wheat exposed to O_3 and PM reduced economic yield by 34%, wheat exposed to PM only by 44%, and plants exposed to O_3 alone by 52%.

A modeling analysis of the effects of PM and O_3 on wheat and rice yields in China revealed similar findings. Their findings showed that anthropogenic aerosols decreased rice and wheat yields by 4 and 7%, respectively, based on current levels of O_3 and aerosols.

This resulted from aerosols' ability to dim photosynthetically active radiation, though there were also positive effects from cooling and nutrient delivery. For rice, the estimated losses from aerosols and O_3 were 11.6% and 11.3%, respectively. The sensitivity to variations in UV radiation is still unknown because the relative contributions of primary and secondary PM were not disclosed.

All things considered, these findings suggest that aerosols and tropospheric O_3 either separately or in conjunction have a negative impact on plants or crop yields. Aerosols cause less loss than O_3, although their effects are additive. The significance to crop plants and human activities is uncertain due to the lack of studies on this interaction, which raises questions about the possibility of O_3 and particulates having additive or synergistic effects. These effects, however, are probably going to be localized, and they might be lessened by tighter controls on air pollutants in the troposphere.

6.7. CHANGES IN ATMOSPHERIC CIRCULATION AND TRANSPORT OF POLLUTANTS

6.7.1. Ozone from the Stratosphere

Traditionally, ozone pollution in the air has been viewed as a local or regional problem. However, O_3 from the stratosphere is also carried to the troposphere, where it serves as a baseline to which locally or regionally generated O_3 is added. At ground level, it contributes a significant but variable fraction of O_3. The phrase "stratospheric-tropospheric exchange" (STE) describes this. Although difficult to measure, the amount of stratospheric O_3 that contributes to tropospheric O_3 is significant. The influence of stratospheric O_3 on tropospheric O_3 was estimated through a comparison of measurements and three different models, highlighting the difficulties in producing reliable results. The amount of O_3 that is near the Earth's surface that is thought to be brought down from the stratosphere was calculated by the study. It was discovered that this varied from 10% throughout the year in the tropics to more than 50% in the winter at mid-to-high latitudes.

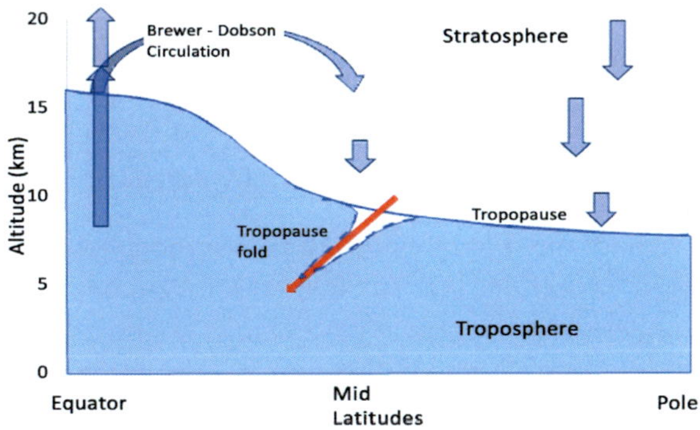

Figure 6.2. *Schematic of the circulation of air into the stratosphere and its return. The general circulation is slow relative to movements in the troposphere. Tropopause folds (shown with a dotted blue line) occur sporadically and inject stratospheric air into the troposphere.*

Source: Madronich, S., Sulzberger, B., Longstreth, J.D. et al. Changes in tropospheric air quality related to the protection of stratospheric ozone in a changing climate. Photochem Photobiol Sci 22, 1129–1176 (2023). https://doi.org/10.1007/s43630-023-00369-6

Both atmospheric models and observations provide inadequate characterizations for STE estimates. According to an evaluation of the amounts of O_3 in the troposphere, estimates of STE for results published in 1995 were approximately 1000 Tg year^{-1} O_3. However, by 2015, models were providing estimates close to 400 Tg year^{-1}, with a multi-model estimate of 535±160 Tg year^{-1} for the year 2000. A value of 628±800 Tg year−1 is reported by the IPCC AR6 assessment for 2010, with a large uncertainty indicating how poorly this value is known. Some more recent estimates are 400±60 Tg year^{-1} (1990–2017) and 347±12 Tg year^{-1} (2007–2010).

Rather than modeling STE transport itself, the magnitude of the STE is typically inferred as the difference between the calculated production and loss of O_3 (termed the residual). These production and loss terms are highly uncertain because they differ by an order of magnitude (approximately 5000 Tg year^{-1}) from the estimated transport. The fact that different definitions of the tropopause, or upper boundary of the troposphere, have been used by models introduces another layer of confusion.

By the year 2000, there will be a noticeable drop in O_3 from the stratosphere, according to modeling of the effect of STE on tropospheric O_3 for the period 1850–2100. ODSs' effect on stratospheric O_3 led to a decrease in the transport of O_3 from the stratosphere to the troposphere, according to a modeling study covering the years 1980–2010. Up until 1994, the model predicted that ODS would cause a 4% drop in global tropospheric O_3 (14 Tg O_3). A different study that employed N_2O measurements to constrain atmospheric modeling calculated an average annual decrease in STE of 30 Tg year^{-1}, with a variation of 5–55 Tg year^{-1}, caused by the Antarctic ozone hole (1990–2017).

On the other hand, other studies found that stratospheric O_3 led to a slight increase over this period in the Northern Hemisphere and no significant change in the Southern Hemisphere. These results contrast with the results of a modeling and observational study of the regional changes in O_3 concentration in the troposphere for the period 1980–1990 through to 2000–2010. Clearly, this is an area that needs more work.

Significant increases in the amount of O_3 transferred from the stratosphere to the troposphere are anticipated for the years 2000–2100. According to estimates made by Abalos et al. using data from seven atmospheric models and with a primary focus on RCP6.0, there will be a 10–16% increase in O_3 from STE in the troposphere during the twenty-first century. They did not get a consistent picture from the models when evaluating the relative importance of changes in GHGs vs. ODSs in driving the changes in STE, even though it seems that the two factors are of similar magnitude. However, there isn't enough consensus among models to measure trends. The total amount of anthropogenic emissions has a significant impact on the net change in O_3 concentration in the troposphere by the year 2100. The anticipated controls on air pollution emissions are what are driving the decline in net chemical production. The RCP8.5 scenario calculations revealed a significant increase in O_3 concentration in the troposphere, with three times more O_3 being transported from the stratosphere than in the RCP6.0 scenario. Since these models report O_3 concentrations averaged over the whole vertical extent of the troposphere and the impact of stratospheric O_3 is much larger in the upper troposphere than at the Earth's surface, it is not possible to infer the magnitude of changes in O_3 at ground level.

Ground-level air quality is directly impacted by tropopause folds. These folds are widespread throughout the Eastern Mediterranean, where they have been linked to elevated ground-level concentrations of O_3 that are higher than the air-quality standards set by the European Union. Their longitudinal distribution is not uniform. Over the Indian and Southern Oceans, the Southern Hemisphere also experiences the same effect. According to a modeling analysis of tropopause folds for the years 1960–2100, which included stratospheric O_3 recovery and emissions as outlined in RCP6.0, folds are expected to rise during this time. Variations in the number of tropopause folds by approximately 3% have been found to be statistically significant in areas that correspond with an estimated rise in O_3 of 6 μg m^{-3} close to the Earth's surface.

Therefore, measuring the amount of O_3 transported from the stratosphere is crucial to comprehend the quality of the troposphere's air, but it is still challenging. The mechanism through which the downward transport takes place contributes to the difficulty in quantifying and modeling the STE of O_3. At the edges of the tropics, there are "folds" where thin stratospheric layers are surrounded (vertically) by tropospheric layers and vice versa, injecting air rich in O_3. Then, these layers combine. Techniques for locating folds in model output are being developed and appear to be fairly consistent across models.

The comparatively small number of measurements of the upper troposphere and lower stratosphere's atmospheric chemistry and physical structure further complicates STE modeling. Therefore, there isn't much data that can be utilized to limit atmospheric models. Currently, attempts are being made to utilize commercial aircraft chemical composition measurements to create a reliable climatology that will aid in modeling. In a similar vein, attempts are being made to enhance the utilization of O_3 measurements from satellites and possibly O_3 sondes and *in situ* measurements in atmospheric modeling. It takes careful consideration to determine whether using observations to constrain models exposes the models to changes in the calibration and quality of the input data. These data should eventually enable more accurate measurement of the variations in tropospheric O_3 resulting from variations in stratospheric O_3.

6.7.2. Effects of Circulation Changes on Extreme Weather Events and Air Quality

Wildfires and other extreme weather conditions have an impact on air quality. The polar stratospheric ozone depletion, which reinforces the stratospheric polar vortex, contributes to both climate change and changes in weather patterns, including extreme weather events. In the Southern Hemisphere, altered weather patterns have been noted as a result of the Antarctic ozone hole. For instance, there is a correlation between the length of the Antarctic Ozone hole and anomalies in rainfall and droughts in the Southern Hemisphere. Australia's weather is influenced not only by the strength of the stratospheric polar vortex but also by the El Niño Southern Oscillation and the Indian Ocean Dipole. Wildfire risk is increased by hot, dry weather. The severe Australian fire season of 2019–2020 resulted in a smoke plume that was seen all over the world and a major deterioration of the country's air quality. Worldwide, there is a growing probability of wildfires, and this trend is anticipated to persist. However, the stability of the Antarctic polar vortex should be weakened by the recovery of stratospheric O_3, which should result in wetter conditions for this region in the Southern Hemisphere soon.

The Arctic Oscillation (AO) shifts to more positive values and the Northern Hemisphere jet stream becomes more zonal due to stratospheric ozone depletion, much like the effects of the Antarctic ozone hole's atmospheric dynamics. As a result, surface temperatures in southern Asia and Southeast Europe are below average, while those in Western Europe, Russia, and northern Asia are above average. For instance, the heat wave and wildfires that accompanied the spring 2020 Arctic stratospheric ozone depletion in Siberia were most likely the result. Given that greenhouse gases (GHGs) influence the Arctic stratosphere through radiative cooling, the likelihood that such events will transpire in the future depends on trends in ODS and GHG emissions. Therefore, changes in atmospheric circulation, polar stratospheric O_3, and the direct effects of climate change all have an impact on the frequency of extreme weather events like droughts and, consequently, wildfires in both hemispheres. Due to the release of PM, CO, and other tropospheric pollutants, wildfires have a negative effect on human health and lower the quality of the troposphere's air (Madronich et al., 2023).

REFERENCES

1. Abalos, M., Orbe, C., Kinnison, D. E., Plummer, D., Oman, L. D., Jöckel, P., Morgenstern, O., Garcia, R. R., Zeng, G., Stone, K. A., & Dameris, M. (2020). Future trends in stratosphere-to-troposphere transport in CCMI models. *Atmospheric Chemistry and Physics, 20*(11), 6883–6901.
2. Ascenso, A., Gama, C., Blanco-Ward, D., Monteiro, A., Silveira, C., Viceto, C., Rodrigues, V., Rocha, A., Borrego, C., & Lopes, M. (2021). Assessing Douro vineyards exposure to tropospheric ozone. *Atmosphere, 12*(2), 200.
3. Benson, S. W., & Axworthy, A. E. (1957). Mechanism of the gas-phase thermal decomposition of ozone. *Journal of Chemical Physics, 26*, 1718–1726.
4. Caprio, V., Insola, A., Lignola, P. G., & Volpicelli, G. (1982). A new attempt for the evaluation of the absorption constant of ozone in water. *Chemical Engineering Science, 37*(1), 122–123.
5. Chang, H. H., Zhou, J., & Fuentes, M. (2010). Impact of climate change on ambient ozone level and mortality in southeastern United States. *International Journal of Environmental Research and Public Health, 7*, 2866–2880. https://doi.org/10.3390/ijerph7072866.
6. Gauderman, W. J., Avol, E., Gilliland, F., Vora, H., Thomas, D., Berhane, K., et al. (2004). The effect of air pollution on lung development from 10 to 18 years of age. *New England Journal of Medicine, 351*, 1057–1067. https://doi.org/10.1056/NEJMoa040610.
7. Jerrett, M., Burnett, R. T., Pope III, C. A., Ito, K., Thurston, G., Krewski, D., et al. (2009). Long-term ozone exposure and mortality. *New England Journal of Medicine, 360*, 1085–1095. https://doi.org/10.1056/NEJMoa0803894.
8. Koman, P. D., & Mancuso, P. (2017). Ozone exposure, cardiopulmonary health, and obesity: A substantive review. *Chemical Research in Toxicology, 30*, 1384–1395. https://doi.org/10.1021/acs.chemrestox.7b00077.
9. Krishna, M. T., Madden, J., Teran, L. M., Biscione, G. L., Lau, L. C., Withers, N. J., et al. (1998). Effects of 0.2 ppm ozone on biomarkers of inflammation in bronchoalveolar lavage fluid and bronchial mucosa of healthy subjects. *European Respiratory Journal, 11*, 1294.
10. Li, K., Jacob, D. J., Liao, H., Shen, L., Zhang, Q., & Bates, K. H. (2019). Anthropogenic drivers of 2013–2017 trends in summer surface ozone in China. *Proceedings of the National Academy of Sciences USA, 116*, 422–427. https://doi.org/10.1073/pnas.1812168116.

11. Madronich, S., Sulzberger, B., Longstreth, J. D., et al. (2023). Changes in tropospheric air quality related to the protection of stratospheric ozone in a changing climate. *Photochemical & Photobiological Sciences, 22*, 1129–1176. https://doi.org/10.1007/s43630-023-00369-6.

12. Malig, B. J., Pearson, D. L., Chang, Y. B., Broadwin, R., Basu, R., Green, R. S., et al. (2016). A time-stratified case-crossover study of ambient ozone exposure and emergency department visits for specific respiratory diagnoses in California (2005–2008). *Environmental Health Perspectives, 124*, 745–753. https://doi.org/10.1289/ehp.1409495.

13. Matrozov, V. I., Kashtanov, S. A., Stepanov, A. M., & Tregubov, B. A. (1975). Ozone solubility in water. *Journal of Applied Chemistry, 48*(8), 1838–1841.

14. McDonald, B. C., de Gouw, J. A., Gilman, J. B., Jathar, S. H., Akherati, A., Cappa, C. D., et al. (2018). Volatile chemical products emerging as largest petrochemical source of urban organic emissions. *Science, 359*, 760–764.

15. Neidell, M., & Kinney, P. L. (2010). Estimates of the association between ozone and asthma hospitalizations that account for behavioral responses to air quality information. *Environmental Science & Policy, 13*, 97–103. https://doi.org/10.1016/j.envsci.2009.12.006.

16. Plass, R., Egan, K., Collazo-Davila, C., Grozea, D., Landree, E., Marks, L. D., & Gajdardziska-Josifbvska, M. (1998). Cyclic ozone identified in magnesium oxide (111) surface reconstructions. *Physical Review Letters, 81*(22), 4891–4894.

17. Schubert, C. C., Schubert, S. J., & Pease, R. N. (1956). The oxidation of lower paraffin hydrocarbons. II. Observations of the role of ozone in the slow combustion of isobutane. *Journal of the American Chemical Society, 78*(21), 5553–5556.

18. Scott, B. F., Spencer, C., Martin, J. W., Barra, R., Bootsma, H. A., Jones, K. C., Johnston, A. E., & Muir, D. C. G. (2005). Comparison of haloacetic acids in the environment of the Northern and Southern Hemispheres. *Environmental Science & Technology, 39*(22), 8664–8670.

19. Solomon, K., Velders, G., Wilson, S., Madronich, S., Longstreth, J., Aucamp, P., & Bornman, J. (2016). Sources, fates, toxicity, and risks of trifluoroacetic acid and its salts: Relevance to substances regulated under the Montreal and Kyoto protocols. *Journal of Toxicology and Environmental Health B, 19*, 289–304.

20. Strickland, M. J., Darrow, L. A., Klein, M., Flanders, W. D., Sarnat, J. A., Waller, L. A., et al. (2010). Short-term associations between ambient air pollutants and pediatric asthma emergency department visits. *American*

Journal of Respiratory and Critical Care Medicine, 182, 307–316. https://doi.org/10.1164/rccm.200908-1201OC.

21. Strosnider, H. M., Chang, H. H., Darrow, L. A., Liu, Y., Vaidyanathan, A., & Strickland, M. J. (2019). Age-specific associations of ozone and PM2.5 with respiratory emergency department visits in the U.S.. *American Journal of Respiratory and Critical Care Medicine, 199*, 882–890. https://doi.org/10.1164/rccm.201806-1147OC.
22. Sun, Q., Wang, W., Chen, C., Ban, J., Xu, D., Zhu, P., et al. (2018). Acute effect of multiple ozone metrics on mortality by season in 34 Chinese counties in 2013–2015. *Journal of Internal Medicine, 283*, 481–488. https://doi.org/10.1111/joim.12724.
23. Tian, Y., Liu, H., Zhao, Z., Xiang, X., Li, M., Juan, J., et al. (2018). Association between ambient air pollution and daily hospital admissions for ischemic stroke: A nationwide time-series analysis. *PLOS Medicine, 15*, e1002668. https://doi.org/10.1371/journal.pmed.1002668.
24. Tomiyasu, H., Fukutomi, H., & Gordon, G. (1985). Kinetics and mechanisms of ozone decomposition in basic aqueous solutions. *Inorganic Chemistry, 24*, 2962–2966.
25. Turner, M. C., Jerrett, M., Pope III, C. A., Krewski, D., Gapstur, S. M., Diver, W. R., et al. (2016). Long-term ozone exposure and mortality in a large prospective study. *American Journal of Respiratory and Critical Care Medicine, 193*, 1134–1142. https://doi.org/10.1164/rccm.201508-1633OC.
26. Wilson, S. R., Madronich, S., Longstreth, J. D., & Solomon, K. R. (2019). Interactive effects of changing stratospheric ozone and climate on tropospheric composition and air quality, and the consequences for human and ecosystem health. *Photochemical & Photobiological Sciences, 18*, 775–803.
27. Xie, G., Zhai, Z., & Zhang, J. (2020). Distribution characteristics of trifluoroacetic acid in the environments surrounding fluorochemical production plants in Jinan, China. *Environmental Science and Pollution Research, 27*, 983–991.

CHAPTER 7

SOURCES OF AIR POLLUTION

Contents

INTRODUCTION

The release of chemicals into the atmosphere that are detrimental to biological systems is what causes air pollution. Natural occurrences like dust storms, forest fires, volcanic eruptions, and biogenics (emissions from soil bacteria and vegetation) can all introduce it. We are unable to significantly lower these kinds of natural emissions, though. Instead, since we have control over it, we will concentrate on pollution that is anthropogenic—that is, caused by human activity. Both stationary and mobile sources can contribute to outdoor pollution. Mobile sources of air pollution emit pollutants as they travel from one location to another. Transport vehicles such as cars, planes, ships, and trains are examples of mobile sources. The location of stationary sources is fixed. Examples include the smokestacks from power plants, industrial operations, burning, heating in homes, building sites, farms, and surface mines, among other things. Point and nonpoint sources are further categories into which stationary sources can be divided. Larger emitters at a fixed location, like an industrial facility, are referred to as point sources. Nonpoint sources, like home heating, are typically more prevalent and of a lesser magnitude (Figure 7.1).

Figure 7.1. *Air pollution from a coking oven.*

Source: By Alfred T. Palmer, Public Domain, https://commons.wikimedia.org/w/index.php?curid=3363860.

Fossil fuel combustion is primarily responsible for the pollution caused by human activity. We use fuel combustion to power a range of vehicles, heat our homes, and produce electricity. When fuels like coal, diesel fuel, fuel oil, gasoline, blended biodiesel, natural gas, and gasoline burn, exhaust gas, also known as flue gas, is released. It plays a significant role in the emissions from stationary internal combustion engines as well as motor vehicles. In certain big cities, motor vehicle emissions play a significant role in the formation of smog and contribute to air pollution. 53,000 premature deaths are attributed to vehicle emissions annually in the United States alone, according to a 2013 MIT study. Unreacted gasoline vapors can be found in exhaust gases in addition to combustion products. Particulate matter is also directly released into the atmosphere by a variety of fuels, including wood and coal.

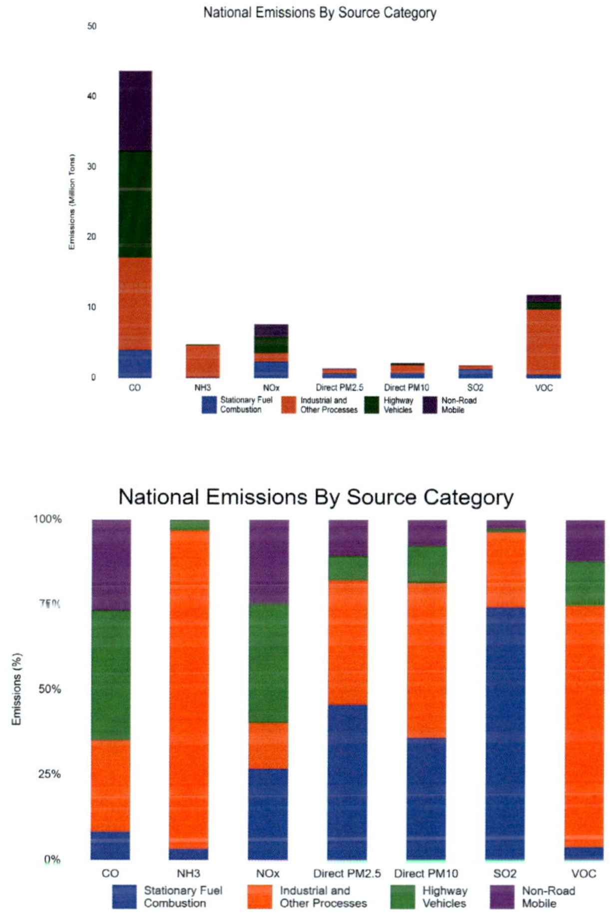

Figure 7.2. *National emissions for the U.S. in 2017. The first figure shows the total amount of pollutant emitted for the year (in millions of tons) along with the source. The second figure expands each bar to show the percent of each of four anthropogenic sources contributing to the emission of that pollutant.*

Source: EPA air report, https://gispub.epa.gov/air/trendsreport/2022/#sources.

Pollution caused by humans can originate from sources other than combustion reactions. While some of these pollutants are the actual chemicals, others are byproducts of industrial processes. Numerous of these are volatile organic compounds (VOCs), which are substances that contain carbon and are frequently found as ingredients in household goods. These substances have the ability to evaporate and so enter the atmosphere. Additional contaminants include substances like asbestos and heavy metals like lead and mercury, which spread as particulate matter and land on surfaces like soils and surface waters where they can be absorbed by plants and go up the food chain. The government tracks the emissions of the most common pollutants (Figure 7.2).

7.1. GENERAL ASPECTS

There are almost as many sources of air pollution as there are sand grains. When the wind entrains them and sends them into the air, the sand grains themselves actually become air pollutants. They would be categorized as natural air pollutants, which suggests that they have always existed. The term "natural sources" refers to sources of air pollution that are not brought on by human activity. Imagine that a layer of exposed soil has been left behind after someone has removed the ground cover. Later, some of this soil is picked up by the wind and carried a great distance before landing at a different location and affecting more people. If there was a sufficient amount of time between the removal of the ground cover and the material becoming airborne, we could classify this as natural pollution rather than anthropogenic pollution. The questions of how long would be long enough are more complicated than they seem at first. This is one of the reasons why decisions about how to control pollution at a given level must first be made after extensive research and analysis.

7.1.1. Natural Sources

Particulate matter is emitted by a volcano in eruption. Additionally released are harmful gases like methane, SO_2, and H_2S. An eruption's emissions could be so strong that they damage the surrounding area a long way from the volcanic source. For extended periods of time, clouds containing gases and volcanic particulate matter have lingered in the atmosphere. The Mount eruption. St. Helens, located in Washington, is a prime illustration of volcanic activity. A picture of Mount is shown in Figure 7.3. St. Helens on May 18, 1980, during the catastrophic eruption. Even though they may have started as a result of human activity, accidental fires in forests and on prairies are typically categorized as natural sources. Lightning starting a fire in a significant area of forestland could only be classified as natural, but foresters frequently purposefully start fires in forestlands to burn off the residue. Witnessing a massive, uncontrolled forest fire can be terrifying. Large amounts of pollutants are released by such a fire in the form of ash, carbon monoxide, carbon dioxide, smoke, and unburned hydrocarbons. Up to 350 kilometers away from the actual fire, forest fires in the Pacific Northwest of the United States have been known to release a plume that reduces visibility and sunlight.

Figure 7.3. *1980 eruption of Mount St. Helens.*

Source: By Austin Post, Public Domain, https://commons.wikimedia.org/w/index.php?curid=3157557.

Large-scale particulate matter-laden dust storms are a frequent natural source of air pollution in many regions of the world. Suspended particulate matter readings one or two orders of magnitude above ambient air quality standards can be obtained even from a relatively small dust storm. Significant dust storms often result in reduced visibility, which can even have an impact on air travel and lead to serious traffic accidents. Dust storms carry particulate matter from the desert to urban areas, where it poses issues for industry, transportation, and residential areas. The majority of the materials that an automobile's air cleaner removes are naturally occurring pollutants like road dust and other entrained material. The world's oceans are a significant natural source of pollution. Aerosols from the ocean constantly release salt particles into the atmosphere, which can corrode metals and paints. Rocks are reduced to sand by the action of waves, which may eventually become airborne. Wave and tidal action erodes even the shells that wash up on the beach, sometimes making them so small that they could fly.

Earth's trees and other plant life constitute a significant source of natural pollutants. Despite being a major contributor to the process of turning carbon dioxide into oxygen through photosynthesis, these green plants are still the planet's main source of hydrocarbons. Another issue with air pollution that can be linked to plant life is the pollen, which causes allergic reactions and respiratory distress in people. The familiar blue haze that hovers over forested areas is almost entirely caused by atmospheric reactions of the volatile organics given off by the forest dot. The environmental effects of other natural sources, like saltwater and alkaline lakes, are typically fairly localized. This also applies to sulfurous gases originating from hot springs; these gases have an overpowering smell near the source that fades a few kilometers distant.

7.1.2. Anthropogenic Sources

7.1.2.1. Industrial Sources

The modern human being's dependence on industry to meet their needs has caused pollution sources to shift from the individual to the industry. Although the combined emissions of all the household soap-making kettles that a soap factory replaces will likely be less than that of the factory itself, soap consumers can still demand that the factory be cleaned up. Manufacturing goods from raw materials—(1) iron and steel from ore, (2) lumber from trees, (3) gasoline and other fuels from crude oil, and (4) stone from quarries—contributes significantly to industrial pollution. Every one of these manufacturing processes results in a product as well as a number of waste products that are referred to as pollutants. Frequently, the polluting material can be recovered in part or in its entirety and transformed into a useful product. Additionally, industries that transform products into new ones—such as those that make furniture out of lumber, paint out of solids and solvents, asphaltic paving out of rock and oil, and automobile bodies made of steel—emit industrial pollution.

Industrial sources release comparatively constant quantities and types of pollutants, and they are all stationary. For example, unless a significant process change is made, a paper mill will remain in the same location and continue to emit the same types and quantities of pollutants. Most of the time, industrial sources can be controlled by using existing technology. The most effective regulatory control is that which is applied uniformly within all segments of industries in a given region, e.g., "Emission from all asphalt plant dryers in this region shall not exceed 230 mg of particulate matter per standard dry cubic meter of air."

7.1.2.2. Utilities

In today's world, utilities play such an integral role in our daily lives that it is difficult to imagine life without them. In addition to producing electricity for our homes' heating and lighting needs, an electric power plant also supplies the energy needed to recharge the batteries in our laptops, cell phones, PDAs, electric toothbrushes, and televisions. In the days before televisions and other entertainment devices, when wood fires warmed our homes, homemade candles were used for lighting, and food was kept in cellars, the combined amount of air pollution produced by all of these individual sources was likely greater than that produced by the contemporary generating stations that provide our energy today. It is simple for residents to blame the utility for air pollution without making the connection between the pollution and their own electricity use.

Transporting and transforming energy between different forms is the business of utilities. A sizable steam generating plant that generates 2000 MW must find a way to dispose of 40 000 kg of ash per hour if it burns a million kilograms of coal with 4% ash per hour. Some will be removed from the furnaces by the ash-handling systems, but some will go up the stack with the flue gases. 200 kg of ash will be released into the atmosphere per hour if 50% of the ash enters the stack and the fly ash collection system operates at a 99% efficiency level.

Gaseous emissions from a typical generating plant would consist of 341 000 kg of sulfur oxides and 185 000 kg of nitrogen oxides per day. If the level of pollution is deemed to be too high, the management may choose to: (1) buy coal with less ash or sulfur; (2) modify the furnace to direct more ash toward the ash pit and less toward the stack; or (3) install more effective air pollution control machinery. In any event, there will be an increase in operating costs, which will be passed on to the customer.

The utilities that manage the waste of contemporary society are another significant source of air pollution. Air pollution can be caused by an overloaded, badly designed, or poorly run sewage treatment plant, which will prompt the public to demand quick action.

Open dumps can still be found in a lot of nations worldwide. These could catch fire and emit toxic smoke and fume plumes. Dust and smoke can still be found in landfills, abandoned mine shaft fires, and industrial waste disposal and reclamation sites, even in more developed economies. Even though the general public may be told that this is the "cheapest" way to dispose of their solid waste, these are undoubtedly causes for concern. In order to have a pollution-free environment, the public has demonstrated a willingness to forbid burning dumps and to pay the extra expense of sufficient waste disposal facilities.

7.1.2.3. Personal Sources

Despite the trend in society toward centralized industries and utilities, there are still a lot of individual sources of air pollution that we are responsible for, including (1) cars; (2) furnaces at home; (3) fireplaces and stoves at home; (4) backyard barbeque grills; and (5) open burning of leaves and trash. The personal emissions of are shown in Figure 7.4. In the U.S., personal sources contribute more to energy release and air pollution emissions than industry and utilities put together.

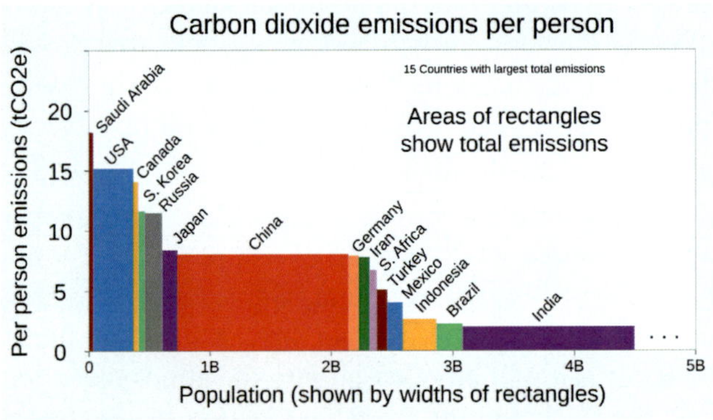

Figure 7.4. *Estimated emissions.*

Source: By RCraig09 – Own work, CC BY-SA 4.0, https://commons.wikimedia.org/w/index.php?curid=107009763.

The quantity of pollutants released by the enormous number of private cars in American cities is greater than that of any other source. Controlling one's own personal sources of pollution can be done in four ways: (1) by enacting regulations (such as limiting the use of fireplaces and stoves to times when atmospheric mixing is favorable); (2) by adopting a different lifestyle (such as selling one's car and using public transportation); (3) by switching from a more polluting to a less polluting source (such as turning a furnace into a natural gas furnace); or (4) by changing the type of pollution (such as composting leaves instead of burning them). Whatever strategy is employed to reduce pollution from personal sources, enforcing it is typically challenging and unpopular. It is challenging to convince the public that the smoking factory stack on the horizon is a bigger source of pollution than their brand-new, lavishly advertised, shiny, free cars. Moreover, pointing out that the factory's visible plume of smoke is not the only source of air pollution that the workers at that plant contribute to; driving their cars to and from work or mowing their lawns on Saturdays are two more ways that they do so (Goldberg, 1998; Schlesinger, 1992).

7.2. COMBUSTION

Any form of open burning puts the public's health and environment at risk. The air we breathe is made dirty by smoke. Ash contaminates our lakes, rivers, streams, groundwater, and soil. Anything that burns outdoors has the potential to start a wildfire. By adhering to state regulations and burning only authorized materials, the likelihood of these negative effects can be reduced.

7.2.1. Smoke

Particles make up the majority of smoke, but it can also contain other gaseous air pollutants like carbon monoxide, nitrogen oxides, and potentially hazardous hydrocarbons. Smoke exposure can have negative health effects and exacerbate pre-existing conditions.

7.2.1.1. Environmental Impacts of Smoke

Burning materials that are forbidden, like plastic, trash, and wood that has been painted or treated, damages the environment by releasing harmful chemicals into the air. Both people and animals can breathe in contaminated air, which can also land on plants, pollute surface waters, and soil. Burning residue contaminates the groundwater and soil, and it can infiltrate the human food chain through livestock and crops. Furthermore, when we eat meat, fish, and dairy products, certain chemicals that are released during burning can build up in the fat of the animals and eventually in humans. Both soot and smoke have a long range. Some people find odors to be bothersome. Odors and smoke residue have the ability to seep into homes or affect items outside of them, such as cars or clothes hanging on clotheslines. In addition to harming building paint, the gases emitted during open burning can corrode metal siding.

7.2.1.2. Impacts of Burning Plastic

Burning plastics releases some of the most hazardous compounds into the air. One such byproduct is dioxin, which is produced when products containing chlorine are burned. Dioxins often stick to the waxy surface of leaves and use this as a pathway into the food chain. Certain plastics (like polyethylene or polypropylene) may not contain chlorine, but other materials burned alongside or adhered to the plastic may contain chlorine. The plastic's unburned parts end up as litter in rivers, lakes, and the ground. Animals that consume the disintegrating plastic may become ill. Larger plastic particles have the potential to harbor bacteria, for example, by retaining water that serves as a mosquito's home.

7.2.1.3. Health Impacts of Smoke

7.2.1.3.1. Impacts of Smoke from Burning Clean Wood and Leaves

Burning domestic waste, such as wood and leaves, releases smoke, which is made up of particulate matter and vapors (solid and liquid droplets suspended in the air). Smoke pollution in the air can harm people's health. Exposure to these air pollutants can cause headaches, breathing difficulties, coughing, and irritation of the nose and eyes. Individuals who suffer from respiratory conditions like emphysema, asthma, or heart disease are particularly vulnerable to air pollution. Allergies, bronchiolitis, pneumonia, and lung infections are among the other medical conditions that burning exacerbates.

7.2.1.3.2. Impacts of Smoke from Burning Trash and Plastic

Burning garbage can have long-term negative effects on health. Nitrogen oxides, sulfur dioxide, volatile organic compounds (VOCs), and polycyclic organic matter (POMs) are among the harmful substances released during burning. Toxic substances like dioxin and heavy metals are also released when plastic and treated wood are burned. Burning plastic also releases other chemicals, such as polyaromatic hydrocarbons (PAHs) and benzo(a) pyrene (BAP), both of which have been linked to cancer. Agricultural bags and containers will also be released into the air if they are tainted with pesticides or other dangerous materials.

7.2.2. Wildfire

The frequency, intensity, and duration of wildfires are all rising globally, making it more important than ever to comprehend the potential health effects of wildfire exposure. Extremely dry conditions, such as drought, heat waves, and strong winds, increase the risk of wildfires. A combination of dangerous air pollutants, including lead, $PM_{2.5}$, NO_2, ozone, and aromatic hydrocarbons, can be found in wildfire smoke. Apart from releasing harmful pollutants into the atmosphere, wildfires also affect the climate by releasing copious amounts of greenhouse gases, such as carbon dioxide, into the atmosphere. The fire season is beginning earlier and ending later due to climate change, which is causing warmer temperatures and drier conditions as well as increased urbanization of rural areas. The number of acres burned, their duration, and their intensity are all increasing during wildfire events, which can also interfere with communications, transportation, the supply of water, power, and gas.

7.2.2.1. Environmental Impacts of Wildfire

The primary cause of wildfires in Wisconsin is debris burning, which results in hundreds of structures being at risk each year as well as thousands of acres of forest land burning accidentally. Unplanned fires are a major risk to property, human safety, and the environment. A wildfire on your land could have disastrous effects. Damage from wildfires can be both physically and visually upsetting. Even after the fire is extinguished, the effects on the environment often continue. Following a wildfire, landowners may encounter hazards such as road instability, invasive plant growth, tree mortality, and erosion.

7.2.2.2. Health Impacts of Wildfire

The sad truth is that certain fires have the potential to endanger or even kill lives. When trying to put out an escape fire, one runs the risk of suffering severe burn injuries or, in extreme circumstances, passing away. Still, most deaths are caused by the fire inadvertently (e.g. cardiac arrest brought on by breathing in too much smoke, or trauma from trying to put out the fire on both a physical and mental level). You should never attempt to put out a fire on your own if it escapes your control. Press the local emergency number right away. Uncontrolled wildfires can pose a threat to emergency responders, who must participate in risky suppression operations that could jeopardize their health and safety, as well as individuals who started the fire or are directly affected by it.

7.2.3. Ash

Fly ash particles are toxic air pollutants. They can trigger heart disease, cancer, respiratory diseases and stroke.

7.2.3.1. Environmental Impacts of Ash

Waste ash will result from any fire. Although wood ash has certain nutrients that plants need for healthy growth, ash is bad for our rivers, lakes, and ponds. Phosphorus, potassium, and trace amounts of iron, manganese, boron, copper, and zinc are among the micronutrients found in ash, and they can all upset the fragile ecosystems found in water bodies. For instance, phosphorus is a potent growth agent that encourages the growth of algae in bodies of water. As essential as some algae are to the natural food chain, an abundance of algae can lead to scum formation, bad smells, low oxygen levels in the water, and offensive viewpoints. Avoid burning near a waterway shoreline when disposing of debris such as brush, leaves, tree limbs, or clean scrap wood. Burning near a shoreline destroys flora and modifies the composition of the soil, which increases soil erosion into the lake.

7.2.3.2. Health Impacts of Ash

The leaching of heavy metals and other potentially hazardous substances into streams, lakes, and rivers, as well as into drinking water supplies and our food chain, can have an adverse effect on human health. This is particularly true if the materials being burned don't consist of dry, flammable trash, yard waste, or untreated or painted wood. These issues can be avoided by disposing of ash waste in an approved landfill (Goldberg, 1998; Schlesinger, 1992).

7.3. STATIONARY SOURCES

Industrial processes produce a wide range of frequently complex emissions, which can be managed by utilizing the best technology available. Similar control technology is typically used in spite of slight variations in emissions between facilities, even though they appear to use similar equipment and processes. For instance, a method used to control emissions at steel mill × may also be applied to control similar emissions at plant Y. If plant × has a system that is functioning satisfactorily, plant Y shouldn't have to spend excessive sums of money on research and development. Finding an identical industrial process with comparable emissions and identifying the kind of control system in use are frequently the first steps in solving a problem. All the same, good engineering demands that every design take the details into account. Thus, even what would seem to be minor variations between two comparable processes (e.g., quantity, flow rates, holding periods, materials, and reactor shape and size) can result in waste products that are significantly different, in addition to unanticipated safety and liability concerns. As such, the talks that follow are only suggestions that need to be customized for every facility.

7.3.1. Chemical and Allied Products

A chemical process's emissions may be connected to that particular process. It is reasonable to assume that a plant producing resin will also release some raw materials and other products that may or may not be related to the resin in addition to the resin itself. It is normal for a plant producing sulfuric acid to release SO_2 and fumes of sulfuric acid. It is reasonable to anticipate a range of smells coming from a soap production facility. Emissions could include any one or a mix of dust, aerosols, fumes, or gases, depending on the process. It's possible that the emissions are poisonous or smell bad. It is possible that some of the primary emissions are harmless at first, but they later react with the atmosphere to create unfavorable secondary pollutants. In any discussion regarding the significance of emissions from a specific process for an area, a number of factors must be taken into account: (1) the percentage of the area's total emissions that the specific process emits; (2) the degree of toxicity of the emissions; and (3) the obvious characteristics of the source (which can be related to either sight or smell). A flowchart and material balance sheet for the specific process are very helpful in understanding and analyzing any process and its emissions dot.

7.3.2. Resins and Plastics

Resins are organic materials that are solid or semisolid, insoluble in water, and have little to no tendency to crystallize. These are the building blocks of plastics and are also applied as coatings on surfaces like paper, particleboard, and other materials that need a particular kind of finish, whether it be decorative, protective, or functional. Resins all share the trait of requiring heat to be applied and manufactured, along with the exhaustion of gases during these processes. A significant amount of the economically recoverable gases is lost to the atmosphere; however some may condense. About 85% of the resin that was purchased was being released into the atmosphere during one process, which involved coating porous paper with resin to create battery separators. This resin left the plant's stacks in a blue haze, and the smell could frequently be smelled more than two kilometers away. The most frequent complaint against any operation using resins is disagreeable odor because the majority of resins and their byproducts have low odor thresholds.

7.3.3. Varnish and Paints

For the formulation and purification processes involved in varnish manufacturing, heat is required. The same might apply to processes used to prepare paints, shellac, inks, and other coatings that are decorative or protective. Some of the gases that are released into the atmosphere have incredibly low odor thresholds. Two potential emissions from varnish cooking operations are acrolein, which has an odor threshold of approximately 4000 μgm^{-3}, and reduced sulfur compounds, which have odor thresholds of 2 $\mu g\ m^{-3}$. While some of the solvents used in paint preparation are regularly condensed for recovery and then returned to the process, the atmospheric emissions from varnish cooking seem to have little or no recovery value. Organic solvents must be recovered, destroyed, or released into the atmosphere if a paint finish is baked in order to harden the surface. The final option—emission into the atmosphere—is undesirable and might be outlawed by the organization in charge of air pollution control.

7.3.4. Acid Manufacture

Acids are basic raw materials used in a wide range of manufacturing and chemical processes. The flow diagram for an acid plant is shown in Figure 7.5. One of the main inorganic chemicals used in contemporary industry is sulfuric acid. One may anticipate gases, such as SO_2, and aerosol mists containing SO_3 and H_2SO_4 in the submicron to 10-μm size range in the atmospheric discharges from a sulfuric acid plant. Paint, plants, metal, and synthetic fibers are among the materials that are most harmed by aerosol mists. It is normal to expect various toxic and nontoxic gases as well as acid mists from processes that produce acids, such as phosphoric, nitric, and acetic acids. A comprehensive analysis of the specific process is necessary to compile a comprehensive inventory of all the emissions.

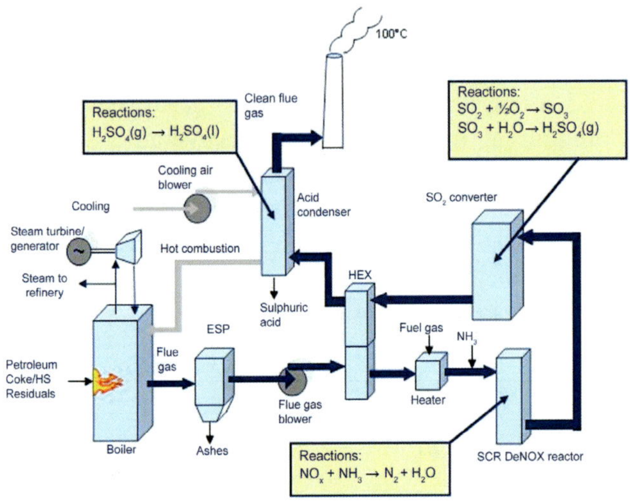

Figure 7.5. *Flow diagram for a sulfuric acid plant.*

Source: By Anne Mette Sørensen – Anne Mette Sørensen, Public Domain, https://commons.wikimedia.org/w/index.php?curid=10251295.

7.3.5. Soaps and Detergents

Oils or fats react with a base to create soap. There are several grades and varieties of soap produced. They could be powdered, granules, liquid, or solid. The smells from chemicals, greases, fats, and oils are the main sources of air pollution in the soap industry, though handling and drying processes can release particulates. Since detergents are made from base stocks that are comparable to those used in petroleum refineries, refineries, and air pollution are related issues.

7.3.6. Phosphate Fertilizers

After phosphate rock has been treated to remove impurities, it is dried and ground to create phosphate fertilizers. Normal superphosphate fertilizer can then be created by reacting sulfuric acid with the PO_4 in the rock. A billion kilograms of phosphate fertilizer are produced annually by more than 100 plants in the U.S.. A typical superphosphate plant's flow diagram, shown in Figure 7.6, lists the pollutants released. Near phosphate fertilizer plants, the emissions of fluoride, both gaseous and particulate, are of the greatest concern.

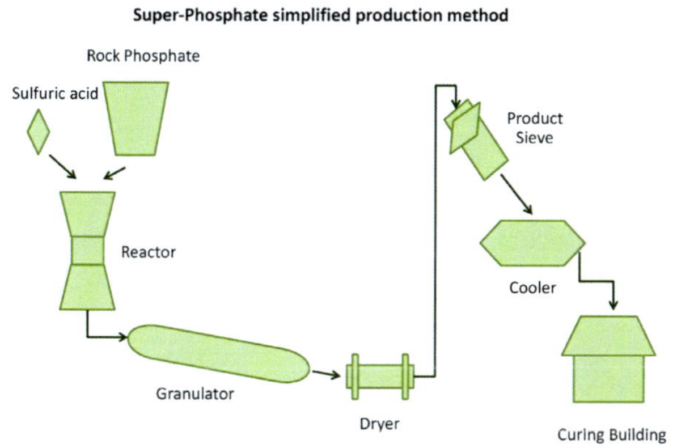

Figure 7.6. *Diagram of the main operations units involved in superphosphate manufacture and further granulation.*

Source: Erro, J., Urrutia, O., Baigorri, R. et al. Incorporation of humic-derived active molecules into compound NPK granulated fertilizers: main technical difficulties and potential solutions. Chem. Biol. Technol. Agric. 3, 18 (2016). https://doi.org/10.1186/s40538-016-0071-7.

7.3.7. Other Inorganic Chemicals

Emissions of air pollutants are an unwanted byproduct of producing the massive amounts of inorganic chemicals required for contemporary industrial processes. Some of the more frequently produced inorganic chemicals are listed in Table 7.1, along with any possible air pollutants that may be released from the particular process.

Table 7.1. *Composition and Sources of Major Air Pollutants*

Category	Pollutants	Main Sources
Gaseous Pollutants	Carbon monoxide (CO), sulfur oxides (SOx), nitrogen oxides (NOx), volatile organic compounds (VOCs), ozone (O_3)	Incomplete combustion of carbon-containing materials, coal-burning power stations, industrial processes, vehicular traffic, biomass burning, photochemical reactions

Persistent Organic Pollutants (POPs)	Polychlorinated dibenzo-p-dioxins (PCDDs), polychlorinated dibenzofurans (PCDFs), hexachlorobenzene (HCB), organophosphates, dichlorodiphenyldichloroethylene (DDE), bisphenol A	Widespread use in controlling pests and diseases, enhancing crop production, and facilitating various industrial processes
Heavy Metals	Lead (Pb), mercury (Hg), cadmium (Cd), chromium (Cr), nickel (Ni), zinc (Zn), manganese (Mn), arsenic (As), and their inorganic salt compounds	Point sources: mines, foundries, smelters, coal-burning power plants; diffuse sources: combustion byproducts from industrial processes, residential heating, waste incineration, vehicle emissions
Particulate Matter (PM)	PM10 (coarse particles, ≤10 μm), PM2.5 (fine particles, ≤2.5 μm), ultrafine PM (UFPs, <0.1 μm); composed of sulfates, nitrates, endotoxins, polycyclic aromatic hydrocarbons (PAHs), heavy metals (iron, nickel, copper, zinc, vanadium)	Primary sources: road traffic exhaust, non-exhaust vehicular emissions (brake wear, tire abrasion, road dust resuspension), industrial processes (combustion, mining, construction), residential solid fuel burning (wood, coal), agricultural activities (crop residue burning, soil tillage); secondary particles: complex chemical reactions in the atmosphere

Source: Olloquequi J, Díaz-Peña R, Verdaguer E, Ettcheto M, Auladell C, Camins A. From Inhalation to Neurodegeneration: Air Pollution as a Modifiable Risk Factor for Alzheimer's Disease. International Journal of Molecular Sciences. 2024; 25(13):6928. https://doi.org/10.3390/ijms25136928.

7.3.8. Petroleum and Coal

Most energy in all industrialized nations comes from coal and petroleum. In addition to hinting at the scope of the air pollution issues related to obtaining the resource, transporting it, refining it, and then transporting it again, this fact indicates the enormous volumes of materials handled.

7.3.8.1. Petroleum

Crude oil is used to make petroleum products. The transportation of crude oil from the ground to the refinery presents numerous opportunities for reduced sulfur gas emissions and hydrocarbon emissions. There is frequently little to no attempt at control because these operations mostly affect individuals who work for the industry and occur in somewhat isolated areas. Huge amounts of air pollution could potentially come from an unregulated petroleum refinery. Refineries all have an odor, to varying degrees depending on how well they are kept up around. As refineries are basically closed processes, emissions are typically not regarded as a component of the process. Emissions from refineries are a possibility, and they do require vents and pressure relief systems. For both regulatory and financial reasons, the majority of refineries employ extremely stringent control measures. The cost of the control equipment can be easily covered by recovering 1% or 2% of the refinery throughput that was previously lost to the atmosphere. The most advanced control technology must be employed to stop catalyst emissions into the atmosphere due to the high cost of the catalyst charge in some crackers and regenerators.

Hydrocarbons from all systems, leaks, loading, and sampling; (2) sulfur oxides from boilers, treaters, and regenerators; (3) carbon monoxide from regenerators and incinerators; (4) nitrogen oxides from combustion sources and regenerators; (5) odors from air and steam blowing, condensers, drains, and vessels; and (6) particulate matter from boilers, crackers, regenerators, coking, and incinerators are examples of possible petroleum refinery air pollutants. Tank trucks and tanker ships must be filled, and loading facilities need to be built to recover all of the vapors that are produced. If not, the atmosphere will lose these vapors. Serious air pollution issues could arise because they have the potential to be both odorous and photochemically reactive. The vapors must be disposed of or returned to the process in some way after being collected. The possibility of an uncontrolled fire at the refinery brought on by an accident or a terrorist act is a growing source of concern. Big and extremely poisonous plumes may originate from this. Refineries therefore require extra security and protection measures in addition to extremely careful operation, maintenance, and monitoring. Every refinery must also have a current backup plan.

7.3.8.2. Coal

One of the main concerns is the air pollution issues related to coal combustion. These issues typically arise outside of the coal mine. In terms of the overall issues with air pollution, the atmospheric emissions caused by the mining, processing, handling, and delivery of coal from the mine to the user are less significant. The handling of coal always results in particulate emissions. Coal dust or inorganic inclusions may be released. If the facilities for the storage and transfer of coal are situated close to residential areas, controlling these emissions may prove to be comparatively costly. This is especially troublesome in underdeveloped countries. For instance, coal combustion has contributed significantly to China's economic and industrial growth over the last 10 years (approximately two-thirds of energy use). This results in three different kinds of issues:

1. Coal is among the worst fuels in terms of production of greenhouse gases, mainly CO_2.
2. Coal often has high sulfur and ash content, so it is a major source of oxides of sulfur and particulate matter, respectively.
3. Coal contains heavy metals, particularly mercury, so it is also a major contributor to air toxics.

7.3.9. Primary Metals Industry

Air pollution from metallurgical equipment has long been known to occur. Since the dusts and fumes from metallurgical furnaces are submicron in size, they are easily seen. Not only are the emissions from related coke ovens visible, but they also have an unpleasant smell.

7.3.9.1. Ferrous Metals

Since the beginning of the industry, the iron and steel sectors have been concerned about emissions from their furnaces and cupolas. Control demands have pushed the companies to reduce their allowable emissions to the point where some of their older operations have closed rather than incur the costs to comply. Instead of going out of business, the majority of the companies in charge of these operations have opened new, controlled plants to replace each old one. The visible aerosol plume is caused by a mixture of metal oxides, smoke, fumes, and dusts released from steelmaking furnaces. They might also contain inorganic and organic gases. When steel scrap is melted, significant amounts of oil, grease, and other combustibles may be present in the charge, which raises the loadings of organic gases and smoke. Both gaseous and particulate fluoride emissions can be very problematic if the used ore has significant fluoride concentrations.

Although foundry cupola emissions are minimal, they are nevertheless noticeable in certain places. Up to 50 kg of dust, fumes, smoke, and oil vapor can be expected to be released hourly from an uncontrolled 2-meter cupola. It's also possible to expect organic gases, nitrogen oxides, and carbon monoxide. Control is feasible, but for a small foundry with only one or two heats per week, the cost of the control might be unaffordable. Coke ovens are frequently connected to the steel industry. Coal that has undergone pyrolysis is known as coke that is heated to between 1000 and 1400°C without oxygen to prevent burning. In order to create nearly pure carbon, this process purposefully releases the gaseous components of the coal. Complex mixtures of gas, liquid, and solid phases are released from coke ovens.

These emissions typically contain a range of approximately 40 PAHs as well as other incomplete combustion products, such as formaldehyde, acrolein, aliphatic aldehydes, ammonia, carbon monoxide, nitrogen oxides, phenol, cadmium, arsenic, and mercury. Near coke plants, more than 60 organic compounds have been found. Up to 635 kg of coke, 90 kg of coke breeze (large coke particulates), 7–9 kg of ammonium sulfate, 27.5–23.4 L of coke oven gas tar, 55–135 L of ammonia liquor, and 8–12.5 L of light oil can be obtained from a metric ton of coal. As gases and vapors, up to 35% of the initial coal charge is released. The majority of these vapors and gases are collected during the manufacturing of coke byproducts. The components of coke oven gas are nitrogen, oxygen, hydrogen sulfide, ammonia, propylene, butylene, acetylene, carbon monoxide, ethylene, and propylene. Coal-tar pitch, pyridine, tar acids, naphthalene, and creosote oil are all components of coke oven gas tar. It is possible to extract solvent naphthas, xylene, toluene, and benzene from the light oil fraction. Between 1880 and the beginning of the 1950s, U.S. coke production rose gradually, reaching a peak of 65 million metric tons in 1951. The United States produced 48 million metric tons of coke in 1976, ranking second in the world that is 14% of global production is produced. The U.S. produced 24 million metric tons in 1990, which dropped them to fourth place globally. Production dropped from 20 million metric tons in 1997 to 15.2 million metric tons in 2002, continuing a steady downward trend. The amount of coke used in relation to the amount of steel produced has decreased by up to 25% as a result of technological advancements, which has also decreased demand for blast furnace coke in recent years.

The volatilized gases are undoubtedly air pollutants. As a matter of fact, thousands of compounds, many of them toxic, come from coke ovens. Coke facilities are using available technologies to recover, separate, and profitably sell these gases because they have realized that many of these gases have economic value as well.

7.3.9.2. Nonferrous Metals

One of the most blatant examples of industrial pollution of the environment around the turn of the century was the total annihilation of vegetation in the vicinity of copper, lead, and zinc smelters. The metallic sulfide ores' smelting was the root cause of this issue. Large amounts of sulfur were oxidized to SO_2, which was toxic to most of the vegetation the plume fumigated, as the metal was released during the smelting process. Although current smelting systems incur significant costs to prevent the uncontrolled release of SO_2, the ecosystem will likely take years or even centuries to recover in many places.

Due to the fluoride emissions from their operations, early aluminum reduction plants were the source of air pollution. Emissions of fluoride have the potential to seriously harm both vegetation and animals that eat it. The end result was a barren area of ground surrounding the plant.

These kinds of scenes remind me of those that exist downstream from some of the unregulated copper smelters. To reduce fluoride emissions, new aluminum reduction plants will need to invest a significant amount of money. The cost of control for some of the older plants is going to be higher than the facility's initial capital investment. Control agencies have created vast sampling networks to track emissions from the plant of concern when the issue is severe.

The main emissions from other nonferrous metal facilities are very small-diameter metal oxides or fumes. The toxic fumes of zinc oxide range in wavelength from 0 to 3 μm. Lead and lead oxide fumes have been thoroughly studied and are very toxic. Numerous metallurgical processes can release trace metals such as arsenic, cadmium, bismuth, and others.

7.3.10. Stone and Clay Products

Particulate matter is released at every stage of operation by the industries that handle and produce different types of stone products.

These particles could include tiny mineral dusts that are large enough to harm the lungs. Mineral fibers, particularly asbestos, may also be released depending on the type of rock. In order to protect workers from incurring incurable illnesses such as lung cancer, mesothelioma, pleural diseases, asbestosis, and silicosis, the threshold values for these types of dusts have been set relatively low. In the process of making chalk, cement, talc, clay, so forth. particulate matter will typically be released during each process.

These procedures, which can be confined and regulated to stop particle emissions, may include drying, grinding, and sieving. The recovered particles can frequently be added back into the process at a profit. There will be a significant amount of dust released during the glass-making process, with the particles averaging 300 μm in size. When handling the raw materials, some dusts may also be released. It is essential to control this dust to avoid creating an annoyance outside the plant. Smoke and gases may escape from the contact of the molten glass with lubricated molds during the blowing or shaping of glass into the final product. Despite being quite dense, these emissions don't last very long.

7.3.11. Forest Products Industry

7.3.11.1. Wood Processing

Trees are considered a renewable resource that are used for sustained yield in most parts of the world. An appropriately maintained forest will perpetually yield wood for chemicals, fiber, and lumber. The process of extracting this resource may produce a lot of dust and other particles. Transportation on unpaved roads generates a lot of dust. Slash burning, a traditional method of burning the leftover material after harvesting timber, is still carried out in some places and is a significant cause of air pollution in the areas downwind from the fire in terms of smoke, gas, and particulate matter. This kind of burning can seriously impair visibility. Sawing, peeling, planing, sanding, and drying are some of the processes that turn harvested timber into the final product. These processes have the potential to release large amounts of wood fiber and smaller amounts of gaseous material into the atmosphere. For a plywood mill or particleboard plant, controlling wood fiber emissions from the pneumatic transport and storage systems can be a significant issue that comes with significant costs.

7.3.11.2. Pulp and Paper

The production of paper and pulp is growing exponentially worldwide. Paper will always be in demand as long as new applications for it are discovered. Paper offers a great use for wood, a renewable resource, as it is primarily made from wood or wood waste. The kraft process, which recovers and regenerates the chemicals, is the most popular pulping method.

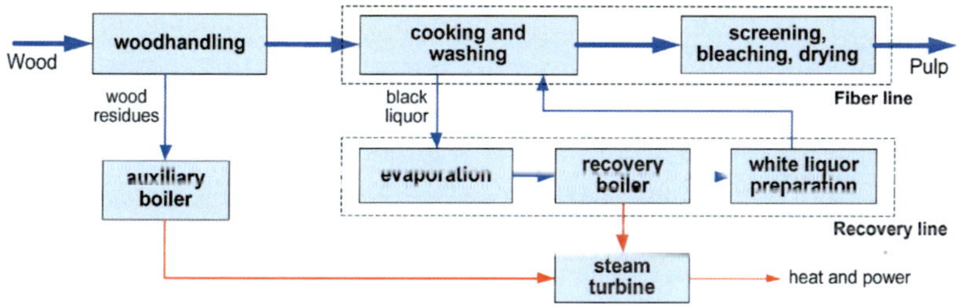

Figure 7.7. *Overview of a kraft pulping process.*

Source: Hamaguchi, M.; Cardoso, M.; Vakkilainen, E. Alternative Technologies for Biofuels Production in Kraft Pulp Mills—Potential and Prospects. Energies 2012, 5, 2288-2309. https://doi.org/10.3390/en5072288.

This takes place in the recovery furnace, which has reducing and oxidizing zones of operation. Particulate matter, sulfur oxides, and very odorous reduced sulfur compounds are among the emissions from these recovery furnaces. Dioxin formation has historically been significantly reduced by bleaching, but many processors have now changed their processes to use significantly less chlorine. The smells and aerosol emissions from the kraft pulp process will spread over a large area if stringent and costly control measures are not taken. Reports of odors have been made more than 100 kilometers away from these plants. Millions of kilograms of completed products can be produced daily from a properly managed and run kraft plant, handling massive volumes of raw materials with little to no complaints about odor or particulate emissions (Figure 7.7).

7.3.12. Noxious Trades

As the name suggests, these operations have the potential to seriously contribute to air pollution if left unchecked. The smells connected to the process are the primary issue. These include the tanning industry, rendering plants, confined animal feeding operations (CAFOs), and numerous food processing facilities, including fish meal plants. Usually, only the smells emanating from these plants are a cause for concern, rather than the emissions of gases and particulates. But there are other common air pollutants as well (e.g., Nitric oxide (NO) and chromium in tanning close to CAFOs. It is no longer appropriate to use these industries as a means of control by forcing them to locate away from residential or commercial areas (Goldberg, 1998; Schlesinger, 1992).

7.4. MOBILE SOURCES

Any source of air pollution that is able to move independently from one location to another is considered mobile. By this definition, a portable asphalt batching plant is not a mobile source; a vehicle is. The internal combustion engine appears to be at the core of the regulatory definition. Though sources like construction equipment, gasoline-powered lawn mowers, and gasoline-powered tools are included in this category, mobile sources generally imply transportation. Therefore, mobile sources include a wide range of vehicle types that are propelled by engines that run on various cycles, are fueled by a variety of products, and release variable amounts of simple and complex pollutants into the air. The automobile, which runs on gasoline and has an Otto cycle (four-stroke engine), is the main source of mobile air pollution in all industrialized nations. In 1990, there were more than 85 million cars on the road in the U.S. The total number of vehicles in the United States surpasses 100 million when the 15 million gasoline-powered trucks and buses and the 4 million motorcycles are taken into account. It is claimed that the engine powering these millions of cars is the most intricately designed device of the century. Given the current reliability, cost, and lifespan of internal combustion engines, it is easy to understand why they have remained so widely used. A contemporary car engine with 100,000 km on it will have roughly 2.5×10^8 power cycles.

A gasoline-powered vehicle emits emissions from a variety of sources. Figure 7.8 shows the expected behavior of a controlled car (1983 or later model) that satisfies federal standards of 1983 and an uncontrolled car (1960 model). Lead emissions are no longer a big deal in the West because the majority of cars on the road today run on unleaded gasoline, but they are still an issue in many other countries.

Figure 7.8. *Emissions and emission sources of light-duty vehicles.*

Source: By Ruben de Rijcke – Own work, CC BY-SA 3.0, https://commons.wikimedia.org/w/index.php?curid=8735834.

7.4.1. Smog and Your Health

Instead of being released into the atmosphere directly, ground level or "bad" ozone is produced chemically when sunlight interacts with nitrogen oxides (NOx) and volatile organic compounds (VOC).

Because of its negative effects on both people and the environment, ground level ozone is considered a harmful air pollutant. It is also the primary component of "smog.." Inhaling ozone can lead to a number of health issues, especially for young people, the elderly, and those with lung conditions like asthma who are of any age. Inhaling ozone can cause:

1. Make it more difficult to breathe deeply and vigorously.
2. Cause shortness of breath, and pain when taking a deep breath.
3. Cause coughing and sore or scratchy throat.
4. Inflame and damage the airways.
5. Aggravate lung diseases such as asthma, emphysema, and chronic bronchitis.
6. Increase the frequency of asthma attacks.
7. Make the lungs more susceptible to infection.
8. Continue to damage the lungs even when the symptoms have disappeared.
9. Cause chronic obstructive pulmonary disease (COPD).

7.4.2. Particle Pollution

A mixture of solid particles and liquid droplets found in the air is referred to as particle pollution (also known as particulate matter or PM). Primary particles are those that are released straight from the source, such as fields, construction sites, fires, smokestacks, and unpaved roads.

Others are created when complex chemical reactions occur in the atmosphere, such as those involving sulfur dioxide and nitrogen oxide emissions from vehicles, factories, and power plants. These particles, also referred to as secondary particles, account for the majority of the nation's fine particle pollution. Particulate matter has ties to the following:

1. Premature death in people with heart or lung disease.
2. Nonfatal heart attacks.
3. Irregular heartbeat.
4. Aggravated asthma.
5. Decreased lung function.
6. Increased respiratory symptoms, such as irritation of the airways, coughing or difficulty breathing.
7. Reduced visibility (haze) in many parts of the U.S., including many of our national parks and wilderness areas.
8. Lake and stream acidification.

7.4.3. Near Roadway Air Pollution and Health

Living, working, or going to school close to a major road seems to be associated with a higher frequency and severity of health issues related to exposure to air pollution from traffic. Those with preexisting cardiopulmonary diseases, children, the elderly, and those from lower socioeconomic backgrounds are among the groups most vulnerable to the negative health effects of air pollution near roads. These dangers consist of:

1. Higher rates of asthma onset and aggravation.
2. Cardiovascular disease.
3. Impaired lung development in children.
4. Pre-term and low-birthweight infants.
5. Childhood leukemia.
6. Premature death.

7.5. AIR TOXIC SOURCES

Because of their dense populations and higher concentration of emission sources, including stationary sources like power plants, urban areas are typically more dangerous for air toxics. Air pollution is a barrier to economic opportunity and security and may be disproportionately experienced by low-income neighborhoods, tribal populations, and communities of color living in urban areas. People exposed to toxic air pollutants at sufficient concentrations and durations may have an increased chance of getting cancer or experiencing other serious health effects, including:

1. Damage to the immune system.
2. Neurological disorders.
3. Reproductive disorders (e.g., reduced fertility).
4. Developmental disorders.
5. Respiratory and other health problems.

Some toxic air pollutants, like mercury, can deposit onto soils or surface waters in addition to being inhaled. From there, they are absorbed by plants, consumed by animals, and eventually amplified up through the food chain. Animals that are exposed to high enough levels of air toxics over time may develop health issues similar to those experienced by humans. Toxic substances like pesticides and metals like mercury can descend from the atmosphere and accumulate in the food chain, affecting fish, birds, wildlife, and even humans in terms of behavior, neurology, and reproduction. Heavy metals like mercury and persistent organic pollutants (POPs) like DDT and pesticides are examples of air toxics. These airborne pollutants are particularly dangerous because they can accumulate in the tissues of living things, have toxic effects, are persistent and do not decompose in the environment. In regions affected by air pollution, pesticides, mercury, and PCBs can find their way into the food chain. Because these and other air toxics become more concentrated at every level of the food chain and have the potential to seriously endanger the health of both humans and wildlife, the National Park Service is concerned about them. While eating fish is the primary way that humans (and wildlife) are exposed to mercury, pesticides, PCBs, and other contaminants, it is also a lean, low-calorie source of protein.

Methylmercury, the toxic form of mercury, hinders a child's and fetus's neurological development. Lower reproductive success, stunted growth and development, aberrant behavior, weakened immunity, illness, and shortened survival are possible side effects. Other air pollutants have different effects on health. Some toxic compounds have been banned from use and production in the U.S. Unfortunately because these toxics take a long time to break down, they remain in the environment and in the fatty tissues of animals. For example, dieldrin, an insecticide banned in the U.S. in 1987, is very carcinogenic and *endocrine-disrupting*. In addition, it causes neurological issues, lowers the success rate of reproduction, and diminishes the effectiveness of the immune system. The pesticide DDT is prohibited in the U.S., which was also identified as an endocrine disruptor in 1972. In addition, it is probably carcinogenic to humans, harms the liver, momentarily impairs the nervous system, lowers the likelihood of successful reproduction, and harms the reproductive system.

The group most at risk of ingesting metals and harmful substances is the seafood industry. However, because of the harmful effects on fetal growth and childhood development, some groups of people are more vulnerable than others. Children and women who are fertile are among these groups. Furthermore, because they depend on fish for their diets, subsistence fish consumers, like some native populations, are more vulnerable. It is food consumption, not airborne exposure, that causes harm to humans from air toxics. This contrasts with worries about other air pollutants, such as ozone and particulate matter, that affect human health.

When fish is consumed occasionally, as during recreational fishing, the chances of negative health effects are relatively low. By adhering to local fish advisories, which list the kinds of fish that may contain higher levels of contaminants locally, risks can be reduced. Sensitive groups may be subject to different restrictions (such as women of childbearing age and children). Fish fillet, or muscle, is more likely to accumulate methylmercury. The amounts of pollutants consumed can vary depending on how fish is prepared. For instance, removing the fish's skin before cooking can reduce the quantity of pollutants. To reduce risk, the following are some general suggestions for consuming and preparing fish:

Verify local fishing laws and consumption advisories before going fishing. If there are no rules, you might want to stick to eating smaller fish. Smaller fish are probably younger and have fewer dangerous pollutants in them. The U.S. Environmental Protection Agency also posts a National Listing of Fish Advisories.

Try consuming less fatty fish or fish that feed on the bottom of lakes or streams if local fish advisories alert you to the presence of contaminants (PCB, pesticides, dioxins, PBDEs) that concentrate in the fatty portion of fish. These fish may have lower levels of these dangerous contaminants.

When preparing fish to eat, remove the skin, fat, and organs. Contaminants are more likely to build up in those parts of the fish.

Prepare fish by grilling, if possible, and letting the fat drip off.

Airborne pollutants can have toxic effects on human and wildlife health, such as pesticides and mercury. Over 275 million annual visitors to U.S. national parks trust the National Park Service to both give them opportunities for recreational activities such as fishing and observing birds and wildlife, as well as take care of all the natural systems within parks. Pollution sources as far away as Europe and Asia as well as close as the local county are the sources of toxic air pollutants that are carried to national parks.

7.6. EMISSION INVENTORY

An emissions inventory is a database that shows how much air pollution was released into the atmosphere over a given year or other period of time, broken down by source. Emission inventories are used by governments to help identify important sources of air pollution and to guide regulatory actions. A crucial component of mathematical models used to estimate air quality is emissions inventories. By using estimated emissions reductions in conjunction with emissions inventory data in air quality models, the impact of potential regulatory actions on air quality can be forecast. Periodic updates to the emissions inventory allow for the establishment of emission trends over time. Moreover, inventories can be used to educate the public about the sources of pollution. Estimates of the emissions from various sources of pollution within a given region are included in an emissions inventory. All the pollutants linked to the local air quality issues should be included. For instance, sources of nitrogen oxides (NOx) and volatile organic compounds (VOC) should be included in an emissions inventory to support the management of ground-level ozone.

The use of emissions factors, short-term emission measurements extrapolated to a longer time period, and continuous monitoring of emissions from a source are some of the methods available for determining emissions. The approach selected is determined by the resources (finance, staff, time, and data) available. A pollutant's quantity released and the activity level linked to its release are attempted to be related using an emissions factor, which is a representative value. Kilograms of particulate matter released per megagram of burned coal is one example of an emissions factor format. In published literature, emissions factors are usually presented as long-term averages for comparable facilities, based on available data of reasonable quality. However, differences between facilities can have a big impact on emissions. These differences include raw material usage, operating conditions, and emission controls. The creation of local, source-specific emission factors is ideal whenever it is feasible (Goldberg, 1998; Schlesinger, 1992).

REFERENCES

1. Askarov, R. A., Askarova, Z. F., & Karelin, A. O. (2019). The evaluation of dynamics of pollution of atmospheric air by stationary sources in the regions of the Republic of Bashkortostan. *Health Care of the Russian Federation, 60*(4), 192–198. https://doi.org/10.18821/0044-197x-2016-60-4-192-198.
2. Gao, G., Zeng, H., & Zhou, Q. (2022). Biomonitoring atmospheric pollution of polycyclic aromatic hydrocarbons using mosses. *Atmosphere, 14*(1), 26. https://doi.org/10.3390/atmos14010026.
3. Goldberg, P. K. (1998). The effects of the corporate average fuel efficiency standards in the U.S.. *The Journal of Industrial Economics, 46*(1), 1–33. https://doi.org/10.1111/1467-6451.00059.
4. Kokkinakis, I. W., & Drikakis, D. (2022). Atmospheric pollution from rockets. *Physics of Fluids, 34*(5), 056107. https://doi.org/10.1063/5.0090017.
5. Nenes, A., Pandis, S., & Pilinis, C. (1998). ISORROPIA: A new thermodynamic equilibrium model for multiphase multicompetent inorganic aerosols. *Aquatic Geochemistry, 4*, 123–152. https://doi.org/10.1023/A:1009604003981.
6. Schlesinger, R. B. (1992). Atmospheric pollution. *Otolaryngology–Head and Neck Surgery, 106*(6), 642–648. https://doi.org/10.1177/019459989210600605.
7. Schroeder, J. R., Crawford, J. H., Fried, A., Walega, J., Weinheimer, A., Wisthaler, A., Müller, M., Mikoviny, T., Chen, G., Shook, M., Blake, D. R., Diskin, G., Estes, M., Thompson, A. M., Lefer, B. L., Long, R., & Mattson, E. (2016). Formaldehyde column density measurements as a suitable pathway to estimate near-surface ozone tendencies from space: O_3-CH_2O relationship. *Journal of Geophysical Research: Atmospheres, 121*, 13,088–13,112. https://doi.org/10.1002/2016JD025419.
8. Simon, H., Baker, K. R., & Phillips, S. (2012). Compilation and interpretation of photochemical model performance statistics published between 2006 and 2012. *Atmospheric Environment, 61*, 124–139. https://doi.org/10.1016/j.atmosenv.2012.07.012.
9. Skarżyńska, K., Polkowska, Ż., & Namieśnik, J. (2006). Sampling of atmospheric precipitation and deposits for analysis of atmospheric pollution. *Journal of Automated Methods and Management in Chemistry, 2006*, 1–19. https://doi.org/10.1155/jammc/2006/26908.
10. U.S. Environmental Protection Agency, Office of Air Quality Planning and Standards. (1999). *Innovative Strategies and Economics Group, OAQPS Economic Analysis Resource Document*. Retrieved from: http://www.epa.gov/ttn/ecas/econdata/Rmanual2/ (accessed on 14 December 2024).

11. Valin, L. C., Fiore, A. M., Chance, K., & González Abad, G. (2016). The role of OH production in interpreting the variability of CH_2O columns in the southeast U.S. *Journal of Geophysical Research: Atmospheres, 121*, 478–493. https://doi.org/10.1002/2015JD024012.
12. Wang, Y., Li, L., Chen, C., Huang, C., Huang, H., Feng, J., Wang, S., Wang, H., Zhang, G., Zhou, M., Cheng, P., Wu, M., Sheng, G., Fu, J., Hu, Y., Russell, A. G., & Wumaer, A. (2014). Source apportionment of fine particulate matter during autumn haze episodes in Shanghai, China. *Journal of Geophysical Research: Atmospheres, 119*, 1903–1914. https://doi.org/10.1002/2013JD019630.
13. Weaving, J. H., & Pouille, J. P. (1990). Atmospheric pollution. In *Internal Combustion Engineering: Science & Technology* (pp. 707–793). Springer Netherlands. https://doi.org/10.1007/978-94-009-0749-2_15.

CHAPTER 8

EFFECT OF AIR POLLUTION

Contents

INTRODUCTION

When substances known as pollutants are present in the atmosphere and endanger the health of people and other living things, or harm the climate or materials, the resultant contamination of the air occurs. The natural characteristics of the atmosphere can also be changed by outside or indoor environments being contaminated by chemical, physical, or biological agents. There are numerous varieties of air pollutants, including particulates (both organic and inorganic), gases (such as ammonia, carbon monoxide, sulfur dioxide, nitrous oxides, methane, and chlorofluorocarbons), and biological molecules. The natural environment (such as habitat degradation, ozone depletion, and climate change) as well as the built environment (such as acid rain) may be harmed by air pollution, which can also kill humans, cause diseases, allergies, and even kill other living things like animals and crops. Human activity and natural events are two sources of air pollution.

Numerous pollution-related illnesses, such as lung cancer, heart disease, chronic obstructive pulmonary disease (COPD), stroke, and respiratory infections are significantly increased by air pollution. There is mounting evidence linking exposure to air pollution to lower IQs, cognitive decline, a higher risk of mental illnesses like depression, and poor perinatal health. While there are many negative effects of poor air quality on human health, the respiratory and cardiovascular systems are the main ones. The kind of pollutant, the amount of exposure, the person's genetic makeup, and their current state of health all influence how each person responds to air pollution.

All things are impacted by air pollution. It damages forests, wildlife, and agriculture in addition to impairing our health. It also affects the environment by obstructing sunlight, decreasing visibility, and producing acid rain. Global warming is caused by greenhouse gas pollution, which has an impact on every part of the earth (Baccarelli et al., 2016; Henneman et al., 2023).

8.1. AIR POLLUTION AND HEALTH

One well-known environmental health risk is air pollution. When a plume emerges from a smokestack, exhaust billows across a busy highway, or a brown haze descends upon a city, we know what we're looking at. Though some air pollution is invisible, you can detect it by its strong odor. It poses a serious risk to the prosperity and health of the world. Over 63.5 million deaths worldwide are attributed to air pollution, in all its forms, a figure that has increased during the last 20 years. Hazardous substances from both natural and man-made sources combine to form air pollution. The main causes of man-made air pollution are automobile emissions, fuel oils and natural gas used to heat homes, waste products from manufacturing and power plants, especially those that burn coal, and fumes from the production of chemicals. Hazardous materials are released into the atmosphere by nature. These include gases like methane released from decomposing organic matter in soils, ash, and smoke from volcanic eruptions, and smoke from wildfires, which are frequently started by humans. Most of the components of air pollution caused by humans are present in traffic-related air pollution (TRAP), which is a mixture of gases and particles. These components include ground-level ozone, different types of carbon, nitrogen oxides, sulfur oxides, volatile organic compounds, polycyclic aromatic hydrocarbons, and fine particulate matter.

At ground level, the atmospheric gas ozone is frequently referred to as smog. It is produced by the chemical reaction of pollutants released by automobiles, power plants, refineries, and other sources with sunlight. As byproducts of industrial processes and constituents of motor vehicle emissions, harmful gases include carbon dioxide, carbon monoxide, nitrogen oxides (NOx), and sulfur oxides (SOx).

Chemicals like carbon, nitrates, sulfates, and mineral dusts make up particulate matter (PM). PM is found in cigarette smoke, burning organic materials like wildfires, and industrial and vehicle emissions from the burning of fossil fuels. 30 times thinner than human hair, fine particulate matter (PM 2.5) is a subset of PM. It may cause major health issues if breathed deeply into the lung tissue. Most health effects associated with air pollution in the United States are caused by PM 2.5. s.

Because volatile organic compounds (VOCs) vaporize at or close to room temperature, they are referred to as volatile. The fact that they contain carbon gives them the name organic. Paints, cleaners, insecticides, certain furniture, and even craft supplies like glue release volatile organic compounds (VOCs). VOCs are emitted during combustion and are primarily found in gasoline and natural gas.

Organic substances with hydrogen and carbon are known as polycyclic aromatic hydrocarbons, or PAHs. The Report on Carcinogens lists 15 of the more than one hundred PAHs that are known to be extensively present in the environment. As a byproduct of numerous industrial processes, including the production of steel, iron, and rubber goods and the production of electricity, PAHs are also produced in addition to combustion. Particulate matter also contains PAHs.

Chemicals or particulates in the air that are harmful to people, animals,' or plants' health are known as air pollution. It harms structures as well. Air pollutants come in a variety of forms.

They could be liquid droplets, solid particles, or gases. Numerous air pollutants have negative effects on the environment and human health that are either known or suspected. These pollutants are mostly the byproducts of combustion from power generation, space heating, or motor vehicle traffic in the majority of Europe.

These sources of pollution have the potential to cause issues not just locally but also over large distances (Baccarelli et al., 2016; Henneman et al., 2023).

Table 8.1 lists the studies selected for the scoping review and details their methodologies for analyzing the relationship between particulate matter and mental health outcomes (stress, anxiety, and depression). Commonly used AQ analysis methods include air quality monitoring station data (inventories, databases, and AQI) and land-use models, such as land use regression (LUR) and the air pollution model (TAPM). For MH, various questionnaires and in-built psychological scales were used: self-reported questionnaires and surveys, Generalized Anxiety Disorder Questionnaire (GAD, GAD-7), Beck Anxiety Inventory (BAI) and Fear Questionnaire (FQ) for anxiety disorder and fear; The structured Clinical Interview for Diagnostic and Statistical Manual of Mental Disorders (SCID-I), Inventory of De-pression Symptomatology (IDS), Centre of Epidemiologic Studies Depression Questionnaire (CES-D) and the Patient Health Questionnaire (PHQ) (PHQ-4, PHQ-8 and PHQ-9) scales for measuring self-reported severity of depression; Altman Self-Rating Mania (ASRM) scale for mania; WHO-5 Wellbeing Index for general wellbeing; Short Warwick-Edinburgh Mental Wellbeing Scale (SWEMWBS), Ecological Momentary Assessment (EMA) and Eysenck Personality Questionnaire-Revised Short Form (EPQR-Short) for social behavior, feelings, and mood; EuroQol-5 dimensions (EQ-5D) index for both anxiety and depression; Chinese version of the Perceived Stress Scale (CPSS) for stress; including medical/healthcare data, and Internet search behavior were also used in selected studies.

For statistical analyses to find the relationship between AQ and MH, linear regression, including multiple, multilevel, Generalized Additive Models (GAMs), generalized linear models (GLMs), and multivariable regressions; logistic regressions, including multiple, multilevel, and multivariable regressions; survival analysis models, including Cox proportional hazard analysis and time-varying Cox proportional hazard analysis; and other statistical approaches, such as two-way FE, negative binomial, GEE, and mix (multiple and linear regression), were incorporated.

The table also indicates which study controlled for meteorological conditions (e.g., temperature, precipitation, wind speed, wind direction, humidity), demographics (e.g., sex, age, family status, occupation, cultural background), and socioeconomic factors (e.g., income, education level, living conditions, resources, etc.) to find the results.

Table 8.1. *Studies Selected for the Scoping Review and Their Methodologies for Analyzing the Relationship Between Particulate Matter and Mental Health Outcomes*

Study	Study Name	References	Country	Methodology—AQ	Methodology—MH	Controlled			Statistical Analysis
						Meteorological	Demographics	Socio-economic Status	
Study 1	Relationship of emergency department visits for suicide attempts with meteorological and air pollution conditions	Miyazaki et al. (2023)	Japan	Monitoring stations	Medical (ED) data	Yes	Yes	No	Multivariate Poisson regression
Study 2	Residential greenspace and anxiety symptoms among Australian women living in major cities: A longitudinal analysis	Mouly et al. (2023)	Australia	Land-use regression (LUR) model	GAD	Yes	Yes	Yes	Generalized estimating equation (GEE)
Study 3	How People's COVID-19-Induced-Worries and Multiple Environmental Exposures Are Associated with Their Depression, Anxiety, and Stress during the Pandemic	Huang et al. (2023)	Hong Kong	Portable air pollutant sensors.	PHQ-4 and self-reported scale of stress symptoms.	No	Yes	Yes	Linear regression models
Study 4	Real-time air pollution and bipolar disorder symptoms: a remote-monitored cross-sectional study	Kandola and Hayes (2023)	UK	Air quality index (AQI)	PHQ-8 and ASRM scale	Yes	Yes	No	Linear regression

Study 5	Long-term Exposure to Multiple Ambient Air Pollutants and Association with Incident Depression and Anxiety	Yang et al. (2023)	UK	LUR	Medical data	No	Yes	Yes	Cox proportional hazards models
Study 6	Multiple environmental exposures along daily mobility paths and depressive symptoms: A smartphone-based tracking study	Roberts and Helbich (2021)	Netherlands	LUR	PHQ-9	No	Yes	Yes	Multiple regression analyses
Study 7	Daily space-time activities, multiple environmental exposures, and anxiety symptoms: A cross-sectional mobile phone-based sensing study	Lan et al. (2022)	Netherlands	LUR	GAD-7	No	Yes	Yes	Multiple linear regressions, including the Random Forest (RF) model
Study 8	Association between traffic-related air pollution and anxiety hospitalizations in a coastal Chinese city: are there potentially susceptible groups?	Ji et al. (2022)	China	Monitoring stations	Medical data	Yes	Yes	No	Generalized additive model (GAM)
Study 9	Spatial Variability of the Relationship between Air Pollution and Well-Being	Li and Managi (2022)	Japan	Database	Survey	No	Yes	Yes	Linear analyses

Study 10	Air pollution and anti-social behavior: Evidence from a randomized lab-in-the-field experiment	Lohmann et al. (2023)	China	Air Quality Index (AQI)	Antisocial behavior: using incentivized economic games. Risk and time preferences: lottery choice task and a convex time budget task. Cognitive ability and well-being: Raven's Matrices, Numerical Stroop Task, and self-reported scales	Yes	Yes	No	Linear regression
Study 11	Do objective and subjective traffic-related pollution, physical activity, and nature exposure affect mental wellbeing? Evidence from Shenzhen, China	Huang, Tian and Yuan (2023)	China	NASA Sedac database	WHO-5 Wellbeing Index Short Warwick-Edinburgh Mental Wellbeing Scale (SWEMWBS) Self-reported survey (single question)	No	Yes	Yes	Multivariate linear regression
Study 12	The effects of long-term exposure to air pollution on incident mental disorders among patients with prediabetes and diabetes: Findings from a large prospective cohort	Feng et al. (2016)	UK	UK Air database	Healthcare data	Yes	Yes	Yes	Time-varying covariates Cox model and generalized propensity score (GPS)

Study 13	The Impacts of Air Pollution on Mental Health: Evidence from the Chinese University Students	Zu et al. (2020)	China	The Air Quality Index (AQI) data	CES-D	No	Yes	Yes	Multi-variable models
Study 14	Associations of co-exposures to air pollution and noise with psychological stress in space and time: A case study in Beijing, China	Tao et al. (2021)	China	The Airbeam sensor	EMA	Yes	Yes	Yes	Multilevel ordinal logistic regression models
Study 15	Effects of vitamin D on associations between air pollution and mental health outcomes in Korean adults: Results from the Korea National Health and Nutrition Examination Survey (KNHANES)	Kim et al. (2023)	S. Korea	Monitoring stations	Self-reported questionnaire	No	Yes	Yes	Multiple logistic regression
Study 16	Relationship between chronic exposure to ambient air pollution and mental health in Korean adult cancer survivors and the general population	Kim et al. (2021)	S. Korea	Monitoring stations	Self-reported questionnaire	No	Yes	Yes	Multiple logistic regression
Study 17	Ambient air pollution exposure and depressive symptoms: Findings from the French CONSTANCES cohort	Sakhvidi et al. (2022)	France	LUR	CES-D	No	Yes	Yes	Negative binomial regressions

Study 18	Associations of $PM_{2.5}$ and road traffic noise with mental health: Evidence from UK Biobank	Hao et al. (2022)	UK	LUR	PHQ and the SCID-I	No	Yes	Yes	Logistic regression model
Study 19	Associations between long term exposures to outdoor air pollution and indoor solid fuel use and depression in China	Zhang et al. (2022)	China	Monitoring stations	PHQ-9	No	Yes	Yes	Multiple logistic regres-sion
Study 20	$PM_{2.5}$ exposure and anxiety in China: evidence from the prefectures	Chen et al. (2021)	China	Air Qual-ity Online Monitor ing and Analysis Platform	Internet search behavior	No	No	Yes	Two-way fixed ef-fect (FE) regression model
Study 21	Short-term effects of traf-fic noise on suicides and emergency hospital ad-missions due to anxiety and depression in Madrid (Spain)	Díaz et al. (2020)	Spain	Monitoring stations	Medical (ED) data	Yes	No	No	Linear models and Poisson regres-sion.
Study 22	The relationship between air pollution and depression in China: Is neighborhood social capital protective?	Wang et al. (2018)	China	Airborne Fine Particulate Matter and Air Quality Index	CES-D	No	Yes	Yes	Multilev-el linear regression analyses
Study 23	Long-term exposure to ambient air pollutants and mental health status: A nationwide population-based cross-sectional study	Shin, Park and Choi (2018)	S. Korea	Monitoring stations	EQ-5D index	Yes	Yes	Yes	Multiple logistic regres-sion

Study 24	Greenspace and mortality in the U.K. Biobank: Longitudinal cohort analysis of socio-economic, environmental, and biomarker pathways	Wan et al. (2022)	UK	LUR	PHQ-4 and EPQR-Short	Yes	Yes	Yes	Cox proportional hazards model
Study 25	Ambient temperature and air pollution associations with suicide and homicide mortality in California: A statewide case-crossover study	Rahman et al. (2023)	U.S.	United States Environmental Protection's (U.S EPA's) AQ System	Healthcare (death) data	Yes	No	No	Conditional logistic regression models
Study 26	Impacts of coal mine fire related $PM_{2.5}$ on the utilization of ambulance and hospital services for mental health conditions	Carroll et al. (2022)	Australia	The Air Pollution Model (TAPM)	Medical (ED) data	Yes	No	No	GAMs
Study 27	Urban Air Pollution and Mental Stress: A Nationwide Study of University Students in China	Zhang et al. (2021)	China	National Bureau of Statistics	CPSS	No	Yes	Yes	Multilevel logistic regression
Study 28	Outdoor light at night, air pollution, and depressive symptoms: A cross-sectional study in the Netherlands	Helbich, Browning and Huss (2020)	Netherlands	LUR	PHQ–9	No	Yes	Yes	GAM

Study 29	Not urbanization level but socioeconomic, physical, and social neighborhood characteristics are associated with the presence and severity of depressive and anxiety disorders	Generaal et al. (2019)	Netherlands	LUR	IDS, BAI, and FQ	No	Yes	Yes	Multilevel logistic and linear regression

Source: Leontjevaite K, Donnelly A, MacIntyre TE. Air Pollution Effects on Mental Health Relationships: Scoping Review on Historically Used Methodologies to Analyze Adult Populations. Air. 2024; 2(3):258-291. https://doi.org/10.3390/air2030016.

8.1.1. Sources of Air Pollution

The Earth's atmosphere is contaminated in a variety of ways. The majority of air pollution is human-caused and comes from vehicles, factories, airplanes, and aerosol cans. Air pollution also includes the smoke from secondhand cigarettes. Anthropogenic sources are those that are caused by humans. Certain forms of air pollution, like volcanic ash and smoke from wildfires, happen naturally. We refer to these as natural sources. Large cities with concentrated emissions from a variety of sources are most likely to experience air pollution. There are situations when high buildings or mountains block the spread of air pollution. This air pollution frequently takes the form of a cloud that taints the air. We refer to it as smog. The terms smoke and fog are combined to form the word smog. Compared to cities in developed countries, large cities in poor and developing nations typically have higher levels of air pollution. The World Health Organization (WHO) lists Lima, Peru; Beijing, China; Cairo, Egypt; Karachi, Pakistan; and New Delhi, India as some of the most polluted cities in the world. However, air pollution is a problem in many developed countries as well. California's Los Angeles is known as "Smog City."

8.1.2. Indoor Air Pollution

Most people associate air pollution with vehicle exhaust or smoke from large factories. However, indoor air pollution comes in a variety of forms as well. The air inside a house can become contaminated when fuels like coal, wood, and kerosene are burned for heating. Smoke and ash can cling to clothing, walls, and food, making breathing difficult. Homes may also accumulate naturally-occurring radon gas, a substance known to cause cancer. The Earth's surface allows radon to escape. Professional installation of inexpensive systems can lower radon levels. A few building supplies, such as insulation, are also harmful to human health. Furthermore, air movement, or ventilation, in houses and rooms can contribute to the growth of toxic mold. In a house, a single colony of mold may reside in a damp, cool space between walls. Spores from the mold get into the air and travel throughout the house. The spores can make people sick to their stomachs.

8.1.3. Effects on Humans

Exposure to air pollution has a wide range of negative health effects on people. There are two categories of effects: short-term and long-term. Illnesses like bronchitis or pneumonia are examples of short-term effects, which are transient. They also include discomforts like skin, eyes, throat, or nose irritation. Air pollution can also result in nausea, headaches, and dizziness. Odors emanating from factories, waste disposal sites, or sewage networks are also regarded as air pollution. Although not as bad, these smells are still disagreeable. Air pollution can have long-term effects that persist for years or even a lifetime. They might even cause someone to pass away. Lung cancer, heart disease, and respiratory conditions like emphysema are among the long-term health effects of air pollution. Long-term exposure to air pollution can harm a person's kidneys, liver, brain, nerves, and other organs. Some scientists believe birth defects are caused by air pollution. Every year, the effects of indoor or outdoor air pollution claim the lives of nearly 2.5 million people worldwide.

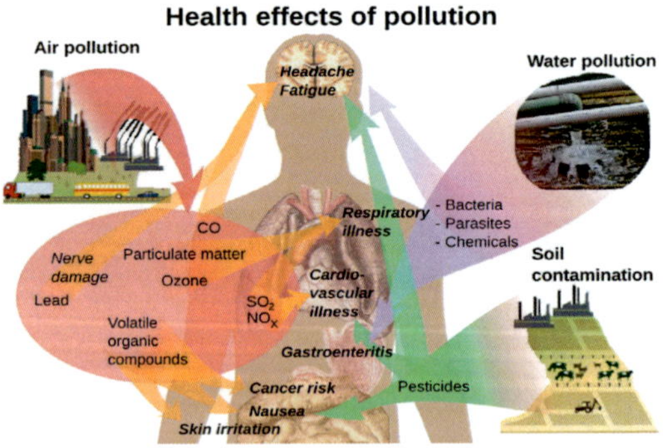

Figure 8.1. *Pollution affects our lungs, making it difficult to breathe.*

Source: https://commons.wikimedia.org/wiki/File:Health_effects_of_pollution.svg.

Individuals respond differently to various forms of air pollution. Due to their weakened immune systems, older adults and young children are frequently more sensitive to pollution. Air pollution exposure can exacerbate conditions like lung disease, heart disease, and asthma. Factors include the duration of exposure as well as the kind and quantity of pollutants. The World Health Organization (link is external) estimates that air pollution claims the lives of 7 million people annually. A severe smog event that struck London in 1952 claimed the lives of over 4000 people in a matter of months. Breathing becomes challenging due to the contraction of lung muscles caused by ground-level ozone. High ozone exposure can lead to lung inflammation, coughing, sore throats, and irreversible lung damage (Figure 8.1).

Long-term exposure is associated with significant illness and disease in several body systems, but short-term exposure usually results in symptoms that go away quickly. Compared to other groups, children, the elderly, and those with chronic illnesses are more susceptible to air pollution. Because of the high levels of pollution in cities, populations living there are also more vulnerable. To find out if you should take precautions like cutting back on or avoiding outdoor activity, check the current air quality in your area.

8.1.4. Effects on the Environment

Entire ecosystems can be negatively impacted by air pollution, just like people, animals, and plants can. Similar to smog, haze is an obvious form of air pollution that masks hues and forms. Pollution from haze can even block out sounds. Particles of air pollution eventually return to Earth. Soil and water surfaces can be directly contaminated by air pollution. This may cause crops to perish or produce less. It can destroy other plants and young trees. When airborne sulfur dioxide and nitrogen oxide particles combine with atmospheric water vapor and oxygen, acid rain can result. The primary sources of these air pollutants are motor vehicles and coal-fired power plants. Acid rain harms crops, erodes the quality of the water in rivers, lakes, and streams, damages plants by altering the composition of the soil, and can hasten the deterioration of structures like buildings and monuments. Animals are susceptible to negative health effects from air pollution, just like humans are. Air pollution has been linked to diseases, reduced reproductive rates, and birth defects.

8.1.4.1. Harming Animals and Plants

Many of the detrimental health effects of air pollution that affect humans can also affect wildlife. Animals most frequently experience damage to their respiratory systems, although neurological issues and skin irritations are also frequent side effects. Crops and plants that are exposed to continuous air pollution grow less. Plants are harmed by ozone pollution because it damages the stomata, which are microscopic pores on the underside of leaves that permit airflow. Certain plant species can defend themselves by making antioxidants or temporarily closing their stomata, but other plant species are more vulnerable to harm. Ozone pollution cost the United States nine billion dollars in lost corn and soybeans between 1980 and 2011. Plants are deprived of the nutrients necessary for growth and survival when the chemical composition of the soil is altered by acid rain, lead poisoning, and exposure to nitrogen oxides. Grasslands, forests, and agriculture are affected by this. In addition to these effects, air pollution also damages the habitats, water supplies, and food sources that are essential for the survival of plants and animals.

8.1.5. Global Warming

Anthropogenic and natural air pollution are the main causes of the environmental phenomenon known as global warming. It alludes to global warming of the atmosphere and seas. The increase of greenhouse gases in the atmosphere is responsible for at least some of this temperature rise. Heat energy is trapped in the Earth's atmosphere by greenhouse gases. Generally speaking, more heat from Earth escapes into space. The greenhouse gas that has most contributed to global warming is carbon dioxide. Burning fossil fuels releases carbon dioxide into the atmosphere (coal, gasoline, and natural gas). Fossil fuels are now widely used by humans to power factories, automobiles, and airplanes as well as to heat homes. Carbon dioxide from doing these things pollutes the air. Methane, nitrous oxide, and fluorinated gases are a few more greenhouse gases released by both natural and artificial sources. An important emission from coal-fired power plants and agricultural practices is methane. Common sources of nitrous oxide emissions include agriculture, industrial factories, and automobiles burning fossil fuels. Industry releases gases that are fluorinated, like hydrofluorocarbons. Many times, gases like chlorofluorocarbons (CFCs) are substituted with fluorinated gases. Because CFCs cause ozone layer thinning, they are prohibited in many places. To combat global warming, numerous nations have implemented policies aimed at reducing or limiting their emissions of greenhouse gases. The Kyoto Protocol is a 1997 agreement between 183 countries to reduce their carbon dioxide emissions. It was first adopted in Kyoto, Japan. That treaty has not been ratified by the U.S. (Baccarelli et al., 2016; Henneman et al., 2023).

8.1.6. Causing Acid Rain

Sulfur and nitrogen oxides are released into the atmosphere when fossil fuels are burned. When sulfur dioxide and nitrogen dioxide combine with atmospheric water droplets to produce sulfuric and nitric acids, acid rain is created. These pollutants can travel thousands of miles on wind before descending to the Earth's surface as acid rain, which damages plant leaves, raises the acidity of water and soil, and is responsible for more than 500 annual deaths. Acid rain affects buildings and other structures as well, resulting in an estimated five billion dollars' worth of property damage annually. Ancient structures and statues carved out of marble and limestone are being destroyed by acid rain, which also dissolves the mortar between bricks and makes stone foundations unstable.

8.2. HEALTH IMPACTS

The modern civilized world is very concerned about air pollution because it poses a serious toxicological risk to both the environment and human health. Although there are many different sources of emissions, the majority of air pollution is caused by motor vehicles and industrial processes. The World Health Organization lists lead, sulfur oxides, nitrogen oxides, carbon monoxide, particle pollution, and ground-level ozone as the six main air pollutants. Humans exposed to air suspended toxicants for short or long periods of time may experience varying toxicological effects, such as respiratory and cardiovascular disorders, neuropsychiatric issues, ocular irritation, skin conditions, and long-term chronic diseases like cancer. Numerous studies have demonstrated a clear correlation between exposure to low-quality air and rising rates of morbidity and mortality, primarily from respiratory and cardiovascular disorders. The incidence and progression of various diseases, including asthma, lung cancer, ventricular hypertrophy, Alzheimer's and Parkinson's disease, psychological complications, autism, retinopathy, fetal growth, and low birth weight, are thought to be primarily influenced by air pollution. Our goal in writing this review article was to go over the toxicology of the main air pollutants, their sources of emission, and how they affect human health.

The effects of air pollution are becoming more common and alarming in a time when urbanization and industrialization are growing. The air we breathe, which was formerly thought to be an endless supply, is now contaminated and poses serious health risks to people. We will examine the complex network of effects that air pollution has on our health in this blog, including its effects on cardiovascular health, respiratory health, and general quality of life. Come along for a journey through the invisible threats that are always present in the air and the urgent need to address this worldwide issue for the sake of everyone's health.

8.2.1. Respiratory Diseases

Animals and plants that are inhabited by humans can suffer from air pollution. Some of the main causes of outdoor air pollution are cigarette smoke, construction activities, vehicle exhaust, factory emissions, and sewer and garbage dump odors. However, poor ventilation, dampness that promotes the growth of mold, rugs, and other things, as well as kitchen fumes, are the main causes of indoor pollution, which is another area of great concern. Particulate matter, which includes tiny dust particles, smoke allergens, and chemicals, irritates the airways and enters the lungs when inhaled. Inflammation may also result from it. Furthermore, carbon monoxide is a dangerous gas that can impede the heart's ability to receive oxygen. The health of humans is impacted by air pollution in both the short and long term. Individuals of all ages are more vulnerable to the health risks associated with pollution. Lung health can be compromised by pollution, increasing the risk of infection. For those who already have a chronic illness, the situation becomes more difficult. Air pollution affects every part of the respiratory system, including the nose, throat, and lungs, having a significant impact on respiratory health.

Here are some key impacts:

1. **Exacerbation of Respiratory Conditions:** Air pollution frequently exacerbates the symptoms of pre-existing respiratory disorders in people, such as asthma or chronic obstructive pulmonary disease (COPD). Reactions to pollutants can start or worsen respiratory attacks.
2. **Increased Respiratory Infections:** Extended exposure to air pollution erodes the protective mechanisms of the respiratory system, increasing susceptibility to respiratory infections such as pneumonia and bronchitis.
3. **Development of Respiratory Diseases:** Lung cancer and other respiratory disorders can arise as a result of prolonged exposure to specific pollutants. Chronic respiratory disorders have been related to particulate matter and hazardous gases.
4. **Reduced Lung Function:** Children who grow up in high-pollution areas may have their lung development stunted. Adults who have been exposed for a long time may experience a reduction in lung function, especially in areas where air quality problems are ongoing.
5. **Aggravation of Allergies:** Airborne pollutants have the potential to exacerbate allergic reactions, increasing a person's sensitivity to common allergens. This may worsen allergy symptoms and cause more discomfort in the respiratory system.

Stricter emission regulations, more green space, and public awareness campaigns to lower exposure and safeguard vulnerable groups are just a few of the coordinated measures needed to address the effects of air pollution on respiratory health. Depending on their size and chemical makeup, air pollutants can either accumulate in or pass through lung tissues, despite the bronchopulmonary tract's many defense mechanisms, including the mucosal cilia and air-blood barrier. Because air pollutants are hydrophilic and hydrophobic, they can either dissolve in bodily fluids or be absorbed by human tissues in the form of vapor, (~10 μm) in size PM_{10} particles can enter the proximal airways and are primarily removed by mucociliary clearance. $PM_{2.5}$, a significant health risk factor, has the ability to penetrate the lungs more deeply. By means of blood circulation, the ultrafine particles have the ability to relocate to distal organs and tissues, including placental tissues during pregnancy and liver tissue for detoxification. Air pollution has been shown to have detrimental effects on the health of people with a variety of respiratory conditions, such as asthma, COPD, lung cancer, respiratory infections, and even heart and stroke disorders when combined. These are a few of the main respiratory illnesses that are directly harmed by air pollution.

8.2.1.1. Respiratory Infections

Respiratory infections become more severe when exposed to air pollution, especially in children. In particular, outdoor pollution in big cities is linked to a high incidence of different acute respiratory infections, which taken together account for almost one-third of pediatric fatalities. In five German cities, exposure to NO_2 and PMs was linked to a rise in laryngo-tracheo-bronchitis cases, the majority of which were caused by influenza viral infections. Similar findings were confirmed by a second study that was carried out in three Finnish cities. However, in Nepal's hilly regions, cooking mothers who do not smoke have high rates of chronic bronchitis due to indoor pollution. This suggests that indoor pollution is more likely to be linked to respiratory infections in developing nations and rural areas. The detrimental effects of air pollution are particularly noticeable in people who already have lung infections or other lung diseases, as they are probably more vulnerable. Children are also more likely to be exposed to air pollution due to their larger lung surfaces and increased outdoor physical activity.

8.2.1.1.1. Asthma and COPD

The majority of asthma emergency room visits are caused by environmental exposure exacerbating the condition. More severe allergic reactions are caused by major indoor and outdoor pollutants such as smoke, dust mite, pollen, O_3, CO, NO_2, SO_2, PM_{10}, and $PM_{2.5}$. In particular, exposure to diesel exhaust particles (DEP) can cause allergic immunoglobulin E (IgE) responses to pollen or ovalbumin, and short-term exposure to nitrogen oxides can increase asthmatic patients' airway responsiveness when they have a house dust mite challenge. In a similar vein, prolonged exposure to indoor air pollution from biomass fuel and secondhand cigarette smoke can cause chronic inflammation that exacerbates COPD, whereas PMs are associated with hospitalization of COPD patients due to acute exacerbations. In general, more epidemiological correlations have been documented connecting air pollution exposure to the onset of asthma and chronic inflammation.

8.2.1.1.2. Lung Cancers

While smoking is widely acknowledged as a primary cause of cancer-related mortality globally, there exist several other risk factors that also play a role in the development of lung cancers, such as indoor and outdoor air pollution. DNA damage, DNA adduct formation, chromosomal aberrations, and methylation modifications are all closely linked to the pathogenesis of lung cancer caused by exposure to air pollution; in fact, some of these processes are being developed as lung cancer biomarkers in relation to exposure to ambient air pollution. The strong correlation between the high incidence rate of lung cancer cases and different air pollutants has been demonstrated, if not with substantial epidemiological evidence.

8.2.2. Cardiovascular Diseases

Conditions affecting the heart or blood vessels are collectively referred to as cardiovascular diseases. The accumulation of plaque in the artery walls, or atherosclerosis, is linked to numerous health issues that accompany heart disease. Blood clots from the accumulation can obstruct blood flow in heart disease patients, increasing the risk of a heart attack or stroke. The majority of deaths from the disease occur in adults over 65, making it the top cause of death in the U.S.. A third of Americans suffer from blood vessel or heart disease. Male sex, advanced age, elevated blood pressure, high total cholesterol, low HDL (high density lipoprotein), and smoking are the conventional risk factors for cardiovascular disease. Furthermore, it has been discovered that exposure to air pollution and diabetes are two additional risk factors that can lead to the development of cardiovascular disease. Numerous scientific studies have demonstrated that air pollution can both cause and worsen cardiovascular disease. The data is especially compelling when it comes to exposure to outdoor particle pollution. A higher risk of cardiovascular events can be associated with fine particulate matter, also known as $PM_{2.5}$, or particulate matter with diameters less than 2.5 μm.

Excessive $PM_{2.5}$ concentrations can cause heart attacks and even death from cardiovascular disease over a period of hours to weeks, according to research conducted by EPA and others. Increased risk of cardiovascular death and shorter life expectancy can result from prolonged exposure. In comparison to many other well-established risk factors that are discussed above, an individual's risk of cardiovascular disease is lower due to particle pollution. Still, it has been demonstrated that both short- and long-term exposure raises hospitalizations for major cardiovascular events in the general population, especially in individuals with established heart disease, such as coronary syndrome, arrhythmia, heart failure, stroke, and sudden cardiac death.

People with chronic heart disease may experience one or more of the following symptoms following exposure to fine particulate matter:

1. Heart palpitations.
2. Unusual fatigue.
3. Lightheadedness.
4. Shortness of breath.
5. Chest tightness or pain in the chest, neck or shoulder.

Air pollution also poses significant risks to cardiovascular function, impacting the heart and circulatory system in several ways:

1. **Increased Risk of Cardiovascular Diseases:** An increased risk of cardiovascular diseases is linked to prolonged exposure to air pollution, particularly nitrogen dioxide (NO_2) and fine particulate matter (PM2.5). Heart attacks, strokes, and high blood pressure are examples of these ailments.
2. **Inflammation and Oxidative Stress:** The body may experience oxidative stress and inflammation as a result of air pollution. Atherosclerosis, a condition where plaque builds up in the arteries, causes them to narrow and harden, is associated with chronic inflammation.
3. **Changes in Heart Rate and Rhythm:** Heart rate variability and rhythm abnormalities have been related to exposure to air pollution, especially fine particles. These changes may heighten the likelihood of unfavorable cardiovascular events and exacerbate cardiac problems.
4. **Aggravation of Existing Cardiovascular Conditions:** Individuals with pre-existing cardiovascular conditions, such as heart failure, may experience worsening symptoms when exposed to air pollution. The added stress on the cardiovascular system can lead to complications.
5. **Impact on Blood Vessels:** Blood vessels that are damaged by air pollutants may find it difficult for them to properly expand and contract. This vascular dysfunction raises the risk of cardiovascular events and is a contributing factor to hypertension.

The significance of addressing air pollution as a public health concern is highlighted by the comprehension of these cardiovascular effects. Globally preserving cardiovascular health depends on putting policies in place to lower emissions and enhance air quality.

8.2.2.1. Cardiovascular Diseases Associated with $PM_{2.5}$

Numerous studies provide evidence of the connection between cardiovascular disease and $PM_{2.5}$. Cardiovascular disease increased significantly from 0% to 1% for every 5–6 μg/m^3 increase in $PM_{2.5}$. They also demonstrated a 69% rise in cardiovascular fatalities following an acute exposure to particulate matter in the air. It's interesting to note that mortality from cardiovascular disease outweighed mortality from respiratory disease in the event of acute exposure to $PM_{2.5}$ (69% cardiovascular versus 28% respiratory). Additionally, a study involving 500,000 adolescents and adults with a 16-year follow-up found that every 10 points5 μg/m^3 of $PM_{2.5}$ increased the risk of ischemic heart disease, heart failure, arrhythmias, and cardiac arrest by 8–18%. Evidence from studies published through March 2009 was included in an updated announcement by the American Heart Association regarding the effects of particulate air pollution on cardiovascular disease. Furthermore, the correlation between cardiovascular disease, morbidity outcomes, and $PM_{2.5}$ has been acknowledged by the Environmental Protection Agency (EPA). Notably, $PM_{2.5}$ levels in developing countries are 10 times higher than in the U.S. Standards for National Ambient Air Quality. A significant correlation between the risk of acute myocardial infarction and NO_2 and $PM_{2.5}$ was found in a Boston emergency room admissions study. A prior extensive investigation from 21 U. S. Cities also reported a link between hospitalization risk and exposure to NO_2 and PM_{10}. According to the Intermountain Heart Collaborative Study, every 10 μg/m^3 increase in $PM_{2.5}$ was associated with a 4.5% increase in coronary artery disease. These findings support $PM_{2.5}$ detrimental effects on public health. Air pollution-related cardiovascular disease is one of the world's most serious health issues.

8.2.2.2. Cardiovascular Diseases Associated with PM_{10}

In the United States, research was done on the link between air pollution and cardiovascular illnesses. Analysis was done on 50 million residents of the 20 biggest cities and metropolitan areas in the United States to determine the rate of cardiovascular disease and the concentration of fine PM_{10}. According to this study, a 10 μg/m^3 increase in PM_{10} the day before death was linked to a 0.68% increase in cardiopulmonary mortality. The Air Pollution and Health European Approach (APHEA-2) study, which examined 43 million people in 29 significant European cities, also produced findings along these lines.

They also demonstrate a strong correlation between PM_{10} and cardiovascular disorders and air pollution. The risk of cardiovascular death increased by 0.76% for every 10 μg/m^3 increase in PM10, which was a higher rate than respiratory disease. Therefore, indicative information from the U.S. and Europe, PM_{10} showed a strong correlation between cardiovascular illnesses and air pollution. Research on PM_{10} also suggested a connection between air pollution and heart disease and stroke, among other cardiovascular conditions. Research indicated that hospitalizations for congestive heart failure (0.8%) and ischemic heart disease (0.7%) increased in tandem with increases in PM_{10} levels of 10 μg/m^3. Furthermore, there has been evidence of a link between air pollution with PM_{10} and stroke. Furthermore, asthma exacerbations and stroke admissions in the United States have been linked to inhalable particles and changes in ambient air conditions, such as warmer versus drier air, grass pollen, upper respiratory infections, and SO_2. In the U.S.A related Danish study also found a link between a higher risk of stroke and particle air pollution (Baccarelli et al., 2016; Henneman et al., 2023).

8.2.3. Impact on Children's Health and Development

For a variety of reasons, children are physically more exposed to air pollution. Compared to adults, young children breathe more quickly and inhale more air per unit of body weight. They breathe air that is closer to the ground and spend a greater amount of time outdoors, which puts them in closer proximity to sources of pollution like dust and car exhaust. Children may be greatly impacted by household air pollution because they spend a large amount of time indoors. Because their brains, lungs, and other organs are still developing, children are physiologically more susceptible to air pollution than adults. Certain air pollutants have the ability to affect developing babies through the placenta. Through adolescence, lung development and function can also be impacted by air pollution. The smallest of children, including neonates, experience morbidity and mortality as a result of air pollution's detrimental effects on numerous bodily systems. Pregnant women who are exposed to polluted air have an increased risk of preterm birth and low birth weight babies. In addition to having a detrimental impact on neurodevelopment and cognitive function, air pollution can cause asthma attacks and pediatric cancer. High air pollution exposure may increase a child's later-life risk of developing chronic illnesses like cardiovascular disease.

Strong evidence suggests that air pollution is closely linked to pneumonia, which causes 22% of all pediatric deaths between the ages of one and five. Strong airway irritants like ground-level ozone can make breathing difficult, especially for kids who already have asthma. Particulate matter from household air pollution is the cause of nearly half of all pediatric deaths from lower respiratory tract infections. Air pollution is a contributing factor in nearly one out of every three lower respiratory infection deaths in children under five in some African and Asian countries. Ozone, which rises in proportion with temperature, is thought to have been a factor in 489,518 deaths worldwide in 2021. Furthermore, air pollution is associated with 34% of preterm births, and the children born into these situations may have health issues, developmental delays, and disabilities for the rest of their lives. Children's middle ear infections, or otitis media, have also been conclusively linked to exposure to ambient air pollution. A higher risk of childhood leukemia is also linked to exposure to air pollution caused by traffic.

8.2.3.1. Children and Air Pollution: Understanding the Problem

One common goal of significant environmental and climate policies, like the zero pollution action plan, is to protect children's health. Despite the fact that the majority of children in the EEA member countries are healthy, there are some environmental risks to their health that should be taken seriously. In certain instances, children, and adolescents may be more exposed than adults to the majority of harmful environmental factors, and they have limited ability to protect themselves or alter the situation. In Europe, air pollution poses the greatest environmental risk to children.

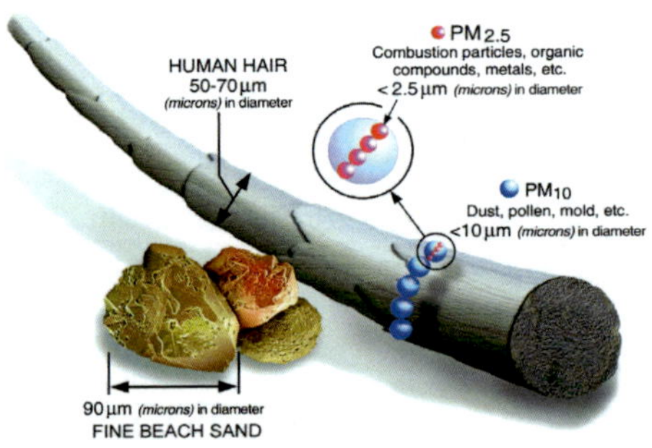

Figure 8.2. *Particulate matter.*

Source: https://en.wikipedia.org/wiki/File:PM_and_a_human_hair.jpg.

This is particularly true. Children and adolescents are particularly vulnerable to air pollution due to a variety of factors. In addition to breathing more quickly than adults do, children also inhale more air per kilogram of body weight. Their lower body height causes them to breathe air that is closer to the ground, where certain pollutants, particularly those from vehicle exhaust, are released and concentrate. Due to their increased physical activity and faster breathing, they also have an increased acquired dose of pollution. In addition, compared to adults, children inhale a higher percentage of air through their mouths. Because the lower respiratory tract is more permeable, pollution enters it deeply as a result of the increased oral breathing. The developing bodies and organs of children, particularly their lungs, further heighten the risk. Moreover, children's immune systems are still developing and are not as strong as adults,' which exacerbates the effects of pollution. The salient features of PM pollution are depicted in Figure 8.2.

8.2.3.2. Ambient Air Pollution and Children: Sources and Exposure

Any substance that alters the natural properties of air, whether indoors or outdoors, is considered to be an air pollutant. It originates from a multitude of natural and, more specifically, artificial sources, such as automobile exhaust and home heating exhausts, as well as factory chimney stacks and numerous other sources. Air pollution can take many different forms and affect both urban and rural areas. It can be caused by ambient air particles, ozone, nitrogen oxides, second-hand smoke (SHS), smoke from burning biomass in homes, mold spores, mites, and allergens, as well as toxic chemicals like formaldehyde. Certain sources and pollutants mentioned earlier are more common or present in higher concentrations indoors.

While the exposure of children and adolescents to air pollutants is similar to that of adults, there are some notable differences. In addition to the air pollution they are exposed to while in their mothers' wombs, children are also commonly exposed to air pollution in child-centric settings such as classrooms where indoor air quality is exacerbated by ambient air pollution, outside school grounds, and during school-related activities such as after-school transportation. Younger children are also exposed to air pollution in lower altitudes, where certain pollutants are more concentrated. Compared to adults, children, and adolescents typically spend more time outside and engage in more physical activity, which may increase their exposure to ambient air pollution. Furthermore, as is the case with other environmental hazards, children from lower-income families are typically more exposed to and impacted by air pollution than children from higher-income families (Baccarelli et al., 2016; Henneman et al., 2023).

8.2.3.3. How Does Air Pollution Affect Children?

From conception until adulthood, children are impacted by ambient air pollution. Air pollution is estimated to be responsible for over 1200 deaths and over 110,000 disability-adjusted life years (DALYs, a measure of the burden of disease caused by a risk factor) among people under the age of 18 each year in EEA member and cooperating countries. Air pollution causes these deaths through a multitude of mechanisms and health consequences, in addition to a substantial burden from non-fatal diseases. Prenatal exposure to ambient air pollution raises the risk of low birth weight, preterm birth, and smaller babies during pregnancy (a condition known as "small for gestational age," or SGA). All of them raise the possibility of developing various health issues in later life. Particulate matter has also been connected to a higher risk of stillbirths and spontaneous abortions, though the exact relationship between the two is less evident. The majority of the prenatal risks associated with air pollution have incompletely understood biological mechanisms, despite strong epidemiological evidence.

After birth, children, and adolescents are more susceptible to a variety of harmful health outcomes due to ambient air pollution. For instance, it raises the risk of upper respiratory infections, pneumonia, acute lower respiratory infections, and otitis media (ear infections) in children. Additionally, brief exposure to air pollution can aggravate allergies, such as children's allergic rhinitis (runny nose), eczema, and conjunctivitis (itchy eyes). Ambient air pollution particularly that of ozone and nitrogen dioxide (NO_2) in the short term and fine particles ($PM_{2.5}$) in the long term, has an impact on children's lung development and function. This effect is evident in children who are healthy as well as those who have asthma, which is made worse by pollution. More than 9% of children in the EU suffer from asthma, which has a significant negative impact on kids, families, and society as a whole. Air pollution exposure over an extended period of time is clearly associated with an increased risk of developing asthma and asthma-like symptoms. This has been noted in laboratory studies where the mechanisms underlying this effect have been investigated and verified, as well as in epidemiological studies. The symptoms of asthma can be anything from moderate to extremely serious, even fatal. Children who are exposed to brief increases in air pollution are more likely to be hospitalized for asthma and to visit the emergency room. There may be a connection between ambient air pollution and other kinds of pediatric health issues. For instance, there is mounting evidence linking air pollution to cognitive decline, altered brain development in children, and the emergence of some forms of autism spectrum disorders. Studies have also found a connection between leukemia in children and air pollution caused by traffic, with benzene being a major contributor.

8.2.4. Long-Term Effects

Long-term exposure to air pollution carries serious health implications that extend beyond immediate respiratory and cardiovascular effects. Here are some enduring consequences:

1. **Chronic Respiratory Conditions:** Long-term exposure to air pollutants, particularly ozone and fine particulate matter, is associated with the onset and aggravation of chronic respiratory diseases like emphysema and chronic bronchitis.
2. **Cardiovascular Diseases:** Chronic cardiovascular diseases, such as atherosclerosis, hypertension, and heart disease, are more likely to develop in people who continue to be exposed to air pollution. Long-term, this increased risk does not go away.
3. **Reduced Lung Function:** People may eventually see a reduction in lung function in places where there are ongoing problems with the quality of the air. This may result in reduced ability to breathe and a higher risk of respiratory infections.
4. **Cancer Risk:** There is a link between a prolonged exposure to specific air pollutants, like formaldehyde and benzene, and a higher risk of lung cancer and other respiratory cancers. Long-term exposure increases this risk.
5. **Neurological Effects:** Recent studies point to a possible connection between neurological disorders brought on by air pollution and neurological conditions such as cognitive decline and a higher risk of neurodegenerative diseases like Parkinson's and Alzheimer's.
6. **Reduced Life Expectancy:** Particularly in areas where pollution levels are persistently high, the cumulative effects of air pollution on different organ systems lead to a decreased life expectancy.

   It will take consistent work to cut emissions, put in place sensible air quality laws, and raise public awareness of the significance of clean air for general health in order to address these long-term health effects.

8.3. ENVIRONMENTAL EFFECTS

Air pollution has terrible consequences for the ecosystem. However, volcanic eruptions and forest fires have caused air pollution long before humans evolved. The only factor that contributed to the current environmental catastrophe was air pollution brought about by human activity.

As soon as coal was first used, greenhouse gases started to build up in the atmosphere. This has a detrimental effect on the earth, which subsequently has an effect on all life. The consequences of air pollution can take on various forms. It can manifest as acid rain or smog. Even passive cigarette smoke is now regarded as air pollution. However, the consequences of this pollution are severe.

1. **Acid Rain:** The pH value of pure water, which includes deionized or distilled water, is 7, while the pH of rainwater ranges from 5 to 5.5, which is slightly acidic. When certain air pollutants like nitrogen dioxide and sulfur dioxide are combined with rainwater, the acidity of the mixture increases significantly. Rainfall with a pH of 4 is usually classified as acid rain because it is 10 times more acidic than rain with a pH of 5. Natural ecosystems are severely harmed by acid rain, especially when it comes to soil. Minerals, including aluminum, make up about 45% of the elements in soil. The aluminum in the soil will dissolve in an acidic environment and obstruct the root's capacity to divide its cells and lengthen. Simultaneously, exposure to acid rain will decrease vital nutrients for plants, like calcium, making it harder for them to absorb water. Additionally, because most aquatic animals operate best at pH values between 6.5 and 8.5, acid rain poses a threat to wildlife.
2. **Eutrophication:** When a body of water becomes more enriched with minerals and nutrients, it can lead to eutrophication, which can eventually cause an excessive amount of algae to grow. Algae take up a lot of oxygen in the water and prevent underwater plants from receiving sunlight, which is essential for photosynthesis. Consequently, the lack of nutrients in the water can cause aquatic plants and animals to perish. While it is true that this phenomenon can happen naturally, there is no denying that human activity has sped up the process. The overabundance of nitrogen dioxide produced during the energy production process is one of the essential nutrients needed for algae to grow. Fertilizer use also adds to eutrophication because nitrogen fertilizer is frequently used in agriculture for a variety of purposes, such as accelerating plant development and preventing yellowing of the leaves, among others. Rainfall will eventually carry excess fertilizer that is not absorbed by plants and ends up in the nearby waterbody.
3. **Climate Change:** The connection between air pollution and climate change is complicated because, despite sharing similar sources and human activity, air pollutants are typically distinguished from greenhouse gases in terms of their effects. While greenhouse gases harm the planet by warming it, endangering the survival of all life on it, air pollutants directly affect human health and the ecosystem. However, there are times when this classification causes the magnitude of the differences between the two groups to be overestimated. Certain air pollutants, like ground-level ozone, have warming properties as well, which means they can retain solar heat in the atmosphere.

It is important to remember, though, that some air pollutants have been found to help protect against climate change. It has been discovered that aerosol, a suspension of airborne particles that is typically created during the burning of fossil fuels, has cooling properties. Aerosol influences the amount of solar radiation that enters and exits the atmosphere as well as its capacity to form clouds, which both contribute to the planet's cooling. Because aerosols are primarily light-colored, solar radiation can be reflected back into space. Aerosols can have very strong cooling properties. For instance, during the 1991 volcanic eruption of Mount Pinatubo in the Philippines, roughly 20 million tons of ash and sulfur dioxide were blasted over a distance of 12 miles into the atmosphere. The planet then underwent two years of consecutive cooling. Its capacity to promote cloud formation is its additional cooling power. Water vapor must condense into water droplets in order to form a cloud. Without airborne particles, which provide surfaces on which water vapor can condense into liquid, condensation is not possible. The formation of aerosols has made more surfaces in the atmosphere available for condensation, which leads to the formation of more clouds. The light color of clouds, which aids in reflecting sunlight into space, is also responsible for their cooling properties. In an effort to slow down climate change, scientists have been experimenting with aerosol manipulation. The problem lies in regulating the amount of particles in the air at a concentration that doesn't endanger human health or the environment.

8.3.1. Acid Rain and Its Impact

Rain or precipitation with high concentrations of hydrogen ions is referred to as "acid rain." Instead of being neutral, the water collected from these types of rains is typically acidic. Sulfur dioxide and nitrogen oxide emissions into the atmosphere are the main causes of acid rain. It is well known that acid rain has a variety of negative effects on human health, infrastructure, animals, plants, and aquatic life.

8.3.1.1. Effects of Acid Rain on Aquatic Animals

The aquatic life that inhabits surface waters may suffer significant harm from acid rain's ability to change the pH of the water. Moreover, acid rain may also be a factor in the elevated concentration of aluminum in surface water. Fish and other aquatic life forms can die from too much aluminum in the water. It is important to remember that fish eggs cannot hatch in an environment where the pH of the surface water falls to 5. The environment will get poisonous enough to kill even adult fish if the pH falls much lower. As a result, acid rain can seriously harm rivers, lakes, and other bodies of water's biodiversity. It should be mentioned that some kinds of lakes with basement rocks made of limestone can withstand acid rain. This is due to the fact that the carbonate-containing basement rocks have buffering qualities and help to keep the water body's pH level stable. But even in these kinds of environments, fish, and insect life can be seriously threatened by acid rain.

8.3.1.2. Effects of Acid Rain on Soil

Acid rains can cause significant disruptions to the chemical and biological processes that take place within soil. For instance, microorganisms that are unable to survive in acidic environments perish because they are unable to adjust to the soil's decreased pH. The enzymes in these microbes are then denatured by the acidic environment. In addition, it is thought that some nutrients—magnesium being the most notable example—are lost from the soil due to the elevated concentration of hydrogen or hydronium ions brought on by acid rain. Certain toxins, aluminum being the most prominent example, can also be mobilized in the soil by increased hydronium ion concentration.

8.3.1.3. What Is Soil Acidification?

The accumulation of hydrogen cations in soil, which is typically accompanied by a decrease in soil pH, is known as soil acidification. It is well known that acid rain directly contributes to soil acidification. The strength of the acid rain and the degree of soil acidification determine how much damage is done to plant life. For instance, mild acidity of the soil does not instantly destroy plant life. But before being carried away, the acidic water that plants growing in such soils absorb dissolves essential nutrients and minerals. As a result, the plant eventually dies from starvation. Highly acidified soils are unsuitable for the development of plants that require a high pH. Such soils cause the plant to die because they wash away the vital nutrients much more quickly. It should be mentioned that plants' waxy leaf cuticles can dry out when acid rain hits their leaves. Acidic water can also interfere with a plant's ability to perform photosynthesis by dissolving the organelles that make up the chloroplasts.

8.3.1.4. Ocean Acidification as a Consequence of Acid Rain

Acid rain can harm oceanic life forms in a number of ways, even though its effects on oceanic ecosystems are not as severe. For example, certain species of coastal life may find it more difficult to form their exoskeletons, which are necessary for their survival, if the pH of the ocean drops (a phenomenon known as ocean acidification). Thus, a decline in biodiversity may be attributed to ocean acidification. It's also critical to remember that the food web or food chain in the ocean may experience significant disruptions as a result of ocean acidification. When acid rain causes biodiversity to disappear, the local food chain is frequently immediately disrupted. For instance, acid rain can dissolve the limestone exoskeletons of corals, which could have a significant impact on the population of those corals in the affected area. Their direct predators' population will be impacted as a result.

8.3.1.5. Effects of Acid Rain on Human Beings

The gases that cause acid rain are dangerous to humans even though acid rain itself does not directly harm people. When sulfur dioxide and nitrogen oxides are present in the atmosphere in significant amounts, it can lead to respiratory issues in humans. In addition, human health issues like asthma and bronchitis can be exacerbated by fine particulate matter in the air. The human-built infrastructure is also impacted by acid rain. For instance, acid rains can harm structures, statues, and historical landmarks composed of limestone, marble, or other rocks containing calcium carbonate. This is due to the fact that the calcium carbonate in the structures reacts with the acids in acid rain to create gypsum, which subsequently flakes off the structure. It should be noted that acid rain can significantly reduce the aesthetic value of monuments and statues.

8.3.2. Impacts of Air Pollution on Ecosystems

Both terrestrial and aquatic ecosystems are negatively impacted by air pollution, which degrades ecosystems and lowers biodiversity. The exposure of vegetation to major air pollutants is examined in this chapter. It uses data from rural monitoring stations as well as estimated total exposure from models. Plants, trees, and agricultural crops are all harmed by ground-level ozone (O_3) because it slows down growth and reduces yields while also damaging biodiversity and ecosystem services. An instance of the impact of ozone on the 2019 wheat harvest is provided. Ecosystems that receive certain air pollutants suffer from degrading deposits on Earth's surface. Excess nitrogen is introduced to land and water bodies as a result of airborne nitrogen oxide molecules (NOX) and ammonia (NH_3) deposition. This leads to eutrophication in water bodies, which is caused by an abundance of nutrients that promote algal blooms and decrease oxygen availability. The exceeding of critical loads for nitrogen deposition in sensitive terrestrial ecosystems, like grasslands, can lead to changes in the structure and function of the ecosystem, the loss of sensitive species, and the increased growth of species that benefit from high nitrogen levels. Acidification—a process that upsets ecosystems and reduces biodiversity—is brought on by the deposition of sulfur dioxide (SO_2), nitrogen oxides (NOX), and nitrous oxide

(NH_3), which alters the chemical composition of soils, lakes, rivers, and marine waters. Surface water and soil acidification has been largely caused by NOX and NH_3, with SO_2 emissions having decreased dramatically in recent decades. Toxic pollutants known as heavy metals are transported over great distances in the atmosphere and settle in ecosystems, whereupon they accumulate in soils and subsequently undergo bioaccumulation and biomagnification in the food chain.

8.3.3. Effects on Agriculture Crop Yields

Exposure of agricultural crops to elevated levels of different air pollutants can cause harm. Damage can take many different forms, from obvious leaf markings to decreased growth and yield to the plant dying too soon. A multitude of other factors influence the development and severity of the injury in addition to the concentration of the specific pollutant. These include the duration of exposure to the pollutant, the type of plant and its developmental stage, and environmental conditions that favor the build-up of the pollutant and the preconditioning of the plant, which determines whether it is vulnerable to harm or not. In general, air pollutants that harm vegetation can be divided into two categories: local and widespread.

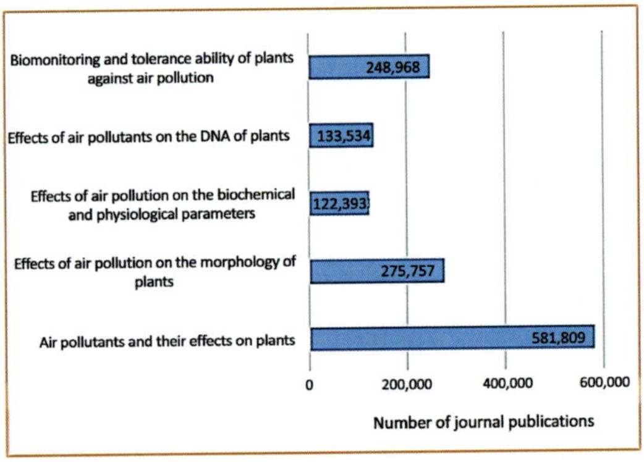

Figure 8.3. *Effects of pollution on plants.*

Source: Mehmood Z, Yang H-H, Awan MUF, Ahmed U, Hasnain A, Luqman M, Muhammad S, Sardar AA, Chan T-Y, Sharjeel A. Effects of Air Pollution on Morphological, Biochemical, DNA, and Tolerance Ability of Roadside Plant Species. Sustainability. *2024; 16(8):3427. https://doi.org/10.3390/su16083427.*

Local pollutants are those that come from a single stationary source and cause damage or contamination to a specific area of vegetation. Particulate matter, ammonia, sulfur dioxide, and fluorides are the most prevalent pollutants in the area. Oxidants make up the majority of widely distributed pollutants. The primary component of oxidants, ozone, is created in the atmosphere through a complicated process involving reactive hydrocarbons and nitrogen oxides, which are byproducts of burning fossil fuels and car exhaust. This process, which only occurs in the presence of sunlight, is known as a photo-chemical reaction. The oxidant build-up in the air can cause vegetation injury over vast rural areas spanning hundreds of square kilometers. Plant damage from air pollution can be seen in many ways. Damage to foliage can manifest itself quickly as necrotic lesions, or dead tissue, or it can take time to manifest as leaf yellowing or chlorosis. There could be a decrease in the growth of different plant parts. Although they can be completely destroyed, plants typically don't die until they have sustained repeated damage (Figure 8.3).

8.3.3.1. Oxidants

The primary pollutant in the oxidant smog complex is ozone. In the Los Angeles region, its effects on plants were initially noted in 1944. Since then, numerous locations across North America, including the southwest and central regions, have reported and documented cases of ozone damage to vegetation. Over the course of the growing season, ozone levels vary significantly, especially in July and August. High ozone periods are linked to regional southerly air flows that travel across the lower Great Lakes after crossing a large number of American cities and industrialized regions. The already high background levels are further enhanced by localized, domestic ozone levels. Because white beans are extremely sensitive, damage levels fluctuate every year, and they are frequently used as a gauge of damage. Cucumber, grape, green bean, lettuce, onion, potato, radish, rutabagas, spinach, sweet corn, tobacco, and tomato are some other species that are delicate. Among the resistant species are apricot, pear, and endive. Ozone symptoms usually show up as flecking, bronzing, or bleaching of the leaf tissues on the upper surface of the affected leaves. Crop loss can happen even in the absence of any visible foliar damage, despite the fact that yield reductions are typically associated with pollutant stress. On the other hand, some crops may suffer from obvious foliar damage without experiencing a negative impact on yield.

Numerous environmental and plant growth factors affect a plant's susceptibility to ozone injury. Susceptibility is increased by high relative humidity, ideal soil nitrogen levels, and water availability. The maturity stage of broad leaves also affects the development of injuries. The newest foliage exhibits resistance. They gradually become more vulnerable at the middle and basal portions as they expand. When the leaves reach full maturity, they once again become resistant.

8.3.3.2. Sulfur Dioxide

Sulfur dioxide is mostly produced by coal-burning businesses, particularly those that provide electricity and space heating. The smelting of ores containing sulfur and the burning of petroleum can also produce sulfur dioxide emissions. The primary way that sulfur dioxide enters leaves is through stomata, which are tiny openings. The damage that results from this can be categorized as acute or chronic. High sulfur dioxide concentrations absorbed in a brief period of time result in acute injury (Figure 8.4). The symptoms manifest as two-sided lesions, or bifacial lesions, which are typically found between the veins and sporadically along the leaf margins. Depending on the season, the plant species impacted, and the weather, the necrotic area's color can change from a light tan or nearly white to an orange-red or brown. The youngest and oldest leaves are usually slightly more resistant to acute sulfur dioxide injury, while recently expanded leaves are typically the most vulnerable.

Figure 8.4. *Acute sulfur dioxide injury to trees.*

Source: By Lovecz – Own work, Public Domain, https://commons.wikimedia.org/w/index.php?curid=938073.

Extended ingestion of sulfur dioxide at levels below the lethal threshold results in chronic harm. In some cases, the symptoms manifest as bronzing on the underside of the leaves, as well as yellowing or chlorosis of the leaf. Plants can differ greatly in how sensitive they are to sulfur dioxide, both among different species and varieties as well as within individual plants. Differences in climate, location, and growth and maturation stages are the causes of these variations. Most agricultural plants are thought to be vulnerable to sulfur dioxide, including tobacco, alfalfa, barley, buckwheat, clover, oats, pumpkin, radish, rhubarb, spinach, squash, and Swiss chard. Cauliflower, celery, cabbage, potato, onion, and celery are examples of resistant crops.

8.3.3.3. Fluoride

When coal is burned, fluorides are released into the atmosphere. They are also released during the manufacturing of steel, aluminum, and ceramics, glass, tile, brick, enamel frit, and fertilizers (Figure 8.6). Leaves absorb fluorides, which are carried to the tips of monocotyledonous leaves (gladiolus) and the margins of broad leaves (grapes). At the location of absorption, there is little damage; however, harmful concentrations accumulate around the edges or tips of the leaves. Beginning as a gray or light-green lesion soaked in water, the injury eventually turns tan to reddish-brown. As the necrotic areas are exposed to more light, they get larger and eventually spread downward on monocotyledonous leaves and inward to the midrib of broad leaves (Figure 8.5).

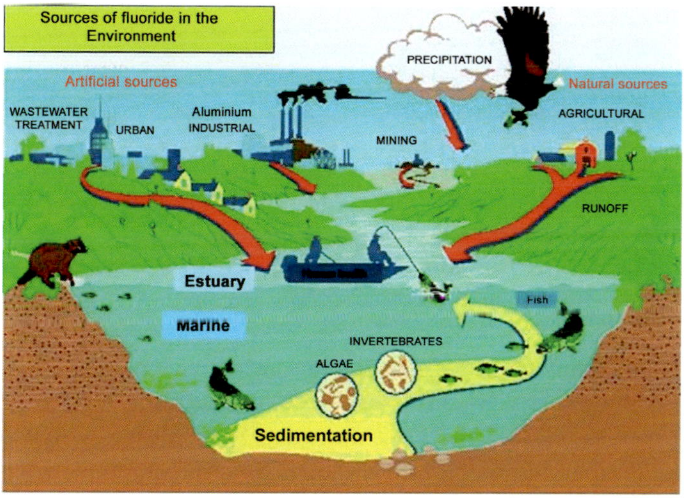

Figure 8.5. *Fluoride sources.*

Source: Shanmugam, T., and Selvaraj, M. (2022). Sources of Human Overexposure to Fluoride, Its Toxicities, and Their Amelioration Using Natural Antioxidants. IntechOpen. doi: 10.5772/intechopen.103714.

According to studies on plant species' fluoride susceptibility, the most sensitive species are tulips, apricot, blueberries, gladiolus, grape, plum, prune, and young barley. Alfalfa, asparagus, bean (snap), cabbage, carrot, cauliflower, celery, cucumber, eggplant, pea, pear, pepper, potato, squash, tobacco, and wheat are examples of plants that are resistant (Figure 8.6).

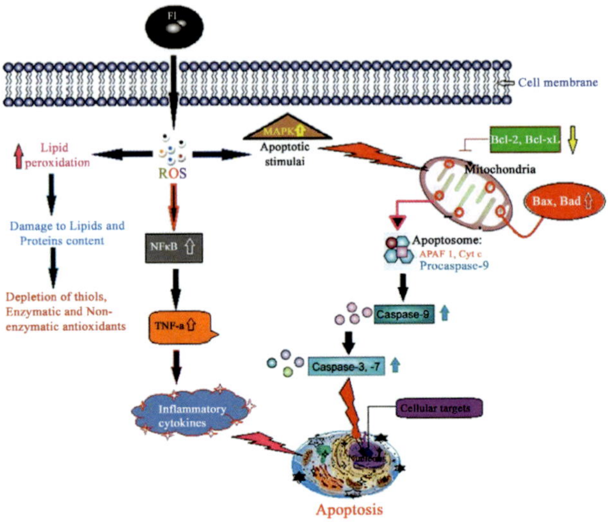

Figure 8.6. *Flouride damage.*

Source: Shanmugam, T., and Selvaraj, M. (2022). Sources of Human Overexposure to Fluoride, Its Toxicities, and Their Amelioration Using Natural Antioxidants. IntechOpen. doi: 10.5772/intechopen.103714.

8.3.3.4. Ammonia

In recent years, accidents involving the storage, transportation, or application of anhydrous and aqua NH_3 fertilizers have resulted in a number of cases of NH_3 injury to vegetation. These events typically result in significant harm to nearby vegetation and the temporary release of large amounts of NH_3 into the atmosphere. On afflicted vegetation, complete system expression typically takes several days to manifest as asymmetrical, bleached, bifacial, necrotic lesions. Grasses frequently exhibit dark discoloration on the upper surface or reddish, interveinal necrotic streaking. Fruit trees that have suffered severe damage may recover by producing new leaves; flowers, fruit, and woody tissues are typically unaffected. Apple, barley, beans, clover, radish, raspberry, and soybean are among the sensitive species. Alfalfa, beet, carrot, corn, cucumber, eggplant, onion, peach, rhubarb, and tomato are resistant species.

8.3.3.5. Particulate Matter

Particulate matter that is deposited on vegetation, such as carbon soot, magnesium-lime dust, and cement dust, can impede the regular processes of photosynthesis and respiration within the leaf. Because of the combination of alkaline toxicity produced by wet weather and a thick crust, cement dust can cause chlorosis and death of leaf tissue. Additionally, the normal action of pesticides and other agricultural chemicals sprayed on foliage may be impacted by the dust coating. Furthermore, the buildup of alkaline dusts in the soil can raise the pH to levels that are harmful to crop growth (Figure 8.7).

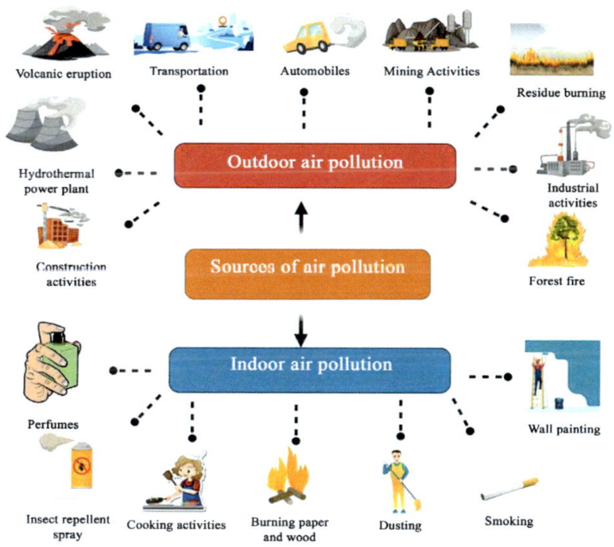

Figure 8.7. *The problems.*

Source: Partap, M., Sharma, D., HN, D., Chandel, A., Thakur, M., VipashaVerma, and Bhargava, B. (2023). Phytoremediation toward Air Pollutants: Latest Status and Current Developments. IntechOpen. doi: 10.5772/intechopen.111559.

8.3.4. Contribution to Climate Change and Global Warming

In addition to having a negative effect on ecosystems and public health worldwide, pollutants also negatively affect the climate. The majority of air pollution control measures provide a win-win solution for the environment and public health. Populations with lower air pollution have better long- and short-term respiratory and cardiovascular health. Reducing home and ambient air pollution can also lower emissions of short-lived climate pollutants like methane and black carbon particles, as well as CO_2. This will help mitigate climate change both in the short and long term. When dangerous concentrations of gases, dust, smoke from fires, or fumes are present in the air, air pollution results. Aerosols are minuscule atmospheric particles suspended in our atmosphere, constituting a subset of air pollution. Aerosols can be liquid or solid. The majority are created by natural phenomena like volcano eruptions, while some are the result of agricultural and industrial human activity. Climate change is measurable as a result of aerosols. In cloud-free air, light-colored aerosol particles can reflect solar energy, while dark-colored particles can absorb it. Global aerosols had the net effect of partially offsetting the rise in global mean surface temperature over the historical period. Aerosols have the ability to alter atmospheric circulation patterns and the amount of energy that clouds reflect.

8.3.4.1. Aerosol Sources, Composition, and Removal Processes

The majority of atmospheric aerosol particles are created naturally throughout the world by processes like biological decay, forest fires, chemical reactions involving atmospheric gases, volcanic injection, and the grinding and erosion of land surfaces that result in dust. However, some particles—those found in industry, agriculture, transportation (including aviation), and construction—have human origins. Depending on where they originate, atmospheric aerosol particles can have a very different composition. They can include organic materials, minerals (like silicon), salts (mostly sulfates), and, most of the time, water. By taking in water vapor and other gases, the particles expand. When water vapor condenses onto these "cloud condensation nuclei" in moist air, clouds are created. After that, these develop into cloud droplets, which ultimately become rain or snow and drop the particles onto land or into the ocean. Aerosol particles in the lower troposphere—the lowest layer of the atmosphere where weather occurs—are typically removed from the atmosphere by settling and precipitating within a few days to weeks of their production, although satellite images show dust plumes from the Sahara and Gobi deserts circling most of the planet. Sulfate particles are created by chemical reactions between gases from volcanoes in the stratosphere, the atmosphere layer above the troposphere, and can linger there for a year or longer, dispersing over a large portion of the planet.

8.3.4.2. Aerosol Particles and Climate

It's common knowledge that human activity causes localized particulate matter to become "air polluted," but less is known about the significant impact that atmospheric particles, both natural and man-made, have on climate. Both outside and inside of clouds, the particles can have significant climatic effects. Tiny aerosol particles interact with the sunbeam in clear air. In essence, "white" particles are those with little to no carbon in them. The air and Earth's surface beneath them are somewhat cooler than they would be otherwise because of their ability to reflect solar radiation. For instance, during the Pinatubo volcanic eruption in 1991, sulfate particles in the stratosphere caused measurable cooling over a large portion of the planet for two years. Conversely, particles with significant concentrations of black carbon (e.g., soot—which is generally created by burning biomass, biofuels, and fossil fuels—warms the area around them by absorbing solar radiation before it hits the ground. Black carbon acts as a shade, causing the ground below to get colder as it reflects incoming sunlight. In the lower troposphere, these microscopic particles also produce cloud droplets. Since ice and water droplets are essentially white, they reflect solar radiation. Condensed water, on the other hand, also emits and traps long-wave radiation, which results in heat production. Hence, depending on the cloud's height and whether its reflecting or trapping properties are stronger, clouds may have either warming or cooling effects on a nearby area. The magnitude of aerosol impacts is one of the major advances in understanding between the fourth and fifth IPCC climate assessments, owing to the numerous unknowns surrounding aerosol particles. The increased confidence about volcanic aerosols and aerosol radiation interactions is especially noteworthy. Currently, NASA operates a number of aerosol monitoring sites around the globe, the data from which is used to improve pollution control as well as our understanding of the climate and air quality.

8.3.4.3. How Is Human-Caused Air Pollution Changing Our Climate?

The effect of particulate matter air pollution caused by humans on our climate is negligible and is probably going to get worse. Human-caused aerosol particles may reduce the overall amount of energy reaching the Earth's surface. Over the past few decades, scientists estimate that particles produced by human activity have resulted in a net loss of solar energy (heat) at the ground by as much as 8% in densely populated areas. This effect, also called "solar dimming," may have obscured part of the global warming caused by heat-trapping gases in the late 20th century. In many places, laws, and contemporary technology have limited human activities that produce aerosol particles that are both reflecting and absorbing. For example, the "pea soup fogs" that plagued London during Sherlock Holmes's time were brought on by particles released when coal was burned incompletely for heat. Thanks to required scrubbers and other cutting-edge combustion techniques, these London "fogs" are now a thing of the past. Changes in the areas where aerosols from human activity have an impact are indicated by trends in both regional and temporal aerosols. While atmospheric soot production from burning fossil fuels for energy production is still very high in many parts of Asia, including large clouds of pollution across much of China, haze clouds seen over urban regions provide dramatic proof of the effects of human-induced particles in the United States.

8.3.4.4. The Primary Cause of Global Warming Is Too Much CO_2

When we burn fossil fuels to make electricity, power our cars, and run our homes, we release an excessive amount of CO_2 and other heat-trapping gases into the atmosphere, which is the main cause of global warming. These gases that trap heat travel the entire planet and stay in the atmosphere for decades or even centuries. As a result, the concentrations of these gases in the atmosphere rise with time as we continue to emit them. Contrarily, atmospheric aerosol particles are mostly concentrated in the vicinity of their sources and do not stay in the atmosphere for very long, meaning that even if we continue to emit them at the current rates, the concentrations of these particles in the atmosphere will not increase significantly over time. The cooling impact of short-lived atmospheric particles will be greatly outweighed by the effects of long-lived global warming emissions.

8.3.4.5. Can Climate Intervention With Aerosols Save U.S. From Global Warming?

A few scientists and engineers have looked into the possibility of using some aerosol particles' reflective properties to temporarily counteract global warming while more advanced non-fossil fuel energy sources are being developed because global warming poses such a serious threat. The use of atmospheric aerosol particles to reflect solar radiation away from Earth is suggested by many climate intervention (also known as "geoengineering") techniques for reducing global warming. Increasing the concentrations of "white" atmospheric aerosol particles above the ocean's surface and/or in the lower stratosphere—above the region where weather occurs—is the theory behind trying to reflect more solar energy away from Earth. We may be able to buy ourselves some time in our attempt to keep warming rates comparatively modest thanks to the nascent field of climate intervention. International governance would face significant challenges from such measures, and there is a risk of large-scale, unintentional, and even dangerous side effects on ecosystems, agriculture, and human health from tinkering with our intricate climate system by sharply raising reflecting aerosol levels. Consequently, before deployment is possible, any deliberate increase in aerosol particle concentration would need to be carefully and thoroughly considered. Primarily, due to the short half-lives of aerosol particles and the long half-lives of global warming gases in the atmosphere, the cumulative effect of heat-trapping gases will outweigh any brief cooling caused by the aerosol particles. Reducing carbon emissions is thought to be more important than intervening in the climate (Baccarelli et al., 2016; Henneman et al., 2023).

8.4. TECHNOLOGICAL AND SCIENTIFIC SOLUTIONS

Redefining air quality through technological innovation is one of the most promising areas of research and development. We have created strategies to stop air pollution both indoors and outdoors, ranging from catalytic converters to low-emitting consumer goods. We're going to talk about six innovations that are actively preventing air pollutants from harming our atmosphere and our health in honor of these technological advancements. Both market-based and regulatory approaches to combating air pollution have advantages and disadvantages that we have previously covered. Since new developments typically don't have the same kind of cost vs. We will talk about each of these innovations separately rather than using technology as a general solution, as these other methods have a beneficial relationship.

1. **Multi-Pollutant Monitoring Devices:** We always enjoy talking about a variety of topics, and this time it's multi-pollutant air quality monitoring. This technological advancement is an essential component of the air pollution control system even though it does not directly lower emissions or eliminate air pollution from our skies. Factories, power plants, and other sources of pollution release multiple pollutants; conversely, certain pollution control technologies manage multiple pollutants. Regulatory agencies can guarantee that emission limits for a wide range of pollutants are being met thanks to multi-pollutant monitoring, which also lowers the expense and time commitment involved in ambient pollution monitoring.
2. **Catalytic Converters:** One of the biggest causes of air pollution in the world is vehicle exhaust. Gasoline and diesel-powered cars were outfitted with catalytic converters, a device that catalyzes a redox reaction that converts harmful air pollutants into less harmful ones, in response to ever-tougher environmental regulations starting in the 1970s. Today's gasoline-powered cars typically have an oxidation-reduction catalytic converter, also known as a three-way converter. These devices function by subjecting exhaust to a catalyst, which is a material that reduces the activation energy required for a chemical reaction. This allows for the conversion of nitrogen oxide to nitrogen and carbon dioxide and carbon monoxide to carbon dioxide. Even though carbon dioxide is a greenhouse gas (GHG), atmospheric nitrogen and carbon dioxide are far less dangerous than nitrogen dioxide and carbon monoxide. For this reason, catalytic converters are an excellent way to lessen the negative effects of vehicle exhaust

on the environment and human health. There are certain drawbacks to catalytic converters. These devices have the potential to limit the exhaust flow, which could lower the vehicle's performance and fuel efficiency. Additionally, there is a warm-up phase for catalytic converters, during which the car releases an unlimited quantity of air pollutants. The process of making these converters is not without its drawbacks, as many of them require rare metals like palladium or platinum, which pollute the refinery during the refining process.

3. **Scrubbers:** Industrial exhaust is cleaned of air pollutants such as sulfur dioxide, chlorine, hydrogen sulfide, and hydrogen chloride using scrubbers, a type of pollution control equipment. We will talk about two primary types of scrubbers: wet scrubbers and dry scrubbers. Wet scrubbers can have different energy levels and work by absorbing particles or gases from an air stream using a liquid, usually water. A spray tower is a typical low-energy wet scrubber that distributes liquid by passing exhaust through an open vessel equipped with sprayers. The liquid in the exhaust absorbs the target gas or picks up floating particles as it passes through the device. Similar in operation, dry scrubbers neutralize gases before they reach the atmosphere by spraying dry reagents into the flue stream rather than a liquid. Scrubbers are an extremely useful instrument for pollution control because they shield nearby communities from the harmful effects of air pollutants, such as those from power plants and water treatment facilities. Furthermore, because these devices don't interfere with production, industry, and the economy can continue without contributing to an increase in air pollution.

4. **Low-Emitting Stoves and Heaters:** Air pollution is not limited to outdoor environments; every year, indoor air pollution from inefficient cooking practices causes 3 point 8 million premature deaths. The majority of these fatalities are caused by burning kerosene or solid fuels in small areas, which raises particulate matter levels to dangerous levels. Cooking is now far more efficient thanks to advancements in stove design and alternative fuels, which have also reduced particulate matter production in homes. These days, biomass stoves can be purchased with fans, insulated combustion chambers, or secondary combustions to burn off leftover fuel and ventilate any leftover pollutants. Although we eventually want to get rid of biomass stoves, they are still useful as stepping stones in low- and middle-income countries without reliable access to natural gas or electricity. Heating also contributes to indoor air pollution, along with cooking. The need for solid fuel is reduced by solar water

heaters and passive solar heating and cooling systems. Additionally, better ventilation designs, such as chimneys and stove hoods, can aid in directing indoor air pollutants outside.

5. **VOC-Free Consumer Products and Building Materials:** Finally, low-emitting building materials and consumer goods are another innovation that helps lower indoor air pollution. Volatile organic compounds (VOCs) are released by many of the products we use on a daily basis, such as paints, cleansers, adhesives, sealants, flooring, and furniture. These VOCs can accumulate in areas with inadequate ventilation. A growing number of consumer goods and building materials are being created and certified as low-emitting or even VOC-free as a result of worries about the effects of these emissions. One can use guidelines provided by ASTM and the California Department of Public Health to ascertain whether a product is low-emitting. Many building standards mandate or give credit for using low-emitting building materials and incorporating low-emitting consumer products into an indoor air quality strategy because these innovations are essential for reducing indoor air pollution.

8.4.1. Advances in Air Quality Monitoring and Data Analysis

Air pollution is a serious problem that needs to be effectively assessed and resolved. Monitoring the quality of the air in real-time has become essential for comprehending and resolving issues related to air pollution.

This article examines the importance of real-time monitoring of air quality and lists its many advantages.

Continuous measurement and analysis of a range of air pollutants, such as particulate matter, VOCs, carbon monoxide, nitrogen dioxide, and ozone, are part of real-time air quality monitoring. We can quickly gather and analyze data by using cutting-edge sensor technologies and data analytics, giving us vital information about the quality of the air we breathe.

8.4.1.1. The Benefits of Real-Time Air Quality Monitoring

1. **Prompt Episode Detection of Air Pollution.** Instantaneous detection and remediation of air pollution incidents are made possible by real-time air quality surveillance. It is possible to quickly identify prolonged periods of poor air quality or abrupt spikes in pollution by regularly monitoring pollution levels. Maintaining public health, minimizing the negative effects of pollution, and quickly putting preventative measures into place all depend on this knowledge.

2. **Knowledge and Empowerment of the Public**. Real-time, accurate, and easily available information on air quality encourages people to take a greater interest in the air they breathe. People can reduce exposure to pollutants and safeguard their health by making educated decisions based on clear and comprehensive data. A greater understanding of air quality issues by the general public can also encourage group initiatives to reduce pollution sources and advance cleaner environments.
3. **Identification of Pollution Sources and Hotspots**. Monitoring in real time makes it easier to locate the sources and hotspots of pollution. Through the examination of data trends and patterns, we are able to identify areas that have consistently low air quality or particular sources of pollution. Urban planning, enforcing emission regulations, and putting targeted interventions into action to reduce pollution in severely affected areas all benefit greatly from this information.
4. **Assessment of Health Effects**. A comprehensive assessment of the health effects of air pollution is made possible by accurate real-time data on air quality. Researchers and policymakers can better understand the direct and indirect effects of air pollution on human health by looking at the relationship between pollution levels and health outcomes. This information serves as the cornerstone for creating public health policies and strategies that effectively safeguard vulnerable groups.
5. **Evaluation of Pollution Prevention Measures**. Monitoring of air quality in real time offers useful data for determining how well pollution control strategies are working. Policymakers can assess their strategies and make well-informed decisions by comparing pollution levels before and after the implementation of interventions, such as regulatory actions and emission reduction technologies. This data-driven review strengthens overall air quality management and pollution control strategies. A vital part of our joint endeavors to lower air pollution and protect public health is real-time monitoring of air quality. Its benefits—which include timely pollution detection, well-informed decision-making, and focused interventions—help to enhance air quality and promote a cleaner, healthier future. We can improve air quality monitoring and make beneficial changes to our surroundings by embracing cutting-edge sensor technologies and data analytics. Let's actively push for the installation of real-time community air quality monitoring systems.

8.4.2. Innovations in Reducing Emissions from Vehicles and Industries

Air pollution and heat-trapping emissions are primarily caused by transportation. Fossil fuel-powered vehicles such as cars, trucks, and buses are a significant source of air pollution. Actually, over half of the nitrogen oxides in our air come from transportation, which also contributes significantly to the U.S.'s heat-trapping emissions. Research has connected car exhaust pollutants to harmful effects on almost all of the body's organ systems. Even worse, not everyone is equally exposed to pollution. Lower-income households, Black people, and Latinos bear a disproportionate share of the damage. Even though there are serious health and environmental risks associated with air pollution, we can drastically cut emissions from our cars and trucks and revolutionize transportation by utilizing clean vehicle and fuel technologies.

8.4.2.1. The Ingredients of Air Pollution

Throughout their entire life cycle, cars, trucks, and buses emit air pollution, including emissions from fuel production and vehicle operation. The production, disposal, and recycling of vehicles, as well as the distribution and refinement of fuels, are the sources of additional emissions.

The following are the major pollutants from motor vehicles:

1. **Particulate matter (PM)**. The soot found in car exhaust is one kind of PM. Because they can enter the lungs deeply, fine particles—less than tenth the diameter of a human hair—pose a serious threat to human health. PM can originate from hydrocarbons, nitrogen oxides, and sulfur dioxides and can act as either a primary or secondary pollutant. One of the main causes of PM pollution is diesel emissions.
2. **Volatile Organic Compounds (VOCs).** When these pollutants are exposed to sunlight, they combine with nitrogen oxides to create ground level ozone, which is a major component of smog. Although this gas is advantageous in the upper atmosphere, it irritates the respiratory system below ground level, leading to reduced lung capacity, coughing, and choking. There is a connection between various forms of cancer and VOCs released by automobiles, trucks, and buses. These pollutants include acetaldehyde, 1,3-butadiene, and benzene.
3. **Nitrogen oxides (NOx).** These contaminants produce secondary PM and ground level ozone. In addition to being a dangerous primary pollutant, NOx can irritate the lungs and impair the body's ability to fight off respiratory illnesses like influenza and pneumonia.

4. **Carbon monoxide (CO).** The combustion of fossil fuels, including gasoline, produces this colorless, odorless, and toxic gas, which is mostly released by cars and trucks. CO depletes the heart, brain, and other critical organs of oxygen when it is inhaled.
5. **Sulfur Dioxide (SO_2).** This pollution is produced when motor vehicles and power plants burn fuels that contain sulfur, such as coal and diesel. Among other air pollutants, sulfur dioxide can react in the atmosphere to form fine particles and is most harmful to young children and asthmatics.
6. **Greenhouse Gases.** Additionally, motor vehicles release pollutants into the atmosphere, primarily carbon dioxide, which fuels global warming. Actually, more than one-fifth of the GHG pollution in the United States comes from the tailpipe emissions of cars, trucks, and buses; the transportation sector, which includes trains, ships, and airplanes, is responsible for about 30% of the nation's total emissions of gases that trap heat.

8.4.2.2. Emissions from Heavy-Duty Vehicles

Every day, thousands of people are transported throughout cities by trucks and buses, which also pick up and deliver packages, carry goods from manufacturers to stores, and clean up trash. However, these vehicles also have a significant impact on public health and global warming. Although making up only 10% of all vehicles on the road, heavy-duty vehicles account for over 25% of emissions related to global warming, 45% of NOx emissions, nearly 60% of direct PM2.5 emissions from on-road vehicles that originate from the transportation sector, and significant amounts of air pollution. Freight-adjacent communities—those close to ports and interstates—are most at risk from this harmful air pollution. Every year, the amount of freight moved in the United States increases, making it more difficult to reduce emissions from this industry. In order to improve air quality and lower heat-trapping emissions in communities across the nation, heavy-duty vehicle pollution must be addressed. Currently, almost all heavy-duty applications can be served by zero-emission vehicles, and while the market for these trucks is growing, it is not doing so quickly enough to have a significant impact on fair access to clean air.

8.4.2.3. Clean Air Benefits of Electric Trucks and Buses

Electric trucks and buses have the potential to completely eliminate tailpipe pollution, much like electric cars are assisting in the reduction of pollution from passenger vehicles. They can also accelerate the development of clean, renewable electricity and hydrogen. Electric trucks and buses not only produce no harmful tailpipe pollution, but they also emit notably less GHG emissions than fossil fuel-powered vehicles. In every region of the nation, a battery-electric bus is the least carbon-intensive option available on the current electrical grid. Cities all over the country are already using battery- and fuel cell-electric trucks and buses, particularly transit buses, and these fleets are growing as new models are made available. The two biggest bus fleets in the nation, those of New York and Los Angeles, are among the many cities pledging to have only 100% zero-emission transit buses in their fleets.

8.4.3. Innovative Technologies for Carbon Emission Reduction

Cutting edge technologies for reducing carbon emissions are becoming indispensable instruments in the unrelenting search for sustainable solutions. These developments, which range from using renewable energy sources to developing novel strategies for carbon capture, are bringing in a new era of environmental responsibility.

1. **Renewable Energy Breakthroughs:-** The fight against carbon emissions is being led by breakthroughs in renewable energy. Renewable energy sources like solar, wind, and hydroelectric power are changing the face of energy production. Future clean energy sources are made possible by these technologies' increased economic viability and decreased costs, which also make them more efficient.
2. **Carbon Capture and Storage (CCS):-** The use of carbon capture and storage technologies is essential to keeping carbon dioxide, one of the main greenhouse gases, out of the atmosphere. Emissions from factories and power plants are captured by CCS and sent to storage locations, which are frequently underground. The goal of ongoing research and development is to improve the cost-effectiveness and efficiency of CCS, making it a vital component of emission-reduction plans.
3. **Direct Air Capture (DAC):-** An innovative method of reducing emissions from the atmosphere directly is Direct Air Capture. These systems offer a special remedy for sectors with few choices for reducing emissions since they capture carbon dioxide using sophisticated chemical processes. With further technological advancements, DAC has great potential to reduce emissions from different sources.
4. **Green Hydrogen Production:-** Green hydrogen is becoming more and more popular as a carbon-neutral energy source. It is created through electrolysis using renewable energy. It has the power to completely transform sectors like long-distance transportation and

heavy manufacturing that now struggle with electrification. Green hydrogen is becoming a more and more practical choice for a variety of applications as its price drops.

5. **Artificial Intelligence for Energy Efficiency:-** Using artificial intelligence (AI) to its full potential is changing the energy efficiency landscape. Energy consumption is optimized by smart grids and AI-powered systems, which aid in balancing supply and demand. These technologies improve the sustainability and resilience of energy infrastructures in addition to reducing waste.

6. **Sustainable Transportation Solutions:-** Transportation innovations are important in the effort to reduce carbon emissions. Renewable energy-powered electric vehicles (EVs) are becoming more popular as a greener option to cars with internal combustion engines. Furthermore, developments in hydrogen fuel cells and biofuels support a more environmentally friendly and sustainable transportation industry.

These cutting-edge technologies represent a sea change in the ever-changing field of reducing carbon emissions and toward a sustainable future. Our perspective on energy is changing as a result of the advancements in renewable energy production, which are both economically and environmentally sound (Yu et al., 2018).

REFERENCES

1. Baccarelli, A. A., Hales, N., Burnett, R. T., Jerrett, M., Mix, C., Dockery, D. W., et al. (2016). Particulate air pollution, exceptional aging, and rates of centenarians: A nationwide analysis of the United States, 1980–2010. *Environmental Health Perspectives, 124*(11), 1744–1750.
2. Henneman, L., Choirat, C., Dedoussi, I., Dominici, F., Roberts, J., & Zigler, C. (2023). Mortality risk from United States coal electricity generation. *Science, 382*(6673), 941–946. https://doi.org/10.1126/science.adf4915.
3. Lucking, A. J., Lundback, M., Mills, N. L., Faratian, D., Barath, S. L., Pourazar, J., et al. (2008). Diesel exhaust inhalation increases thrombus formation in man. *European Heart Journal, 29*(24), 3043–3051. https://doi.org/10.1093/eurheartj/ehn464.
4. Mohai, P., Lantz, P., Morenoff, J., House, J., & Mero, R. (2009). Racial and socioeconomic disparities in residential proximity. *American Journal of Public Health, 99*(3), S649–S656. https://doi.org/10.2105/ajph.2007.131383.
5. Singh, R., Kumar, S., Karmakar, S., Siddiqui, A. J., Mathur, A., Adnan, M., et al. (2021). Causes, consequences, and control of persistent organic pollutants. In N. Kumar & V. Shukla (Eds.), *Persistent Organic Pollutants in the Environment: Origin and Role* (pp. 31–54). CRC Press. https://doi.org/10.1201/9781003053170.
6. Smith, A. (2021). Pollution on other planets could help us find aliens, NASA says. *The Independent*.
7. Yu, T., Wang, W., Ciren, P., & Sun, R. (2018). An assessment of air-quality monitoring station locations based on satellite observations. *International Journal of Remote Sensing, 39*(20), 6463–6478. https://doi.org/10.1080/01431161.2018.1460505.

CHAPTER 9

RAIN WATER QUALITY AND ACIDITY

Contents

INTRODUCTION

An important environmental indicator, rainwater quality affects ecosystems, human health, and is a reflection of the condition of our atmosphere. Acidity is a crucial component of rainwater quality, and it has a big impact on both natural systems and man-made buildings. Rainwater contains dissolved carbon dioxide, which reacts to form carbonic acid, making it naturally somewhat acidic. Nonetheless, this naturally occurring acidity is usually not very strong, with pH values between 5.0 and 5.6.

However, the acidity of rainwater can be increased by the introduction of pollutants into the atmosphere, resulting in what is commonly referred to as acid rain. Acid rain is created when water vapor in the atmosphere reacts with sulfur dioxide (SO_2) and nitrogen oxides (NOx) released from burning fossil fuels to produce sulfuric acid (H_2SO_4) and nitric acid (HNO_3). Rainwater becomes more acidic as a result of these acids being precipitated down to the earth, which lowers the pH of the water considerably. Rainwater is generally regarded as acidic if its pH is less than 5.

There are several environmental and health risks associated with acidic rainwater. Because acid rain can lower the pH of water bodies and cause toxic metals like aluminum to leach into waterways, ecosystems are particularly vulnerable. Fish and other aquatic organisms' life cycles may be disturbed by these changes, which could result in a decline in biodiversity. Because acidic rain can harm leaves, decrease the availability of nutrients in the soil, and hinder plant growth, terrestrial plants are also impacted. Acid rain's effects are not unaffected by man-made structures. Acidic weather can hasten the decay of structures, particularly those composed of concrete, marble, and limestone, as the acid reacts with the calcium carbonate in the materials to cause erosion and structural damage. Repairs and maintenance may become very expensive as a result. Resolving the acidity of rainwater necessitates addressing the underlying causes of air pollution. It is essential to reduce sulfur dioxide and nitrogen oxide emissions by using cleaner energy sources, better industrial processes, and more stringent regulations. Additionally, to lessen the negative effects of acid rain and encourage more sustainable practices, rainwater quality monitoring and awareness-raising regarding its effects on infrastructure and the environment are important (Satake, 2012; Seinfeld and Pandis, 1998).

9.1. OVERVIEW OF RAIN WATER QUALITY AND ACIDITY

When evaluating the sustainability and health of the environment, rainwater quality is a crucial consideration. Since rainwater condenses from atmospheric moisture, it is naturally quite pure. Nonetheless, airborne contaminants can have a big impact on its quality. Particulate matter, gases, and chemicals are just a few of the contaminants that can enter the environment through human activity, vehicle exhaust, and industrial emissions. Rain can gather these contaminants as it falls, changing the water's quality. Rainwater, for example, may contain contaminants such as heavy metals, nitrogen oxides, and sulfur dioxide, which may compromise its safety for use or direct consumption. Pollution levels too high can cause problems like acid rain, which harms water bodies, ecosystems, and infrastructure. On the other hand, appropriately collected and treated rainwater can be a useful resource for irrigation, drinking, and other purposes. In order to ensure that rainwater remains a safe and sustainable resource while also minimizing potential negative effects on the environment and human health, monitoring, and managing the quality of the water is crucial.

Rainwater acidity is closely linked to the chemistry of air pollutants, specifically NOx and SO_2. These substances, which are frequently emitted by burning fossil fuels, cars, and industrial processes, are essential in the development of acid rain. One major factor contributing to the acidity of rainwater is sulfur dioxide. Sulfurous acid (H_2SO_4) is created when SO_2 is released into the atmosphere and combines with oxygen and water vapor. Rainwater's pH can be lowered by sulfuric acid (H_2SO4), a stronger acid that is created when this acid undergoes additional oxidation. In a similar manner, water vapor and nitrogen oxides—which are the result of high-temperature combustion processes—react to form HNO3. Another strong acid that adds to the general acidity of rain is nitric acid. There are multiple steps involved in the reactions that cause acid rain. Firstly, winds carry the emissions of SO_2 and NOx into the atmosphere. These gases go through chemical reactions when they come into contact with moisture and oxygen.

Sulfurous acid is created when sulfur dioxide and water vapor first react:

$$SO_2+H_2O \rightarrow H_2SO_3$$

Sulfurous acid can further oxidize in the atmosphere to form sulfuric acid:

$$2H_2SO_3+O_2\rightarrow 2H_2SO_4$$

For nitrogen oxides, the reaction with water vapor forms nitric acid:

$$2NO_2+H_2O\rightarrow HNO_3+HNO_2$$

Nitrous acid (HNO_2) is a minor product that can further react to form more nitric acid:

$$2\ HNO_2+O_2\rightarrow 2HNO_3$$

The precipitation that carries these acids to the ground causes rainwater to have a lower pH, a phenomenon known as acid rain. Rainfall with higher acidity can be detrimental to vegetation, aquatic systems, and man-made structures. It can also accelerate the erosion of building materials and degrade soil. Developing strategies to reduce emissions and mitigate the effects of acid rain on the environment and infrastructure requires an understanding of the chemistry of these air pollutants and their role in the acidity of rainwater.

9.1.1. What Is Acid Rain?

In its most basic form, acid rain is simply the precipitation of acid, as the name implies. Acid rain is produced when air pollutants such as sulfur and nitrogen oxides combine with precipitation and descend with it. Rain or any other type of precipitation with an unusually high concentration of hydrogen ions (low pH) is referred to as acid rain. The neutral pH of most water, including drinking water, is between 6.5 and 8.5; acid rain, on the other hand, typically has a pH between 4 and 5. The pH of acid rain decreases with increasing acidity. Infrastructure, aquatic life, and plants can all suffer from acid rain's negative effects. Emissions of nitrogen oxide and sulfur dioxide combine with the water molecules in the atmosphere to form acids, which is what causes acid rain. It has been demonstrated that acid rain negatively affects freshwater ecosystems, soils, microbes, insects, and aquatic life forms. Constant acid rain weakens the resilience of tree bark in ecosystems, making flora more vulnerable to environmental stresses like pest infestation, heat waves, and drought. Additionally, acid rain can harm the composition of soil by depleting it of nutrients like calcium and magnesium, which are essential for plant growth and the upkeep of healthy soil. Acid rain affects human health in addition to causing paint to peel, steel structures like bridges to corrode, stone buildings and statues to weather, and other aspects of human infrastructure.

Since the 1970s, several governments—including those in North America and Europe—have worked to reduce air pollution by enacting laws that limit the amount of nitrogen oxide and sulfur dioxide that are released into the atmosphere. Because of the extensive research on acid rain that began in the 1960s and the publicized information about its detrimental effects, these efforts have yielded positive results. Although sulfur and nitrogen compounds that cause acid rain are mostly caused by human activity, sulfur dioxide and nitrogen oxides can also be produced naturally by volcanic eruptions and lightning strikes.

9.1.2. Definition of Acid Rain

Because of air emissions, particularly the excessive amounts of nitrogen and sulfur released by manufacturing processes and automobiles, acid rain is composed of extremely acidic water droplets. Although this idea encompasses a variety of acidic precipitation types, it is frequently referred to as "acid rain.." Wet and dry methods are both used to cause acidic deposition. Acids that are extracted from the atmosphere and deposited on the earth's surface are referred to as wet deposition. Polluting gases and particles adhere to the ground through smoke and dust when there is no precipitation.

"Acid rain" is a popular term referring to the deposition of a mixture of wet (rain, snow, sleet, fog, cloud water, and dew) and dry (acidifying particles and gases) acidic components. Distilled water, once carbon dioxide is removed, has a neutral pH of 7. Liquids with a pH less than 7 are acidic, and those with a pH greater than 7 are alkaline. "Clean" or unpolluted rain has an acidic pH, but usually no lower than 5.7, because carbon dioxide and water in the air react together to form carbonic acid, a weak acid according to the following reaction:

$H_2O\ (l) + CO_2\ (g) \rightleftharpoons H_2CO_3\ (aq)$

Carbonic acid then can ionize in water forming low concentrations of carbonate and hydronium ions:

$H_2O\ (l) + H_2CO_3\ (aq) \rightleftharpoons HCO-3\ (aq) + H_3O^+\ (aq)$

Other chemicals may also be present in unpolluted rain, affecting its pH (acidity level). Nitric acid, which is frequently produced by electric discharges in the atmosphere, such as lightning, is one example. In addition to H_2CO_3, other acids would also be included in acid deposition, which is covered in more detail later in the article. In industrialized areas, pH readings in rain and fog water have occasionally been reported to be well below 2.4. The primary causes of the SO_2 and NO_x pollution that results in acid rain are the processes of industrial manufacturing, oil refining, burning fossil fuels for electricity generation, and internal combustion engine power.

9.1.3. Causes of Acid Rain

Particles of nitrogen and sulfur that combine with the rain's liquid components are what cause acid rain. There are two sources of sulfur and nitrogen particles that mix with water: either man made or natural.

These include the emissions released by industries or by natural events, such as volcanic eruptions that release sulfur oxide and lightning strikes that release nitrogen oxides into the atmosphere. Scottish chemist Robert Angus Smith is credited by the Royal Society of Chemistry, which calls him the "father of acid rain," with coining the term in 1852. Smith came up with the term while researching the chemistry of rainwater in the vicinity of industrial towns in Scotland and England. Even though it's not clean, we regularly receive clean rain. When carbon dioxide and water combine, weak carbonic acid is created that is generally not very dangerous on its own.

The response taking place is:.

$H_2O\ (l) + CO_2\ (g) \rightleftharpoons H_2CO_3\ (aq)$

The pH value of regular rainwater is around 5.7, giving it an acidic nature

The wind carries the dust particles and releases the oxides of sulfur and nitrogen. Once they descend as precipitation, they settle on the surface of the earth. In essence, acid rain is a result of human activity that releases sulfur and nitrogen oxides into the atmosphere. Fossil fuel combustion and unethical waste emission disposal methods are two examples. After being subjected to oxidation, sulfur dioxide and nitrogen dioxide combine with water to form sulfuric acid and nitric acid, respectively. The acid formation reaction will be made clear by the following reaction:

$$2SO_2\,(g) + O_2\,(g) + 2H_2O\,(l) \rightarrow 2H_2SO_4\,(aq)$$

$$4NO_2\,(g) + O_2\,(g) + 2H_2O\,(l) \rightarrow 4HNO_3\,(aq)$$

9.1.4. Effects of Acid Rain

The natural and artificial environments are severely harmed by acid rain. It lowers the pH of water bodies in aquatic systems, causing acidification that is harmful to fish and other aquatic life. Particularly vulnerable species include salmon and trout, whose reproductive cycles may be disturbed by acidification, ultimately resulting in a reduction in biodiversity. On land, acid rain weakens trees and leaches vital nutrients from the soil, leaving them more vulnerable to pests and disease.

It also modifies soil chemistry and decreases nutrient availability, which have an impact on crop yields. Man-made structures also suffer from acid rain, which hastens the disintegration of infrastructure and buildings, particularly those composed of concrete, marble, and limestone. These materials' calcium carbonate reacts with the acidic environment, causing erosion and higher maintenance expenses. Furthermore, there are health risks when drinking water sources are contaminated by acid rain.

All things considered, the effects of acid rain highlight the necessity of efficient pollution control strategies to safeguard human resources as well as natural ecosystems:

1. Acid rain is very harmful to agriculture, plants, and animals. It washes away all nutrients which are required for the growth and survival of plants. Acid rain affects agriculture by the way it alters the composition of the soil.
2. It causes respiratory issues in animals and humans.
3. When acid rain falls down and flows into the rivers and ponds it affects the aquatic ecosystem. It alters the chemical composition of the water, to a form which is actually harmful to the aquatic ecosystem to survive and causes water pollution.

4. Acid rain also causes the corrosion of water pipes, which further results in leaching of heavy metals such as iron, lead, and copper into drinking water.
5. It damages the buildings and monuments made up of stones and metals.

9.1.4.1. Adverse Effects

Research has demonstrated that acid rain can have detrimental effects on freshwater ecosystems, forests, and soils. It can also kill insects and aquatic life forms, damage structures, and have negative health effects on people.

9.1.4.2. Surface Waters and Aquatic Animals

Fish and other aquatic animals may suffer harm from the lower pH and increased aluminum concentrations in surface water brought on by acid rain. Most fish eggs won't hatch at pH values lower than 5, and lower pH can even kill adult fish. The amount of biodiversity decreases as rivers and lakes get more acidic. Insect life and some fish species, such as brook trout in certain lakes, streams, and creeks in environmentally sensitive areas like the U.S. Adirondack Mountains, have been eradicated by acid rain.

The degree to which acid rain, however, directly or indirectly increases the acidity of lakes and rivers through catchment runoff (i.e., contingent upon the features of the surrounding watershed) is changeable. According to the U.S. Environmental Protection Agency (EPA) website, 75% of the acidic lakes and roughly 50% of the acidic streams that were studied had acidity due to acid rain. Lakes found in silicate basement rocks have a higher acidity level compared to lakes found in limestone or other basement rocks that have a carbonate composition, like marble); even with the same quantity of acid rain, this is because carbonate minerals act as a buffer.

Concentrations of sulfur dioxide and nitrous oxide have a variety of effects on aquatic ecosystems, such as changes in acidity, elevated levels of nitrogen and aluminum, and modifications to biogeochemical processes. Usually, exposure to sulfur dioxide and nitrous oxide does not cause direct physiological effects; instead, these gases build up in the environment and cause long-term effects that alter the chemistry of soil and water.

9.1.4.3. Soils

Acid rain can cause significant harm to the biology and chemistry of soil. Certain microbes die because they can't withstand low pH changes. The acid denatures the enzymes in these microorganisms, causing them to lose their ability to function. In addition to mobilizing toxins like aluminum, the hydronium ions in acid rain also deplete vital nutrients and minerals like magnesium.

Soil biology and chemistry can be seriously impacted by acid rain. Some microbes perish because their resistance to low pH changes is compromised. These microorganisms' enzymes become denatured by the acid, which renders them inoperable. The hydronium ions in acid rain not only mobilize toxins like aluminum but also deplete essential nutrients and minerals like magnesium. Acidic water and soil acidification can have either minor or, more often, significant effects on plants. The reason for the majority of minor cases that do not cause plant death is that the plants are less vulnerable to acidic environments and/or acid rain has less force. But because the acidic water lowers the plant's natural pH, the plant will eventually die even in mild cases. Acidic water seeps into the plant, dissolving, and carrying away vital plant minerals until the plant dies from a shortage of minerals for nourishment. The loss of vital minerals happens in major cases, which are more severe, in a manner similar to that of minor cases but much more quickly. Similarly, acid rain that falls on soil and leaves dries out the waxy cuticles on leaves, which leads to a rapid loss of water from the plant to the atmosphere and ultimately to the plant's death (Figure 9.1).

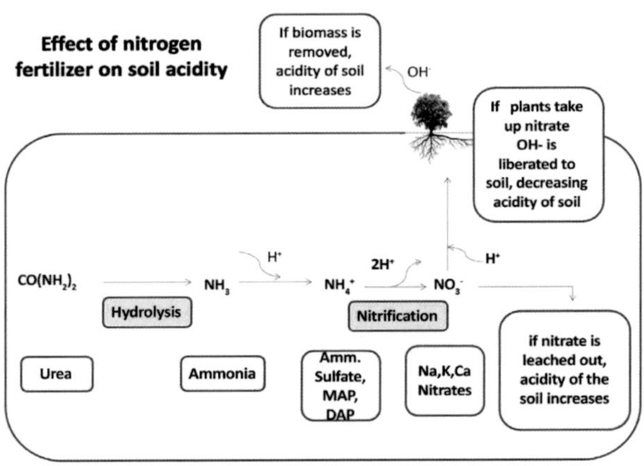

Figure 9.1. *Diagram of nutrient leaching in soil with high levels of soil acidity.*

Source: https://upload.wikimedia.org/wikipedia/commons/e/ed/Effect_of_nitrogen_fertilizer_on_soil_acidity.jpg.

Due to a change in pH brought on by soil acidification, soil microbes may decline, which could be detrimental to plants because they rely on them for nutrient uptake. It is possible to determine whether a plant is impacted by soil acidification by closely observing the leaves of the plant. If the leaves appear healthy and green, the pH of the soil is normal, which is suitable for plant life. However, if there is yellowing between the veins on the leaves, the plant is unhealthy and experiencing acidification. Furthermore, a plant experiencing soil acidification is unable to photosynthesize, and the process of the plant drying out due to acid-water induced damage can damage chloroplast organelles. Without the ability to photosynthesize, a plant cannot produce the nutrients it needs to survive or the oxygen it needs to survive aerobic organisms. This affects the majority of Earth's species and, in the end, renders the plant useless.

9.1.4.4. Forests and Other Vegetation

Indirect effects of acid rain include the effects of the acid on soil (discussed above) and high concentrations of gaseous precursors to acid rain. Because the clouds and fog that surround high altitude forests are more acidic than rain, they are particularly vulnerable. Plants can adjust to the effects of acid rain. Plant species were observed on Chongqing's Jinyun Mountain adjusting to their new surroundings. Both positive and negative effects were experienced by the species. The biochemical and physiological traits of plant seedlings were improved by natural rainfall or slightly acidic rainfall. Acid rain loses its beneficial effects and starts to have negative effects when the pH rises to a threshold of 3.5 (Figure 9.2).

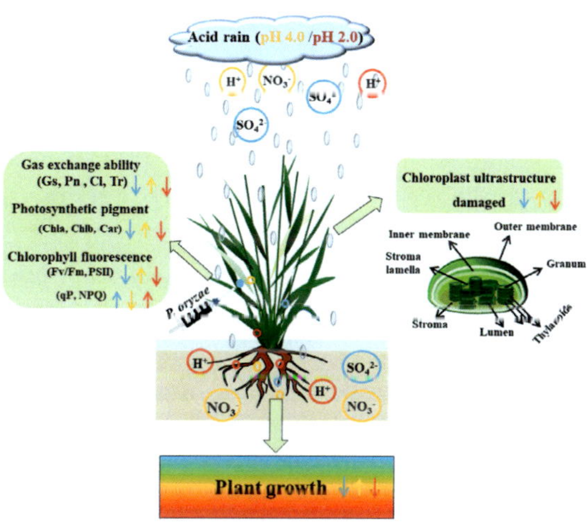

Figure 9.2. *Acid rain can have severe effects on vegetation.*

Source: Li H, Xu Q, Li C, Zhang J, Wang Q, Xiang H, Liu Y, Wei H, Qin Z. Dual Role of Acid Rain and Pyricularia oryzae on Growth, Photosynthesis, and Chloroplast Ultrastructure in Rice Seedlings. Agronomy. *2022; 12(3):567. https://doi.org/10.3390/agronomy12030567.*

When leaves are exposed to a lower pH, photosynthesis is impacted because of the decrease in chlorophyll. This is why acid rain can have a negative effect on photosynthesis in plant leaves. At the cellular level, acid rain can also deform leaves. Examples of this include alterations to the stomatal, epidermis, and mesophyll cells as well as tissue scarring. One of the other effects of acid rain is a reduction in the thickness of the cuticle on the surface of leaves. Acid rain directly affects a plant's capacity to have a robust canopy cover; a reduction in canopy cover can make plants more susceptible to disease.

Acid rain-affected areas are often ted with dead or dying trees. Aluminum leaks out of the soil due to acid rain, endangering plant and animal life. Moreover, it depletes the soil of vital minerals and nutrients required for the growth of trees. Higher altitudes can see an acidic fog and clouds deplete tree foliage's nutrients, resulting in dead or discolored leaves and needles. Because of this reduction in sunlight absorption, the trees become weaker and are less able to withstand cold weather. Although acid rain can harm other plants as well, the impact on food crops is lessened by the use of lime and fertilizers to replenish lost nutrients. Limestone can also be added to soil in cultivated areas to help the soil maintain a stable pH, but this method is not very effective in wilderness areas. Red spruce trees that have had their calcium dissolved from their needles are less able to withstand cold, and they may even die in the winter. By causing necrosis or altering the nutrients in the soil, acid rain can also reduce crop productivity by keeping plants from maturing.

9.1.4.5. Ocean Acidification

Globally, acid rain is far less harmful to oceans than it is to coastal waters, where its effects are amplified in shallower waters. Acid rain can result in ocean acidification, a drop in pH that makes it harder for various coastal species to build the exoskeletons they require to survive. As a component of the ocean's food chain, these coastal species are interconnected, and more marine life will perish if they are not a source for other marine species to eat. Because calcium carbonate, a vital component of the limestone skeleton, dissolves in acidic (low pH) solutions, coral's limestone skeleton is especially vulnerable to pH decreases. Excess nitrogen inputs from the atmosphere not only cause acidification but also encourage the growth of phytoplankton and other marine plants. This can lead to more frequent harmful algal blooms and eutrophication, or the formation of oxygen-depleted dead zones, in some areas of the ocean.

9.1.4.6. Human Health Effects

Human health may be adversely affected by acid rain, particularly if inhaled particles are present. The impact of acid rain on human health is multifaceted and can manifest in various forms, including respiratory complications from extended exposure and secondary exposure via tainted food and water sources.

9.1.4.7. Nitrogen Dioxide Effects

Respiratory health may suffer from exposure to air pollutants linked to acid rain, such as nitrogen dioxide (NO_2). Nitrogen dioxide that is soluble in water builds up in the small airways, where it is converted to nitric and hydrous acids. Pulmonary edema is the direct result of nitric acid-induced pneumonia, which harms the epithelial cells lining the airways. Nitrogen dioxide exposure also suppresses the immune response by preventing alveolar macrophages from producing inflammatory cytokines in response to bacterial infection. According to research on animals, pollution weakens the immune system's capacity to fight off respiratory infections by lowering mucociliary clearance in the lower respiratory tract.

9.1.4.8. Sulfur Trioxide Effects

Because sulfuric acid and sulfur trioxide both produce sulfuric acid when they come into contact with wet surfaces of the skin or respiratory system, their effects are similar. Compared to the amount of SO_3 breath through the nose, more SO_3 breath passes through the mouth. Depending on the size of the particles, tiny droplets of sulfuric acid will form inside the body when humans breathe in sulfur trioxide and travel through the respiratory system and into the lungs. People with asthma symptoms experience difficulty breathing due to the effects of SO_3 on the respiratory system. When exposed directly to a particular concentration or over an extended period of time, sulfur trioxide can also be extremely corrosive and irritate the skin, eyes, and digestive systems. It has been documented that consuming concentrated sulfuric acid can result in death, burn the mouth and throat, erode a hole in the stomach, burn skin when it comes into contact with skin, and cause tears to well up in the eyes.

9.1.4.9. Federal Government's Recommendation

1. **Nitrogen Dioxides:** A 25 parts per million (ppm) maximum for nitric oxide in working air has been set by the Occupational Safety and Health Administration (OSHA) for an 8-hour workday and a 40-hour workweek. Additionally, OSHA has established a 5-ppm nitrogen dioxide exposure limit for 15 minutes in the workplace.
2. **Sulfur Trioxide:** Regulations that recommend not-to-exceed limits in the air, water, soil, or food are frequently based on levels that affect animals before being changed to help protect humans. Depending on whether they use various animal experiments, vary in the duration

of exposure (e.g., an 8-hour workday versus a 24-hour day), or for other causes, these values that should not be exceeded can differ amongst federal agencies. The EPA sets a limit on the quantity of sulfur dioxide that can be released into the atmosphere. As a result, there is less sulfur dioxide in the atmosphere to convert to sulfuric acid and sulfur trioxide. OSHA limits the amount of sulfuric acid in workroom air to 1 mg/m^3. Furthermore, a time-weighted average limit of 1 mg/m^3 is recommended by NIOSH. If you are concerned about your exposure to NO_2 and SO_3, you should ask those around you, especially the kids, and speak with your doctor.

3. **Other Adverse Effects:** Buildings, historical sites, and statues are susceptible to damage from acid rain, particularly if they are made of calcium carbonate-rich rocks like marble and limestone. The calcium compounds in the stones combine with the acids in the rain to form gypsum, which flakes off.

   $CaCO_3\ (s) + H_2SO_4\ (aq) \rightleftharpoons CaSO_4\ (s) + CO_2\ (g) + H_2O\ (l)$

   The effects of this are commonly seen on old gravestones, where acid rain can cause the inscriptions to become completely illegible. Acid rain also increases the corrosion rate of metals, in particular iron, steel, copper, and bronze (Satake, 2012; Seinfeld and Pandis, 1998).

9.2. WATER QUALITY

The phrase "water quality" refers to the ability of the water to support different processes or uses. Specific requirements for the physical, chemical, or biological properties of water will apply to any given use; for example, limits on the concentrations of hazardous substances for drinking water use, or limitations on pH and temperature ranges for water that supports invertebrate communities. As a result, a variety of factors that restrict water use can be used to define water quality. Every use will have different requirements and influences on the quality of the water, even though many of them share some common requirements for particular variables. Different users will not always have compatible needs in terms of quantity and quality, and one user's actions may limit another's ability to do certain things. This can happen when one user uses water at a lower quality than the other or when one demands water that is outside of their acceptable range. The quality and quantity requirements of various users are frequently compromised in the process of trying to maintain or improve a particular water quality. The idea that natural ecosystems should be taken into account when evaluating options for managing water quality is becoming more widely acknowledged. This is due to their inherent worth as well as the fact that they are sensitive markers of modifications or declines in the general quality of the water, offering a helpful supplement to other data such as physical and chemical.

Surface and subsurface water composition is influenced by a variety of natural elements in the drainage basin, including topography, geology, weather, hydrology, and biology. Seasonal variations in runoff volumes, weather patterns, and water levels also affect this composition. Therefore, even in situations where only one watercourse is involved, significant natural variations in water quality may be seen. Water quality is significantly impacted by human activity as well. A portion of these impacts are the outcome of hydrological modifications, like the construction of dams, the removal of wetlands, and the redirection of flow. Polluting activities are more obvious; examples include the purposeful or inadvertent release of domestic, industrial, urban, and other wastewater into watercourses, as well as the dispersal of chemicals onto agricultural land within the drainage basin.

Numerous natural and human factors can impact the quality of water. Geological, hydrological, and climatic factors are the most significant natural influences because they have an impact on the amount and quality of water that is available. Their impact is typically greatest in situations where there is a shortage of water and every effort must be taken to utilize it to the fullest; excessive salinity, for instance, is a common issue in coastal and arid regions. Although desalinating saltwater or saline groundwater is possible if the necessary financial and technical resources are available, this is often not practical. As a result, even though there may be enough water available, its poor quality restricts the uses that can be made of it. Even though the natural ecosystem and the natural water quality coexist, any major changes to the water quality will typically have an adverse effect on the ecosystem.

The degree to which human activities disturb the ecosystem and/or limit water use has a wide range of effects on the quality of water. For instance, there is only one source of human feces-related water pollution, but there are several causes for this kind of pollution, its effects on water quality, and the appropriate corrective or preventive actions. The lack of community waste disposal facilities, inadequate or mismanaged collection and treatment facilities, or on-site sanitation facilities (like latrines) that drain straight into aquifers are all potential causes of fecal pollution. Fecal pollution has a variety of effects. In developed countries, eutrophication, and organic load (in the rivers into which sewage or effluent is discharged and in the sea into which rivers flow or sewage sludge is dumped) may be of greater concern than intestinal disease, which is the main issue in developing countries. Thus, just as a problem may have multiple contributing factors, a single influence may result in multiple water quality issues. Eutrophication can arise from both diffuse and point sources, including runoff from livestock feedlots and agricultural land fertilized with both organic and inorganic fertilizers. Point sources include wastewater discharges with high nutrient loads, primarily nitrogen and phosphorus. It is particularly challenging to control pollution from diffuse sources, like agricultural runoff, or from numerous small inputs over a large area, like fecal pollution from unsewered settlements.

A water's concentration and state (dissolved or particulate) of any or all of its organic and inorganic material can be used to characterize its quality, along with some of its physical attributes. *In situ* measurements and on-site or laboratory analysis of water samples are used to determine it. Thus, on-site measurements, the gathering and examination of water samples, the examination and assessment of the analytical outcomes, and the dissemination of the results constitute the principal components of water quality monitoring. Analyses done on a single water sample are only valid for the specific place and time that the sample was collected. Thus, one goal of a monitoring program is to collect enough information (through frequent or thorough sampling and analysis) to evaluate temporal and/or spatial variations in water quality.

The quality of the aquatic environment is a broader issue which can be described in terms of:

1. Water quality.
2. The composition and state of the biological life present in the water body.
3. The nature of the particulate matter present.

The physical description of the water body (hydrology, dimensions, nature of lake bottom or river bed, etc).

Therefore, a thorough evaluation of the aquatic environment's quality necessitates an investigation and assessment of the water body's physical features, biological life, particulate matter, and quality.

This is attainable by means:

1. Chemical analyses of water, particulate matter and aquatic organisms (such as planktonic algae and selected parts of organisms such as fish muscle).
2. Biological tests, such as toxicity tests and measurements of enzyme activities.
3. Descriptions of aquatic organisms, including their occurrence, density, biomass, physiology, and diversity (from which, for example, a biotic index may be developed or microbiological characteristics determined).

Physical measurements of water temperature, pH, conductivity, light penetration, particle size of suspended and deposited material, dimensions of the water body, flow velocity, hydrological balance, so forth.

Pollution of the aquatic environment, as defined by GESAMP (1988), occurs when humans introduce, either by direct discharge to water or indirectly (for example through atmospheric pollution or water management practices), substances or energy that result in deleterious effects such as:

1. Hazards to human health.
2. Harm to living resources.
3. Hindrance to aquatic activities such as fishing.
4. Impairment of water quality with respect to its use in agriculture, industry or other economic activities, or reduction of amenity value.

The importance attached to quality will depend on the actual and planned use or uses of the water (e.g., water that is to be used for drinking should not contain any chemicals or microorganisms that could be hazardous to health) (Satake, 2012; Seinfeld and Pandis, 1998).

9.2.1. Characteristics of Surface Waters

These are the following characteristics of surface waters:

9.2.1.1. Hydrological Characteristics

There are many different kinds of continental water bodies, such as lakes, ground waters, flowing water, and reservoirs. The hydrological cycle links them all together with numerous intermediary water bodies, both man-made and natural. Hydrologically, wetlands are in between rivers, lakes, and ground waters. Examples of wetlands (Figure 9.3) include floodplains, marshes, and alluvial aquifers. Marshes and wetlands are particularly significant to biology.

A lake's theoretical residence time can be calculated by dividing its total volume by its total outflow rate ($V/\Sigma Q$). Because residence time is related to the amount of time needed to recover from a pollution incident, it is a crucial concept for studies on water pollution.

For instance, quick dispersion and transport of waterborne pollutants in a river facilitates the aquatic system's recovery from a pollution input. Long residence times—like those found in deep lakes and aquifers—often lead to extremely sluggish recovery from pollution inputs because it can take years or even decades for waterborne pollutants to be transported away from their source. Even though the water body's water residence time is brief, pollutants trapped in sediments take a long time to be eliminated from the aquatic system.

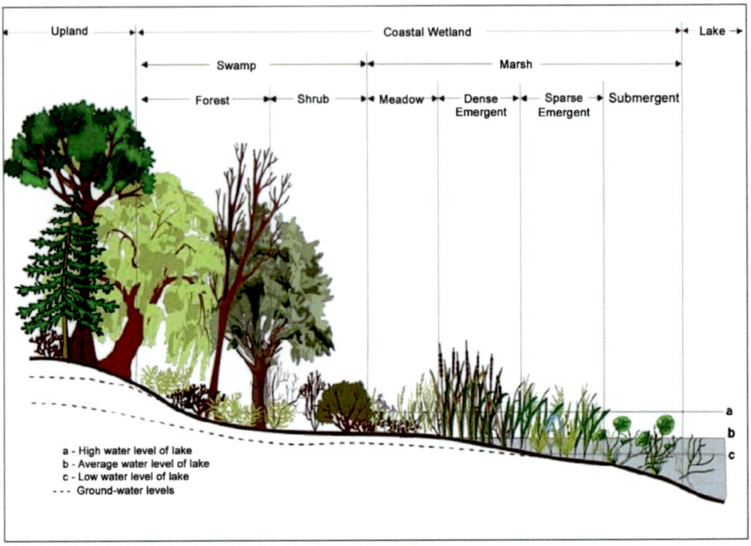

Figure 9.3. *Wetlands.*

Source: By Douglas A. Wilcox, Todd A. Thompson, Robert K. Booth, and J.R. Nicholas – USGS Document, Circular 1311, Lake-Level Variability and Water Availability in the Great Lakes, Public Domain, https://commons.wikimedia.org/w/index.php?curid=19748553.

River flow is unidirectional and frequently exhibits good lateral and vertical mixing, although it can vary greatly depending on the drainage pattern, weather, and climate. Periods of alternating stratification and vertical mixing are characteristics of still surface waters, such as deep lakes and reservoirs. Water currents can also flow in multiple directions and move far more slowly than they do in rivers. Furthermore, the movement of the upper layers of lake and reservoir water is significantly influenced by wind. Water in lakes often stays there for longer than six months and sometimes even for several hundred years. Reservoir residence times, on the other hand, are typically shorter than a year.

9.2.1.2. Lakes and Reservoirs

Stratification is a significant factor affecting the quality of the water in deep, relatively still bodies of water, like lakes and reservoirs. Stratification is the process by which the water in a reservoir or lake behaves as two distinct, denser bodies that float on top of one another. The most frequent cause is temperature variations, which results in variations in density (water has a maximum density of 4°C). However, variations in solute concentrations can also occasionally cause it. Different factors also affect the water quality in the two bodies of water. For instance, the lower layer is physically isolated from the atmosphere, a source of gases like oxygen, and may come into contact with decomposing sediments, which place an oxygen demand on it, whereas the surface layer receives more sunlight. It is typical for the lower layer to have a much lower oxygen concentration than the upper layer as a result of these influences. Different compounds may increase in interstitial waters (through dissolution or reduction) and diffuse from the sediments into the lower water layer when anoxic conditions occur in bottom sediments. Ammonia, nitrate, phosphate, sulfide, silicate, iron, and manganese compounds are among the substances that are produced in this manner.

9.2.1.2.1. Temperate Lakes

For many years, researchers have examined thermal stratification in temperate climates where the water's surface layers warm up and lose density in the spring and summer. They float on top of the denser, colder layer beneath, and vertical mixing is resisted. The colder water trapped below is referred to as the hypolimnion, and the warmer surface layer is called the epilimnion. The temperature of the epilimnion doesn't change much with depth and can be mixed by surface currents and wind. There is a shallow area in between the layers known as the metalimnion or thermocline, where the temperature drops from the epilimnion to the hypolimnion. An "overturn" occurs when the wind causes vertical circulation and mixing in the lake water due to a decrease in surface layer temperature and a reduction in the density difference between the two layers. This can happen very fast. The climate (temperature, insolation, wind) and the features of the lake and its environs (depth and wind exposure) are the main determinants of the frequency of overturning and mixing.

Lakes may be classified according to the frequency of overturn as follows (Figure 9.4):

1. **Monomictic:** Once a year – temperate lakes that do not freeze.
2. **Dimictic:** Twice a year – temperate lakes that do freeze.
3. **Polymictic:** Several times a year – shallow, temperate or tropical lakes.
4. **Amictic:** No mixing – arctic or high altitude lakes with permanent ice cover and underground lakes.
5. **Oligomictic:** Poor mixing – deep tropical lakes.
6. **Meromictic:** Incomplete mixing – mainly oligomictic lakes but sometimes deep monomictic and dimictic lakes.

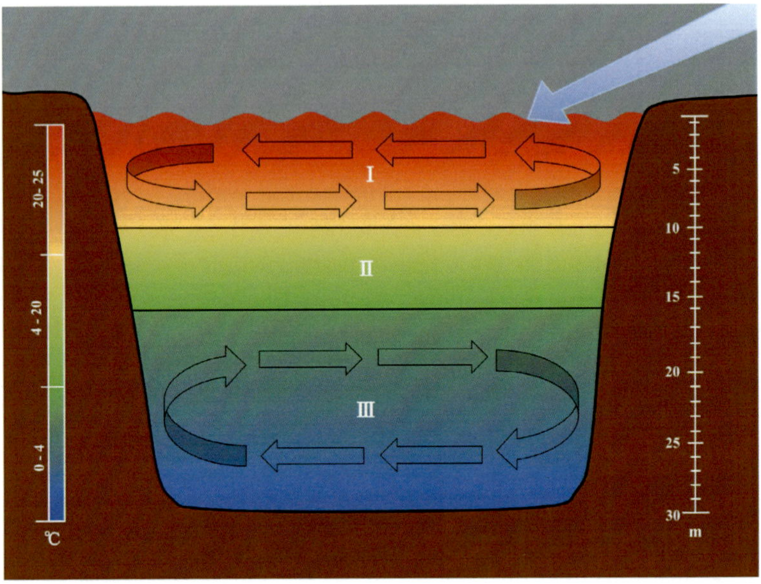

Figure 9.4. *The classification of lakes according to thermal stratification and mixing in the water column: I. The Epilimnion II. The Metalimnion III. The Hypolimnion.*

Source: By Mbrookings19 – Own work, CC BY-SA 4.0, https://commons.wikimedia.org/w/index.php?curid=69783069.

Lakes that are shallower than approximately 10 m typically do not experience thermal stratification because surface wind and water flow promote mixing. Several times a year, shallow tropical lakes can undergo complete mixing. But even in tropical and equatorial areas, stratification can last the entire year in very deep lakes. The result of this ongoing stratification is "meromixis," a persistent anoxia of bottom waters that occurs naturally.

9.2.1.2.2. Tropical Lakes

Due to comparatively constant solar radiation, tropical lakes share the physical trait of having little seasonal variations in water temperature. The average temperature of the water is high, but it gets colder as you go higher. Because the water temperature is nearly constant, density differences are negligible. The annual temperature range of the surface is only 2–3°C, and it is even less at depths greater than 30 m. Mixing is greatly influenced by winds and precipitation, both of which have a tendency to be seasonal. In tropical lakes, the very small seasonal temperature variation also leads to a correspondingly small annual heat budget.

However, because the peak value of heat storage could be the consequence of a single meteorological event, there could be a significant relative variation in the heat budget for several years. The significant variations in rainfall between the wet and dry seasons can cause several-meter fluctuations in water levels in certain tropical lakes. These changes have a significant impact on nutrient availability and dilution, which in turn influences fish spawning, zooplankton reproduction, and algal blooms. Wind speeds are typically higher during the dry season than they are during other seasons, and evaporation rates are at their highest. Mixing is facilitated by wind-induced turbulence and the ensuing heat losses. Tropical lakes are generally not classified according to seasonal temperature variations at different depths.

A classification that takes into account physical attributes like size, depth, and the following is more pertinent:

1. Large, deep lakes all have a seasonal thermocline in addition to a deep permanent thermocline over an anoxic water mass. Recirculation of the deep water may occur but the responsible mechanism is not clear.
2. Large, shallow lakes have a distinct diurnal temperature variation. Temperature is uniform in the morning, stratification develops in the afternoon and is destroyed during the night. The fluctuation in water level may be considerable relative to lake volume and the large flood plain that results will have profound effects on the productivity of biological life in the water.
3. Crater lakes generally have a small surface area relative to their great depth and are often stratified. Despite such lakes being in sheltered positions, special weather conditions can cause complete mixing of lake contents.
4. High-altitude lakes in climates where there is only a small diurnal temperature difference are unstable and experience frequent overturns. Where temperature differences are larger, a more distinct pattern of stratification can be identified. There may also be substantial losses of water by evaporation during the night.
5. River lakes are created when areas of land are flooded by rivers in spate. When the water level in the river goes down, the lake water flows back towards the river. This annual or semiannual water exchange affects the biological and chemical quality of the water.
6. Solar lakes. In saline, dark-bottomed lakes an anomalous stratification can develop. A lower, strongly saline water layer may be intensely heated by solar radiation, especially if it is well isolated from the atmosphere by the upper layer of lighter brine. Temperatures as high as 50 °C have been recorded in the lower levels of solar lakes.

7. Temporary lakes occur in locations where the fluctuations of water level cause a shallow lake basin to dry up completely. In regions where there are pronounced wet and dry seasons this can occur annually, while in other regions the frequency of occurrence may be medium to long term. Temporary lakes often have an accumulation of salts on the lake bottom.

9.2.1.3. Rivers

Understanding a river's discharge regime is crucial for interpreting measurements of water quality, particularly those that involve suspended sediment or are meant to calculate the flux of contaminants or sediment. A river's discharge is influenced by the characteristics of its catchment, specifically the geology, geography, and climate.

9.2.1.3.1. Tropical Rivers

The yearly cycle of wet and dry seasons greatly influences a tropical river's regime. These are a few regimes and some of the environmental and climatic factors that influence them.

Equatorial rivers with one flow peak, resulting from heavy annual precipitation (1750–2500 mm) in areas with no marked dry season.

Equatorial rivers with two flow peaks, produced by precipitation total ling more than 200 mm monthly and well over 1750 mm annually. Equatorial forest predominates in the catchment area.

Rivers in the moist savannah of tropical wet and dry lowlands exhibit pronounced seasonal effects of rainfall patterns. In these areas, the dry season persists for at least three months.

In some areas of the tropical wet and dry highlands, the length of the dry season varies significantly. River basins in such areas are covered by woodland and relatively moist savannah.

In the relatively drier regions of the tropical wet and dry highlands, river basins are located in the marginal parts of the dry climate zones. Precipitation rarely exceeds 500–700 mm annually, which is typical of semi-desert regions, and vegetation in the river basins is predominantly dry savannah.

In areas where the dry season is prolonged, ecologists may divide the associated vegetation into wooded steppe and grass steppe. River regimes, however, do not differ between the two types of regions, and flow is likely to be intermittent. In desert regions, where annual rainfall is less than 200 mm, river basins are covered with sand, desert grass or shrubs, and rivers are of the wadi type. The drainage network is poor and where it is traceable, it is likely to have developed before the area reached its present stage of aridity. Many of the rivers of tropical mountain regions have drainage basins of very limited size.

9.2.2. Characteristics of Groundwater

Groundwater is stored within the voids of sediments like sand or gravel, or in the cracks of broken rock such as crystalline rock and limestone. This encompassing mass of rock or sediment that retains the water is referred to as an aquifer, and the uppermost level of water saturation within this aquifer is known as the water table. Generally, groundwater moves in a consistent manner. Its speed is predominantly influenced by the porosity and permeability of the material it flows through, and is typically significantly slower than that of surface waters, leading to limited mixing.

Aquifer media, which can be either rock or sediment, are distinguished by their porosity and permeability. The volume of pores and fissures divided by the media's total volume is known as porosity. It indicates the media's water or storage volume and is expressed as a percentage of voids. The relative rate at which water or other fluids move through media under specific circumstances is indicated by permeability, which is a measure of how easily fluids may pass through the media generally under potential gradients. It is known as hydraulic conductivity for water.

9.2.2.1. Types of Aquifers

Hard crystalline rocks, consolidated sedimentary formations, and unconsolidated sediments are the three main categories of underground formations. Along with some volcanic rock varieties like basalts and dolerites, the hard crystalline rocks comprise granites, gneisses, schists, and quartzites. Aquifers rely on rock fissures and fractures to provide porosity and pathways for groundwater movement because these formations typically had little to no porosity at first. While weathering frequently enhances them, aquifers in hard rocks are typically small, localized, and low-yielding.

Fluoride and boron concentrations in groundwater in volcanic formations in "recent" volcanic activity regions are often too high for some applications. Sedimentary formations that have consolidated are frequently large, thick, and occasionally artesian. Some of the world's largest, most significant, and highest-yielding aquifers are formed by limestone and sandstone formations, which can also be extremely porous and permeable. These formations' permeability is mostly caused by fissures (fractures, faults, and bedding planes). For some pollutants to move and store, porosity is also important. The rock's dissolution may cause the permeability to rise. The formation of karst aquifers, which can contain sizable subterranean caverns and channels yielding significant amounts of water, is specifically caused by the dissolution of carbonates. Unconsolidated sediments can be found in thick sequences in the main river or lake basins or as thin, superficial deposits over other rock types. Grain size has an impact on permeability and porosity. Silts and clays are less productive than sand and gravel deposits, which can produce significant and high-yielding aquifers. Thick sedimentary deposits that have accumulated over a long period of time in the largest river basins can have multiple layers of various materials, creating significant multi-aquifer sequences. Aquifers can be either restricted or unrestricted (Figure 9.5). An impermeable layer that shields a confined aquifer from surface water or rainfall recharge (and contamination) occurs on top of it. Where permeable rock emerges at or close to the surface, which may be some distance from the area of exploitation, is where confined aquifers recharge. This characteristic might make pollution and quality control more challenging. Some aquifers are referred to as semi-confined or leaky because they are not completely contained.

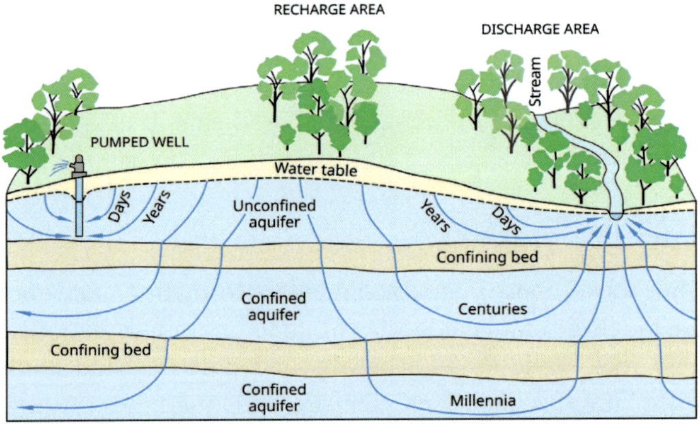

Figure 9.5. *Confined and unconfined aquifers.*

Source: By T.C. Winter, J.W. Harvey, O.L. Franke, and W.M. Alley – Ground Water And Surface Water A Single Resource. U.S. Geological Survey Circular 1139, Figure 9.3., Public Domain, https://commons.wikimedia.org/w/index.php?curid=25976976.

A permeable, unsaturated zone covers unconfined aquifers, allowing surface water to seep down to the water table. As a result, they are frequently shallow with a propensity to interact with surface water and are typically recharged over a large area. Because contaminants and surface water cannot seep into confined aquifers, they are less susceptible to pollution outside of their recharge zone than unconfined aquifers. However, since confined aquifers are typically deep and there are few locations where contaminated water may be pumped out, contamination is frequently difficult to remediate when it does occur. Contaminants may also become more concentrated in small aquifers due to the limited outflow, which could limit water abstraction. Unconfined aquifers are more susceptible to contamination due to their larger recharge area, which increases the potential for contamination, and their increased interaction with contaminated surface water bodies, which increases the risk of contaminant movement into groundwater. The depth of the unsaturated layer that lies on top, the rate at which it permeates the water table, and the land use in the vicinity of groundwater sources all affect the likelihood of contamination.

9.2.2.2. Water Quality

The composition of recharge water, how water interacts with soil, soil-gas, and rocks in the unsaturated zone, as well as the length of time it spends there and the processes that occur there, all affect the quality of groundwater. Therefore, even within a given general area, there can be significant variation, particularly in areas where rocks with varying compositions and solubility are present. The main processes that affect the quality of water in aquifers are geochemical (complexation, acid-base reactions, oxidation-reduction, precipitation-solution, and adsorption-desorption), biochemical (microbial respiration and decay, cell synthesis), and physical (dispersion/dilution, filtration, and gas movement). Because rainwater infiltrates the surface to recharge groundwater, human activities that cause pollution at the land surface have an impact on groundwater quality. Because of human activity, the acidity of the rainwater itself might also be higher. In addition to acting as a storage area for large amounts of pollutants like nitrates that may eventually be released, the unsaturated zone can help lower the concentrations of some pollutants—especially microorganisms—entering groundwater. Certain pollutants infiltrate ground waters straight from closed-off wells, quarries, mines, and underground sewage pipes that avoid the unsaturated zone (and, consequently, may undergo some natural decontamination processes).

9.2.2.3. Contamination

Artificial sources of groundwater pollution can come from both diffuse and point sources. Domestic sewage and latrines, municipal solid waste, agricultural wastes and manure, and industrial wastes (including tipping, direct injection, spillage, and leakage) are a few of the more frequent sources. Groundwater contamination can occur via a convoluted process. Large areas of the aquifer recharge area may contain contaminants, like agricultural chemicals, which may take decades to enter the groundwater and even longer to leave after use has stopped. Large unintentional spills and other point sources of pollution can quickly contaminate a small area before spreading throughout the aquifer. Pollutants that are completely soluble in water and have a similar density—like sewage water tainted with chloride—will permeate the aquifer at a pace determined by the groundwater flow velocity. Less dense pollutants have a tendency to gather at the water table and move along the surface. Dense substances, like chlorinated solvents, sink vertically and collect at an aquifer's bottom.

Due to the unsaturated zone's slow movement and the aquifer's slow flow, pollution incidents typically occur after the contamination is discovered at the point of water abstraction. Similar to this, it takes a while to "flush out" a pollutant, and the deterioration of groundwater quality may be irreversible in some circumstances. Land use around boreholes and in recharging aquifers needs to be closely observed as part of a pollution control program. The amount of human activity and land use in places where rainfall or surface water may seep into the aquifer will determine how vulnerable the aquifer is to pollution. In these regions, groundwater pollution is likely to result from contamination of surface water or the unsaturated layer above an aquifer (Satake, 2012; Seinfeld and Pandis, 1998).

9.2.3. Natural Processes Affecting Water Quality

While human activity is almost always the cause of water quality degradation, there are some natural phenomena that can cause water quality to drop below what is needed for specific uses. Natural occurrences like hurricanes and torrential rains cause landslides and excessive erosion, which raise the amount of suspended material in impacted rivers and lakes. In certain lakes, seasonal water overturning can raise water to the surface that contains little to no dissolved oxygen. Such occurrences of nature could be regular or sporadic. In certain places, permanent natural conditions may render water unfit for human consumption or certain applications, like irrigation. Common instances of this include the high salt content of some ground waters under specific geological conditions and the salinization of surface waters through evaporation in arid and semi-arid regions. Because many ground waters have high levels of carbonates (hardness) by nature, they must be treated before being used in some industrial applications. Certain ions (like fluoride) and hazardous substances (like arsenic and selenium) can be found in groundwater in some areas in quantities that are hazardous to health, while other regions have substances or elements that result in different kinds of issues (like iron and manganese discoloring sanitary fixtures). Various natural processes can alter the composition and concentration of chemical elements and compounds in a freshwater system, that is hydrological, chemical, biological, and physical (Figure 9.6).

These are the main environmental factors:

1. **Distance from the Ocean:** Extent of sea spray rich in Na^+, Cl^-, Mg^{2+}, SO_4^{2-} and other ions.
2. **Climate and Vegetation:** Regulation of erosion and mineral weathering; concentration of dissolved material through evaporation and evapotranspiration.
3. **Rock Composition (Lithology):** The susceptibility of rocks to weathering ranges from 1 for granite to 12 for limestone; it is much greater for more highly soluble rocks (for example, 80 for rock salt).
4. **Terrestrial Vegetation:** The production of terrestrial plants and the way in which plant tissue is decomposed in soil affect the amount of organic carbon and nitrogenous compounds found in water.
5. **Aquatic Vegetation:** Growth, death, and decomposition of aquatic plants and algae will affect the concentration of nitrogenous and phosphorous nutrients, pH, carbonates, dissolved oxygen and other chemicals sensitive to oxidation/reduction conditions. Aquatic vegetation has a profound effect on the chemistry of lake water and a less pronounced, but possibly significant effect, on river water.

Seasonal variations in the concentrations of various chemicals in river water can be attributed to the influence of these primary environmental factors. One factor can have multiple orders of magnitude of variation in small watersheds (less than 100 km^2). Watersheds larger than 100,000 km^2 typically have more consistent water quality, with most measured variables showing variation within one order of magnitude. The range of total dissolved solids in different water bodies reflects these natural variations. The total of the major ions in the water plus the silica makes up the total dissolved solids. Electrical conductivity can be measured to get a close approximation of its value.

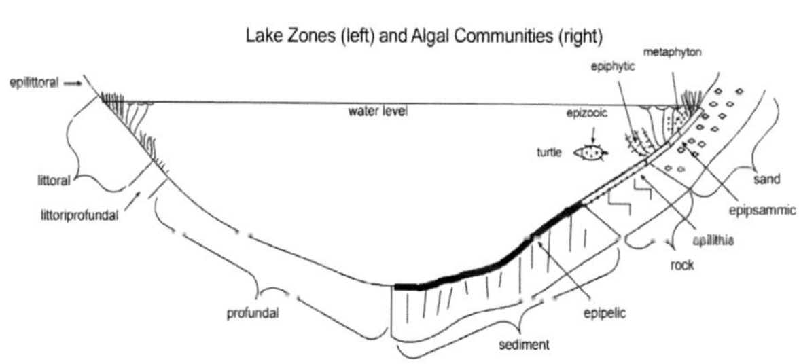

Figure 9.6. *Lake dynamics.*

Source: By Pdietry – Own work, CC BY-SA 4.0, https://commons.wikimedia.org/w/index.php?curid=97678384.

Certain chemical elements may only be found in trace amounts in solution due to their strong affinity for PM and the processes of precipitation/dissolution and adsorption/desorption. Conversely, some elements are extremely soluble and are hardly ever, if ever, found in water as PM. The Soluble Transport Index represents a chemical's propensity to be present in its soluble form as opposed to being linked to particles. The process of separating PM from material in solution involves passing a water sample through a filter that has a pore size of either 0 or 50 μm. Although it is acknowledged that some fine colloids, such as iron, aluminum, and manganese hydroxides, may pass through the filter, using finer filters is very expensive and necessitates extended filtration times.

Compared to dissolved concentrations, the elemental composition of river particulates varies less. This is less true for the trace elements arsenic, cadmium, cobalt, chromium, copper, mercury, nickel, lead, selenium, tin, zinc, so forth. but it is true for the major elements aluminum, iron, silicon, calcium, magnesium, sodium, potassium, manganese, titanium, and phosphorus. With the exception of aluminum, iron, calcium, and sodium, the major element composition of particulates in tropical and temperate and cold rivers is remarkably similar. Local geological factors can cause significant differences in the concentration of trace elements in particulates within small watersheds. Water quality can vary over time and space within a single body of water.

There are five categories of time-related differences:

1. Minute-to-minute and day-to-day differences resulting from water mixing and fluctuations in inputs, usually as a result of meteorological conditions. These differences are most evident in small water bodies.
2. Diurnal (24-hour) variations resulting from biological cycles and daylight/darkness cycles which cause changes in, for example, dissolved oxygen and pH. Diurnal patterns also result from the cyclic nature of waste discharges from domestic and industrial sources.
3. Irregular patterns. Irregular sources of pollution include fertilizers, pesticides, and herbicides, present in the run-off from agricultural land, and wastes discharged from food processing plants. The resultant variations in water quality may be apparent over a matter of days or months.
4. Seasonal biological and hydrological cycles.
5. Year-to-year trends, usually as a result of increased human activities in the watershed.

Variations in the quality of water can arise from both internal and external factors. Internal processes typically follow cycles, occurring on a daily or seasonal basis, and are unrelated to the volume of the water body. Large water bodies and extended water residence times may act as buffers against external processes like the addition of pollutants, depending on flow regimes. A very large lake's average composition most likely doesn't change much from year to year. Similar to this, a stream's water quality will vary considerably more throughout the year than a large river's.

This indicates that the sampling frequency required for a stream to accurately describe its average water quality is typically much higher than that of a river, and it is typically much lower for lakes. Variations in water quality between locations are primarily caused by the body of water's homogeneity rather than its size. For instance, one sample taken from close to the lake's center may be sufficient to characterize the water in a circular lake. Long, thin lakes and lakes with lots of bays and inlets will need more samples; three is the minimum, and 10 or more is the ideal amount. It is especially crucial to collect samples from various depths in lakes that are stratified or deeper than roughly 30 m to create a vertical profile. Up to five samples should be taken at depths determined by field measurements of conditions, depending on the lake's depth. Any given river's flow pattern, or regime, is the result of extremely specific circumstances. While it is comparable to rivers in the surrounding area, altitude, the slope's wind exposure, and rainfall variations will all have a significant impact (Satake, 2012; Seinfeld and Pandis, 1998).

9.3. WATER QUALITY STANDARDS

Water quality standards (WQS) are clauses in federal, state, territorial, authorized tribal, or other laws that have been approved by the EPA. They specify the ideal state of a body of water and the procedures that will be followed to protect or attain that state. Water features can be utilized for leisure activities (e.g., They are also home to a variety of aquatic life and are great for fishing, boating, swimming, and enjoying the scenery. States, territories, and recognized tribal nations establish WQS in order to safeguard both aquatic life and public health in these waters. WQS provide a legal foundation for limiting the amount of pollutants that can enter U.S. waters.

9.3.1. Core Components of WQS

Water quality standards consist of three core components. This includes designated uses of a water body, criteria to protect designated uses, and anti-degradation requirements to protect existing uses and high quality/high value waters.

1. Designated Uses.
2. Criteria.
3. Antidegradation Requirements.

1. **Designated Uses:** The WQS regulation requires states, territories, and authorized tribes to specify goals and expectations for how each water body is used. Typical designated uses include:

   i. Protection and propagation of fish, shellfish, and wildlife.

   ii. Recreation.

   iii. Public drinking water supply.

   iv. Agricultural, industrial, navigational, and other purposes.

2. **Criteria:** Water bodies' designated uses are safeguarded by the adoption of water quality standards by states, territories, and authorized tribes. Criteria for water quality can be numerical (e.g., the highest levels of pollution concentration allowed in a body of water) or story (e.g., a standard that outlines the ideal circumstances for a body of water to be "free from" specific unfavorable conditions. Usually, states, territories, and recognized tribes use both narrative and numerical standards.

3. **Antidegradation Requirements:** The preservation of the chemical, physical, and biological integrity of the country's waters is one of the main goals of the Clean Water Act. The framework for preserving and safeguarding water quality that has already been attained is provided by antidegradation requirements. The main tools used by states and authorized tribes to accomplish the objectives and goals of the Clean Water Act are designated uses and water quality criteria. Antidegradation requirements serve as a framework for preserving current uses, safeguarding waters of a higher quality than is required to support the goals of the Clean Water Act, and safeguarding waters designated as Outstanding National Resource Waters (ONRWs) by states and authorized tribes.

9.3.2. Additional Components of WQS

States, territories, and authorized tribes also have the choice of including additional components in their water quality standards such as general policies and WQS variances.

1. General Policies.
2. WQS Variances.

1. **General Policies:** Policies and provisions that generally impact the application and implementation of water quality standards, such as low-flow, mixing-zone, and WQS variance policies and procedures, may be adopted by states, territories, and recognized tribes. These regulations must be reviewed and approved by the EPA.

2. **Water Quality Standards Variances:** States, territories, and recognized tribes can use a variance from the WQS as a tool to gradually improve the quality of their water. These time-limited WQS strike a compromise between accountability measures to reassure the public that progress will be made and allowing states, territories, authorized tribes, and dischargers the time and flexibility to make incremental improvements in water quality that reflect the best that can be achieved in a given amount of time. When states, territories, and recognized tribes are willing to continue their efforts to improve the quality of their water and it is unclear how much progress can be made, a WQS variance may be utilized.

9.3.3. How Are Water Quality Standards Developed?

Adopting water quality standards is subject to the legal and administrative processes unique to each state, territory, and recognized tribe. Generally, work groups or unofficial public meetings are used to develop standards, which are then presented for public comment. Following that, states, territories, and recognized tribes start a public engagement process that includes holding public hearings on the suggested standards. The public is given access to the proposed water quality standards and the supporting data before the hearing. To protect designated uses, states, territories, and recognized tribes must adopt water quality standards that are both sufficiently broad and stringent.

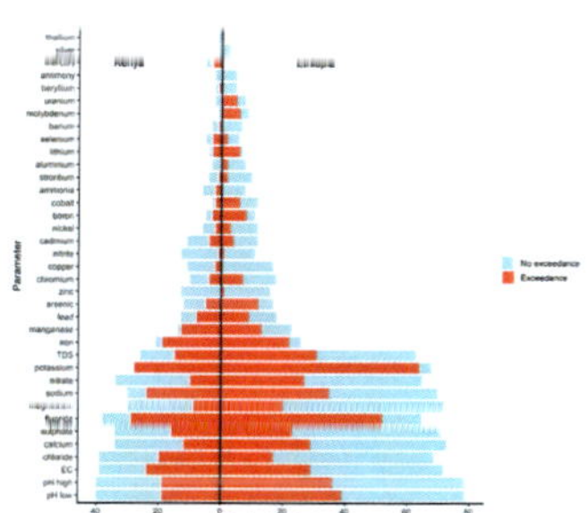

Figure 9.7. *Example of comparison of drinking water quality parameters with thresholds (standards) in Kenya and Ethiopia. The chosen thresholds are from the country's standards or WHO health guideline or East Africa Standard (EAS).*

Source: By Saskia Nowicki a, Behailu Birhanu b, Florence Tanui c d, May N. Sule e, Katrina Charles a, Daniel Olago c d, Seifu Kebede b f – https://www.sciencedirect.com/science/article/pii/S0048969723055547?via%3Dihub#f0025Water chemistry poses health risks as reliance on groundwater increases: A systematic review of hydrogeochemistry research from Ethiopia and Kenya, CC BY-SA 4.0, https://commons.wikimedia.org/w/index.php?curid=139625463.

States, territories, and recognized tribes may adopt the following standards to safeguard the designated use:

1. Adopt the EPA recommended criteria.
2. Adopt unique criteria to reflect site-specific conditions.
3. Use other scientifically defensible methods to develop their own criteria.

The process used by the state, territory or authorized tribes to propose water quality standards may look similar to the steps described in the diagram (Figure 9.7):

Each state, territory or authorized tribe must include the following key elements in their water quality standards to meet EPA approval:

- Designated uses that are consistent with the Clean Water Act.
- Criteria are sufficient to protect designated uses.
- Antidegradation requirements.
- General policies affecting the application and implementation of the standards.
- In addition, the state, territory or authorized tribe must provide the methods and analyses that were used to develop the standards.

9.4. WATER QUALITY CHALLENGES AND IMPACTS

The quality of surface waters and groundwater is influenced by both human activity and natural processes. Naturally occurring dissolved materials, non-dissolved PM, and living organisms can all be found in water; in fact, these elements are essential to clean water because they support healthy biogeochemical cycles. Rarely do naturally occurring substances cause problems with water quality that are harmful to human health. For instance, up to 77 million people in Bangladesh, where almost 90% of the population relies on groundwater as their main source of freshwater, may have been exposed to arsenic in recent decades.

However, a variety of factors, including mining, forestry practices, power generation, agricultural, and industrial production, and domestic use, can change the chemical, biological, and physical characteristics of water in ways that could jeopardize ecosystem integrity and human health. The main causes of water pollution are industrial and agricultural processes, as well as human settlements. Negative aspects of these activities include improper agricultural practices, unsafe solid waste discharge, improper management and treatment of industrial residues, and the unkempt disposal and inadequate handling of human and animal wastes. For instance, more than 80% of sewage in developing nations is dumped untreated into bodies of water. An estimated 300–400 million tons of heavy metals, solvents, toxic sludge, and other waste are dumped into waters annually by industry. The most prevalent chemical contaminant in groundwater aquifers worldwide is nitrate from agriculture. Manure and fertilizer runoff from agriculture is the leading cause of water pollution in the United States; croplands alone are responsible for nearly 40% of nitrogen pollution and 30% of phosphorus pollution. Furthermore, there has been a surge in the use of pesticides, with some regions reporting growth rates as high as 4 to 5% in recent times.

Water salinization is a corollary issue in nearly all nations where land salinization is a major issue. Over 16 million hectares of irrigated land in Central Asia are salinized; these countries include Sudan, China, India, Argentina, and many others. Another significant factor contributing to the salinization of coastal areas' water resources is the intrusion of salty seawater into aquifers, especially in cases where excessive groundwater extraction is occurring. Increased acidity, higher concentrations of nutrients, sediments, salts, trace metals, chemicals, and other toxins, as well as dangerous pathogenic organisms that might flourish in warmer waters, are some of the ways that pollution and contamination from such sources present themselves. One of the most pervasive issues with water quality nowadays is nutrient enrichment, which is seriously harming freshwater and coastal ecosystems. Furthermore, an increasing amount of novel pollutants are found in the world's waterways. These include substances found in contaminated water, such as pharmaceuticals, steroids, hormones, industrial additives and agents, and additives found in gasoline. Due to a lack of information regarding the ecotoxicology and related risks of these contaminants, managing water quality is now faced with a new challenge. Furthermore, their synergistic interactions with already-existing pollutants and contaminants may create complex mixtures that are challenging to treat and have unknown long-term effects.

Among other things, declining water quality harms ecosystems. The biodiversity of freshwater ecosystems has declined more than that of any other ecosystem, including tropical rainforests, according to the Millennium Ecosystem Assessment. For instance, about 40% of freshwater fish species and 40% of amphibian species in the U.S. are in danger of going extinct. Over 40% of the freshwater fish species found in Europe are in immediate danger of going extinct. Furthermore, almost two thirds of freshwater species in South Africa are regarded as endangered or threatened. One of the biggest challenges to managing ocean and coastal resources is the majority of dirty freshwater ends up in the ocean, severely harming numerous coastal areas and fisheries. Aquatic environments are vital for livelihoods in many parts of the world, and while they were once thought of as merely sources of water supply and waste disposal, to be consumed or contaminated without consideration for the consequences, there is a growing appreciation of the vital goods and services these environments offer. Humans are directly impacted by the destruction of fisheries and biodiversity caused by contaminated water, which jeopardizes food production and other benefits to humanity.

Those who live close to polluted rivers and those without other access to clean water or better sanitation are the most affected, 42% of the world's population, or 2 and a half billion people, did not have access to better sanitation in 2008. Toxins and chemicals can either directly harm human development and neurological health, or they can bioaccumulate in fish and other organisms that humans eat. One in eight million people die from diarrheal illness each year, which can be linked to contaminated water or inadequate hygiene and sanitation. Of these, over 1.5 million are children under five, who are more vulnerable to diarrhea than to malaria, HIV, or even all forms of deadly injuries put together. Other startling figures include the fact that the World Health Organization (WHO) received cholera reports from more than 100 countries in 2009 and that 200 million people worldwide are afflicted with the crippling parasitic disease schistosomiasis. The entire cost of illness brought on by water contaminated by discharges from homes, businesses, and farms cannot be accurately estimated. Since many of these effects are long-lasting, it can be challenging to pinpoint their exact cause.

Investments in drinking water and sanitation have high rates of return in terms of money saved, lives saved, illness, and medical costs decreased, more productive workdays, better education, and reduced mortality. In order to eradicate poverty and enhance the lives of billions of people globally, the international community has pledged through the Millennium Development Goals to cut in half the percentage of the population lacking sustainable access to clean water and basic sanitation by the year 2015. In July 2010, the United Nations General Assembly declared that access to clean, safe water for human consumption was a fundamental human right. Ensuring water quality is essential to achieving this right.

Water quantity is directly impacted by poor water quality in a variety of ways. There is less usable water in a given area when it is unfit for drinking, bathing, industry, or agriculture due to pollution. However, using low-quality water (such as brackish or saline waters) can have a significant and direct impact on productive water use, like irrigated agriculture, which can have a significant impact on crop production and land degradation, as well as on rural income and food security. Water scarcity is a condition in which there are not enough water resources to meet the average long-term needs of 700 million people across 43 countries. Two thirds of the world's population may be living under water stress by 2025, with 11.8 billion people expected to reside in nations or regions where there is an absolute shortage of water. Nearly half of the world's population—between 75 million and 250 million people in Africa—will reside in regions with severe water stress by 2030 under the current climate change scenario.

When managing water resources, water quality should be given equal weight with water quantity. However, considering the complexity and scope of the issues raised above, addressing water quality issues will require coordinated action from all relevant parties, including the public and commercial sectors as well as civil society.

9.4.1. Strategies to Combat Water Quality Problems

There are four fundamental strategies to combat water quality problems that can form the basis of policy solutions for improving water quality:

1. Prevention of pollution.
2. Treatment of polluted water.
3. Safe use of wastewater.
4. Restoration and protection of ecosystems.

9.4.1.1. Prevention of Pollution

Reducing or eliminating waste at its source is the main goal of pollution prevention strategies. Most people agree that the cheapest, simplest, and most efficient way to preserve water quality is through prevention. Furthermore, since waste generation, particularly from industrial and agricultural processes, is a sign of inefficient use of materials and resources, preventing or reducing pollution may have enormous financial benefits in addition to environmental ones. Reformulating products to reduce pollution and manufacturing resource requirements, including water, is one way that industry is finding solutions. Reducing the use of toxic chemicals in agriculture for nutrient application, pest control, and general water use can help cut down on pollution. Crop rotation, cover crops, conservation agriculture, better irrigation control, and integrated pest management are a few examples of what is meant by good agricultural practices. The most evident ways to improve sanitation in human settlements include expanding the coverage of improved sanitation, taking settlement design factors into account (like the kinds of materials used in construction), strategically placing industries, managing storm water, and minimizing wastewater output.

9.4.1.2. Treatment of Polluted Water

Before being discharged, wastewater that contains contaminants from industrial, agricultural, or domestic sources needs to be treated. Water that has been contaminated can be treated using a variety of techniques, from high-tech, energy-intensive methods to low-tech, low-energy methods that emphasize biology and ecology. If there are already effective water distribution and treatment systems in place, ongoing work is required to keep them operating as efficiently as possible. However, a lack of personnel, equipment, and funding for upkeep and operation has left many wastewater treatment plants inoperable. When systems are not properly maintained or operated, even high-quality water can deteriorate before it is put to use. The circumstances and intended use determine the best course of action. For example, larger settlements usually need the supply of drinking water from modern, multi-stage water treatment plants. Solutions at the community level could be solar stills and smaller-scale plants for filtration or disinfection. Options for the home include solar pasteurization, chlorine treatment, solar water disinfection, boiling, and basic water filtration. A number of the options mentioned may also be suitable for treating water for industrial use, but in certain cases, agricultural source water may be of much lower quality, particularly in the absence of harmful contaminants that can accumulate in soil and crops.

9.4.1.3. Safe Use of Wastewater

Usually, wastewater is dumped into water bodies after being treated to make it safe for the environment. However, under conditions where the effects on human health and the environment are well understood and every effort is made to reduce risks, it can be used safely—and in some cases, even untreated. Safe wastewater use, for instance in agriculture, can supplement water supply in water-scarce and semi-arid zones, as well as in rapidly expanding pen-urban settings, and lessen the strain that human activity places on freshwater resources.

Moreover, wastewater has the potential to be a valuable resource for some agricultural applications and a source of nutrients when managed appropriately, which could eliminate the need for pricey chemical fertilizers. Moreover, agriculture can remove nutrients from water that would otherwise contaminate watercourses, serving as a biological treatment method.

Reusing treated human wastewater as a source of water can be feasible in both rural and peri-urban areas. For instance, ecological sanitation is a cheap approach to managing human waste that is supported by numerous development organizations. Urine and feces must be separated; sterile urine can be sprayed directly on plants, while feces must be composted until it is suitable for spreading on land. Numerous nations and areas, including China, India, Burkina Faso, Kenya, Niger, Sweden, and portions of Eastern Europe, have adopted this strategy. There are jobs created for the local population and market opportunities for the provision of locally produced fertilizers and soil conditioners for agriculture through the recycling of water and the use of dry pre-stored human wastes. Certain industries, like the food and processing sector, use a lot of water and frequently release a lot of wastewater into the environment. Under such circumstances, safe wastewater reuse as part of cleaner production applications can enhance energy and water efficiency and produce positive environmental effects. For instance, industry can use wastewater from some processes again in other applications where high-quality water is not needed, or it can process wastewater using the right technologies for processes that need higher-quality water. Examples come from Namibia and Singapore, where treated wastewater is added to freshwater resources for industrial and human consumption.

9.4.1.4. Restoration and Protection of Ecosystems

Water filtration is one of the benefits of healthy ecosystems for improving water quality, and it frequently happens at a much lower cost than subsequent engineered attempts to clean contaminated water. Remedial strategies are crucial to restore systemic health and functions when low water quality negatively affects water systems, including watersheds. The process of aiding an ecosystem that has been harmed, destroyed, or degraded is known as ecological restoration. Restoring rivers and wetlands and tearing down upstream dams are simple approaches to freshwater restoration. Ecohydrology is one of the proven strategies that can be applied to address pollution originating from both point and non-point sources. An ecohydrological approach promotes the role of ecosystem processes in improving water quality and is predicated on an understanding of the interactions between ecological processes and the water cycle in a particular catchment. System solutions incorporating social and environmental aspects are based on connecting this approach with the social and economic capacities of a region by involving all users through a watershed management council. Threats relating to water can be addressed through ecohydrology. For example, flood risk can be decreased by enhancing the landscape's ability to retain water or by establishing wetlands that keep pollutants out of streams. There are ecohydrological approach examples all over the world, such as in Iraq, Japan, and Poland. Preventive measures are often the most economical and practical ways to safeguard degraded groundwater systems, as restoration can be costly and challenging. Additionally, there is often a delay between the release of pollutants and the effects they have on groundwater reserves. Because groundwater systems are incapable of effectively purifying themselves, treating contaminated groundwater on-site is challenging and costly. It is crucial to concentrate on adequate monitoring of groundwater supply and quality because conditions in groundwater cannot be easily observed (Hodgson and Manus, 2006).

REFERENCES

1. Canencia, O. P., Dalugdug, M. D., Emano, A. M., Mendoza, R., & Walag, A. M. P. (2016). Slaughter waste effluents and river catchment watershed contamination in Cagayan de Oro City, Philippines. *ResearchGate, 9*(2). https://doi.org/10.2220/2220-6663.
2. Dickens, C. W. S., & Graham, P. M. (2002). The Southern Africa scoring system (SASS) version 5 rapid bioassessment for rivers. *African Journal of Aquatic Science, 27*, 1–10.
3. Diersing, N. (2009). Water quality: Frequently asked questions. Florida Brooks National Marine Sanctuary, Key West, FL.
4. Floor, G. H., Calabrese, S., Román-Ross, G., D'Alessandro, W., & Aiuppa, A. (2011). Selenium mobilization in soils due to volcanic-derived acid rain: An example from Mt Etna volcano, Sicily. *Chemical Geology, 289*(3, 4), 235–244. https://doi.org/10.1016/j.chemgeo.2011.06.003.
5. Furusawa, T., Maki, N., & Suzuki, S. (2008). Bacterial contamination of drinking water and nutritional quality of diet in the areas of the western Solomon Islands devastated by the April 2, 2007 earthquake/tsunami. *Tropical Medicine and Health, 36*(2), 65–74. https://doi.org/10.2149/tmh.2007-63.
6. Hanaor, D. A. H., & Sorrell, C. C. (2014). Sand supported mixed-phase TiO_2 photocatalysts for water decontamination applications. *Advanced Engineering Materials, 16*(2), 248–254. https://doi.org/10.1002/adem.201300259.
7. Hodgson, K., & Manus, L. (2006). A drinking water quality framework for South Africa. *Water SA, 32*(5), 673–678.
8. Kjellstrom, T., Lodh, M., McMichael, T., Ranmuthugala, G., Shrestha, R., & Kingsland, S. (2006). Air and water pollution: Burden and strategies for control. In D. T. Jamison, J. G. Breman, A. R. Measham, & G. Alleyne (Eds.), *Disease Control Priorities in Developing Countries* (2nd ed.). World Bank. PMID: 21250344
9. Rafferty, J. P. (2012). What happened to acid rain? In K. Satake (Ed.), *Acid Rain 2000: Proceedings from the 6th International Conference on Acidic Deposition: Looking Back to the Past and Thinking of the Future* (p. 20). Springer.
10. Seinfeld, J. H., & Pandis, S. N. (1998). *Atmospheric Chemistry and Physics–From Air Pollution to Climate Change*. John Wiley and Sons, Inc. ISBN: 978-0-471-17816-3.
11. Weathers, K. C., & Likens, G. E. (2006). Acid rain. In W. N. Rom & S. Markowitz (Eds.), *Environmental and Occupational Medicine* (4th ed., pp. 1549–1561). Lippincott-Raven Publishers. ISBN: 0-7817-6299-5.

CHAPTER 10

AIR POLLUTION ANALYSIS AND CONTROL

Contents

INTRODUCTION

One of the most important environmental issues of our day is air pollution, which puts human health, ecosystems, and the climate at serious risk. The release of harmful pollutants into the atmosphere has increased due to global urbanization and industrial activities, which has resulted in a decline in air quality in many regions. In order to lessen its effects and safeguard the environment and public health, scientists, legislators, and environmental organizations must now prioritize the analysis and control of air pollution. Analyzing air pollution entails methodically tracking and evaluating air pollutants. Particulate matter (PM), sulfur dioxide (SO_2), nitrogen oxides (NOx), carbon monoxide (CO), volatile organic compounds (VOCs), and ground-level ozone (O_3) are some of these pollutants. These pollutants' concentration and distribution in the atmosphere are measured using sophisticated analytical methods and monitoring systems. An improved understanding of the factors causing air pollution is made possible by data gathered from satellites, mobile sensors, and air quality monitoring stations. These sources of information offer insights into pollution patterns, trends, and sources.

Effective control strategies require an understanding of the sources of air pollution. The main causes of air pollution are emissions from automobiles, manufacturing, electricity production, and agriculture. Particularly in developing nations with fast-expanding populations and underdeveloped infrastructure, traffic emissions are a substantial source of pollution in urban areas. By using air pollution analysis to identify and quantify these sources, targeted interventions to lower emissions and enhance air quality can be developed.

Reducing air pollution necessitates a multidimensional strategy that includes both technological advancements and legislative actions. Technological advancements in emission control technologies, like scrubbers, electrostatic precipitators, and catalytic converters, have significantly decreased the amount of pollutants released by automobiles and industrial sources. Furthermore, adopting energy-efficient practices and switching to greener energy sources, like renewable energy, help to reduce air pollution.

10.1. AIR POLLUTION MONITORING

The term "air pollution" describes the degradation of the natural environment or the release of physical, chemical, or biological substances into the atmosphere that could endanger or discomfort humans or other living things. Our health and quality of life are significantly impacted by the composition of the outside and indoor air. Significant effects of air pollution and greenhouse gas emissions can be seen in the environment, encompassing more general worldwide environmental problems like climate change and stratospheric ozone depletion.

Primary and secondary air pollutants are the two broad categories into which they fall. Primary air pollutants are typically defined as materials that are directly released into the atmosphere as a result of either natural or man-made processes. Examples of these include sulfur dioxide (SO_2) emitted from factories, CO (CO) gas from motor vehicle exhaust, and ash from volcanic eruptions. Primary pollutants do not, however, cause every negative consequence of air pollution on their own. Primary pollutants and atmospheric constituents may undergo chemical reactions, as Figure 10.1 illustrates. Subsequently, a wide variety of novel chemical compounds are created when gaseous pollutants react with one another and with airborne particles. Secondary air pollutants are air pollutants that are not emitted directly but instead form in the air. They cause smog, haze, eye irritation, and damage to materials and vegetation, among other negative effects of air pollution.

Obtaining a comprehensive picture of the existing pollution situation through air pollution monitoring is the first step towards putting policies against air pollution in place. A solid scientific foundation for managing air quality and controlling the sources of air pollution is provided by measuring air quality and comprehending its effects. Therefore, monitoring air quality or pollution levels is essential for formulating policies and strategies, gauging adherence to recommended values, and tracking advancement toward environmental objectives.

Thus, a great deal of work needs to go into the systematic measurement of air pollution levels at various scales, from the local to the global. Monitoring air pollution and the data it yields are not goals unto themselves. Rather, it provides the most effective means of comprehending air pollution issues, evaluating, and reevaluating environmental actions to enable pollution issues to be successfully addressed at the local, state, and federal levels (Bower and Mucke, 2004).

The ultimate goal of air pollution monitoring is to gather trustworthy data that scientists, decision-makers, and planners can use to help them manage and enhance the general quality of the environment (Bower and Mucke, 2004). Monitoring data can be utilized to create policies that are specifically designed to address the most pressing issues of concern, in addition to emission inventories and well-defined targets. Presenting monitoring data in an eye-catching manner also aids in informing nearby communities about air quality concerns and guarantees that the effects of management choices can be fairly evaluated. Any effective air quality management strategy must include high-quality monitoring data that track changes in pollution levels and signal potential health or environmental effects (Mucke et al., 1995). All monitoring programs have limitations, though, and using analytical data from one method should therefore be used in conjunction with the findings of other methods of assessment, such as modeling, emission inventories, interpolation, and mapping.

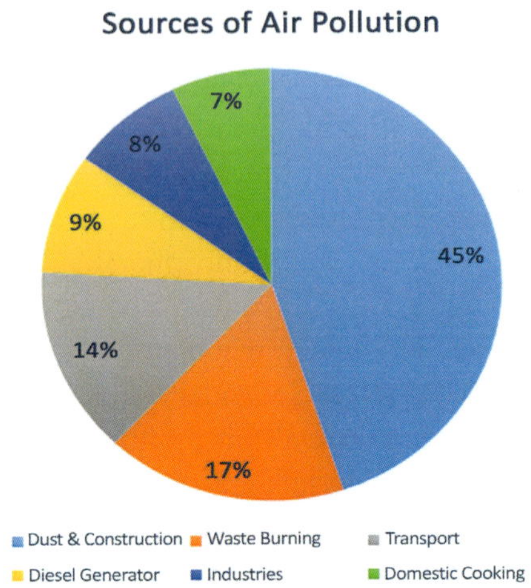

Figure 10.1. *Air pollution: an overview.*

Source: By Laira Aggarwal – http://indpaedia.com/ind/index.php/Air_pollution:_India, CC BY-SA 4.0, https://commons.wikimedia.org/w/index.php?curid=77198080.

10.1.1. Objectives of Air Pollution Monitoring

Any air pollution monitoring program generally has the following broad objectives (Ahmet and Dijk, 1994; Fisher et al., 1995):

1. To provide a sound scientific basis for the development of cost-effective control policies and solutions to abate air pollution.
2. To assess how far air quality standards, limit values, and objectives are being met.
3. To evaluate the potential impacts of air pollution on population health and welfare.
4. To determine the impact of air pollution on ecosystems and our natural environment.
5. To provide the public with reliable and up-to-date information on air pollution.
6. To fulfill statutory air quality reporting requirements.

Identifying the areas in the region where monitoring is necessary because the air quality there is likely to deteriorate or be at risk is a crucial first step in creating a successful monitoring program. The areas at risk can be determined by considering the following (Harrison and Perry, 1986):

1. Emissions sources (e.g., domestic, industrial, transport, agricultural, and natural) and contaminants emitted, and also location and magnitude.
2. Meteorology (areas prone to temperature inversions, etc).
3. Topography.
4. Geography.
5. Population centers (especially where domestic fires and traffic emissions occur).
6. Historical monitoring data (if available).
7. Areas with high natural environmental value (e.g., in, and around natural parks, forests, wilderness, and wetlands).
8. Areas vulnerable to air pollution plumes from other areas.
9. Areas planned for development (e.g., to get a picture of background concentrations).
10. Any public complaints or issues of concern relating to air pollution.
11. Any epidemiological studies on air quality effects on health already carried out.

10.1.2. Sample Collection

10.1.2.1. Selection of Contaminants for Air Quality Monitoring

It is impossible to monitor all contaminants in the air; hence, monitoring programs usually measure only a small group of contaminants that act as indicators of air quality. Particular contaminants are selected on the basis that:

1. They cause significant adverse health or environmental effects.
2. They are commonly discharged from known or suspected sources within the area.
3. They provide a good indication of the overall quality of air.

Which pollutants to keep an eye on can be effectively determined by the characteristics of the sources in the area. For instance, monitoring should be done for nitrogen dioxide (NO_2), carbon monoxide (CO), and potentially benzene as well as fine particles ($PM_{2.5}$) if automobiles are the main source of pollution. If wood-burning domestic fire emissions are present in the area, it is important to look for particles, CO, and polycyclic aromatic hydrocarbons (PAHs). SO_2 levels should be kept an eye on in areas where coal is frequently used for home heating or by industry. It's crucial to take into account any guidelines or standards when choosing which contaminants to monitor and whether the results can be compared to them. It is possible to determine whether a contaminant is likely to have a negative impact on human health or the environment by comparing monitoring data with recommended values. Typically, reference is made to regional, national, and occasionally international guideline values for this purpose. Both highly qualitative and quantitative guidelines are possible.

10.1.2.2. Selection of Monitoring Sites

The choice of site is crucial for any program that monitors air pollution. The area that the data would represent is directly correlated with the location of a monitoring station. The goals of the monitoring program determine where the sample will be taken. It is typically inappropriate to place a monitoring site near a strong localized source, like a factory or garage vent, for the purpose of monitoring air pollution within an urban airshed (Boubel et al., 1994; Harrison and Perry, 1986). High contaminant values that typically do not cause issues or accurately reflect the general air quality in an area may be detected at these sites. However, it is appropriate to position the monitoring equipment close to the source for certain projects, such as figuring out the relationship between CO concentrations at a roadside and vehicle emissions, or between an industrial discharge and its effects on air quality. A site that has either shielding or funneling must be carefully chosen so that local airflows do not significantly impact it. The environment being observed, or potentially being observed, should be reflected in the monitoring location. To put it another way, the monitor ought to be placed where people could potentially be impacted, both now and in the future if negative effects on human health are a worry. Similarly, if ecosystem impacts are the problem, the monitor ought to be placed close to a delicate ecosystem that might be impacted.

It is imperative to consider the pertinent averaging period for the guideline of the contaminant of concern when choosing a site. People who are exposed to CO levels above the recommended levels for either one hour or eight hours, for instance, may experience health issues. Therefore, it is appropriate to place the CO equipment in areas where people may be exposed for one or more hours at a time. If domestic emissions account for a sizable portion of CO emissions, these sites are situated within broader airsheds, close to busy traffic intersections, and along traffic corridors. However, the health effects of benzene are typically assessed over the course of a year of exposure. In order to determine the annual total exposure of the majority of urban residents, it is, therefore, appropriate to place a benzene monitor in an urban residential area that is not directly impacted by a busy road. This is because people rarely spend a significant amount of time directly next to a busy road. Still, placing the monitor close to the road might make sense if a more thorough personal exposure assessment is being conducted. Furthermore, information from all the major locations where a person may be exposed should be gathered to provide data for epidemiological studies. This may apply to air inside buildings or air inside cars. Site selection should also take into account atmospheric reactions affecting the formation and destruction of the contaminant. For example, ozone is formed over time, as pollutants in the atmosphere react with each other in the presence of sunlight. It is therefore appropriate to monitor ozone (O_3) where it is most likely to be present, namely downwind from the major sources of its precursors (e.g., NOx, VOCs, etc.). Similarly, NO_2 is another contaminant that is not discharged into the air in significant concentrations but rather formed over time from the NO released from combustion sources. Therefore a monitor close to a busy road, where NO has not had time to convert to NO_2, may not be the best location for determining NO_2 concentrations.

10.1.2.3. Equipment Maintenance and Inspection

Accurate and trustworthy monitoring results depend on proper equipment maintenance and inspection. The majority of data loggers and other air pollution monitoring devices need to be calibrated on a regular basis. Drift and instrumental bias are frequent occurrences. Based on the results of the calibration, data should be updated. It is necessary to keep a record of the calibration processes and history, which should be accessible upon request. Simple visual inspections of the functionality of the equipment can be used as a kind of calibration, as can multipoint measurements on intricate contaminant mixtures and in-depth analyses of individual parts. Checking if the monitoring apparatus measures a contaminant's concentration accurately in a sample with a known concentration is a common calibration step. When feasible, data loggers should be used to automatically record air quality data at the highest possible temporal resolution (WHO, 1994c).

It is best to test the accuracy and variability of various equipment/method combinations by comparing their performance in different labs or at the same location while analyzing duplicate samples. This is especially helpful in distinguishing between different particulate sampling techniques. When there is uncertainty about the accuracy or dependability of a specific method, or when alternative, less costly methods are being considered, such comparisons ought to be conducted.

10.1.2.4. Data Validation and Interpretation

In addition to gathering data, monitoring aims to generate knowledge that will benefit technical, policy, and general public end users. There is very little use for raw data by itself. Before being compiled, these data must be verified and validated to create a trustworthy dataset (CRC, 1997; Kotlik and Vondruskova, 1996; WHO, 1994a,b,c).

The site situation and its implications should be considered when interpreting data from any air pollution monitoring site. Along with any nearby sources of air pollution, a general description of the site's features should be provided. In addition to recording meteorological data like wind speed, direction, and temperature, it is important to include a brief description of any other meteorological conditions that might have an impact on the site's air quality, such as inversions, the direction of the prevailing wind, and elevation above sea level (Larsen and Schaug, 1995; USEPA, 1987; WMO, 1992).

10.1.3. Selection of Monitoring Methods

There exists a vast array of techniques for quantifying pollutants in ambient air, varying correspondingly in terms of precision and cost. The goals and purposes of the monitoring program, the available funds, and the necessity of adhering to any national recommendations or standard procedures should all be taken into account when selecting the specific monitoring techniques (Larsen et al., 1999). The following categories apply to air monitoring techniques.

10.1.3.1. High-Precision Instrumental Methods

These techniques have a high degree of measurement precision and offer continuous records of pollutant levels for prolonged periods of time (weeks or months) with little assistance from the operator. These systems' detection levels are frequently at least one order of magnitude lower than the average background levels found in cities. These are the priciest monitoring techniques, and proper operation and calibration are needed to guarantee high precision. When conducting research studies or other targeted investigations, high-precision instrumental techniques like differential optical absorption spectroscopy (DOAS) are typically employed to better understand how pollutant levels fluctuate over brief time periods (hours or days).

10.1.3.2. Lower-Level Instrumental Methods

High-precision methods are more expensive than instrumental methods, such as automated monitors used for routine air quality monitoring. Nonetheless, they typically call for more frequent operational inspections and modifications, and the measurement accuracy is only slightly below average background levels.

10.1.3.3. Manual Particulate Methods

Manual filter-based particulate collection techniques yield time-averaged PM results, usually over a 24-hour period, and usually necessitate manual sampling filter changes between samples; however, a variety of semiautomated systems are also accessible. Most manual particulate methods only achieve precisions of 10–20% of average background levels.

10.1.3.4. Manual Wet-Chemical Methods

Typically, manual wet-chemical methods are employed for a broad range of gaseous pollutants, including SO_2 and NO_2. These methods yield time-averaged results, usually spanning over 24 hours, and necessitate frequent operator intervention. Most wet-chemical methods have a precision that is only slightly lower than average background levels, and other pollutants can occasionally cause interference with the methods. The lower-level instrumental monitors' operational costs are strikingly similar to ours (Larsen et al., 1999).

10.1.3.5. Passive Monitoring Methods

Typically, long-term (often monthly) sampling is used to survey an area or categorize it as having low, medium, or high pollution levels. Typically, they don't offer any information regarding short-term pollution levels and don't permit accurate measurement of the real pollutant concentrations without field calibration through simultaneous real-time measurements with active analyzers. Due to its lack of need for regular maintenance and electricity, this is the least expensive monitoring option (WHO, 1994a). When measuring trace gases like O_3, SO_2, and NO_2 in rural or isolated locations, passive samplers are frequently employed.

10.1.4. Common Monitoring Methods for SO_X, NO_X, and PM

10.1.4.1. Monitoring of Gaseous Sulfur and Nitrogen Compounds

The corresponding oxides, hydrides, and organic compounds of sulfur (S) and nitrogen (N) are the gaseous compounds that are of concern in atmospheric pollution. Severe oxides (SO_2) and sulfur trioxide (SO_3) are the most significant pollutants among them. When fossil fuels containing sulfur are burned, sulfur is oxidized to SO_2 and, to a lesser extent, SO_3. This process is the primary source of SO_2. Nitric oxide (NO) and NO_2 are the nitrogen oxides that are most concerning. The quantification of these pollutants can be done using both chemical and physical techniques. Acidimetric, colorimetric, chromatographic, and coulometric chemical techniques are used to analyze gaseous sulfur and nitrogen pollutants; specifics of these techniques are available in the pertinent literature (Harrison and Perry, 1986; Lodge, 1989a,b; Quirit et al, 2007; Vadjic et al., 1992). Furthermore, physical methods that use optical techniques like fluorescence, chemiluminescence, and absorption spectroscopy are used to measure gaseous S and N compounds.

10.1.4.2. Measurement of PM

An array of morphological, chemical, physical, and thermodynamic properties characterize ambient aerosol, which is a complex mixture of both organic and inorganic, volatile, and semi volatile, water-soluble, and -insoluble matter. Particles produced by combustion, such as fly ash, diesel, and soot; particles produced by photochemistry, such as those found in urban haze; salt particles produced by sea spray; and crustal particles from resuspended dust are all included in atmospheric PM. Certain particles might contain moisture that is bound to them and be hygroscopic. The relative humidity of the surrounding air and the makeup of the particles determine how much moisture is present. Generally speaking, water makes up more than half of the mass of fine air particles at relative humidity levels above 80%. Airborne PM measurements are frequently performed for many purposes, such as determining the distribution of sources, assessing the efficacy of control measures, and examining the connection between air quality and health. Measurements of PM10 (PM with aerodynamic diameter ≤10 μm) and $PM_{2.5}$ (PM with aerodynamic diameter ≤2.5 μm) are problematic due to the highly variable nature of airborne PM.

Accurate quantification of minute amounts of mass is necessary for measuring the mass of PM suspended in air, whether it is coal dust in mines or ambient air PM. In general, there are two methods for going about this measurement: indirect methods that estimate the mass based on other properties of the particles, and direct micro weighing techniques that actually determine the mass.

10.1.4.3. Direct Mass Measurement Techniques

The gravimetric technique is one of the direct mass measurement methods. The easiest and most straightforward method of sampling is to gather particles on filters. The PM mass is obtained by calculating the difference between the pre- and post-sample weights using a laboratory gravimetric balance. Determining the PM mass concentration is made possible by knowing the volume of air that passes through the filter. This traditional technique of measuring the mass deposited on a filter by weighing it is frequently regarded as the gold standard (Federal Register, 1997; Noble et al., 2001).

Soluble aerosol particles can be separated using water, and the concentrations of cations and anions can be measured using an ion chromatograph. For elemental composition analysis, insoluble particles are commonly subjected to instrumental neutron activation analysis (INAA), proton-induced x-ray emission (PIXE), or inductively coupled plasma mass spectrometry (ICP-MS). Gas chromatography is usually used to analyze semi volatile components. An electron capture detector or one combined with a mass spectrometer is usually used (Parungo et al., 1992). Particulate monitors collect individual particles, and the bulk composition of these particles varies greatly, reflecting the variety of the particles' origins and atmospheric processing. The relationship between the mass and composition of particles gathered on a filter and the actual composition of PM in the atmosphere has been observed to be fraught with uncertainty. Chuen-Jinn and Hsin-Ying (1995) reported that sampling artifacts, such as the loss of semivolatile material, can be found for a variety of filter-based measurement systems. Due to their high alkalinities, glass fiber filters are known to frequently adsorb SO_2 and HNO_3 gases, forming extraneous sulfate and nitrate in collected particle samples. This phenomenon is observed to cause an overestimation of both substances and may have an impact on the mass of PM that is recorded. This could happen in the deposited PM as well as the filter material. During the sampling process, evaporation, and/or chemical reactions may result in negative artifacts. One instance is the nitrate loss brought on by the reactions of ammonium nitrate that has accumulated on the filter media with acidic aerosols, like particulate H_2SO_4. Nitric acid gas evaporates as a result, and ammonium nitrate is lost. Furthermore, as air passes through aerosol sampling systems, the pressure drop causes a drop in the gas phase concentration of some species. This results in another evaporative loss of semivolatile material from the mass gathered on the filter by creating a concentration gradient between the air stream and the particulate mass. The extent of these processes varies with location, based on meteorological conditions, aerosol mass concentration, and composition. There have also been reports of artifacts related to transportation, storage, and sampling. One instance is volatile compound loss, which can occur following sampling but before weighing or chemical analysis. Additionally, the dust-loaded filter's temperature and humidity history have a significant impact on the filter's actual mass.

10.1.4.4. Indirect Mass Measurement Techniques

The beta gauge and the tapered element oscillating microbalance (TEOM) are two instances of equivalent indirect methods that can yield measurements of ambient particulate concentrations almost continuously. Filter pressure drop methods and optical methods are two more instances of indirect techniques. Since there is no consistent physical relationship between other particle properties and particle mass, indirect methods are inherently problematic.

1. **Tapered Element Oscillating Microbalances:** This real mass measurement method, which is filter-based and has high time-resolution capability, is a significant advancement in PM monitoring. For a variety of demanding applications, inertial mass detectors, or TEOMs, have been developed over the past few decades to provide this capability. TEOM mass detectors were first created for space-

related applications. Today, they are utilized in instruments that examine the chemical kinetics of catalysts as well as in near-real-time PM monitors for ambient air, diesel exhaust, stack emissions, and fly ash carbon concentration (Allen et al., 1997; Ayers et al, 1999; Hering et al, 2004,).

The TEOM monitor draws a sample through a sampling inlet, filters it, and weighs it continuously to provide a filter-based, direct mass measurement of PM in ambient air. Systems that do not directly measure mass have uncertainty in their mass calibration, which is eliminated by TEOM technology. Furthermore, filter-based mass measurements can be obtained in almost real time with manual, gravimetric methods, which also eliminate the possibility of filter handling errors (both at the sampling site and in the laboratory). All TEOM-based monitoring instrumentation's engineering design offers automated, real-time flow control, sample conditioning, and PM mass quantification. The TEOM can be used to measure total suspended particles (TSPs) or size-fractionated PM (e.g., PM_{10}, $PM_{2.5}$, and PM_1) mass concentration when used in conjunction with a size-selective inlet.

Furthermore, a number of well-known research projects that examine semivolatile particles, fine, and coarse particles, and particle behavior use the TEOM as their central mass measurement system. The total nonvolatile and volatile particle mass, the retention of particular semivolatile particle constituents, and the optimization of the TEOM to remove particle-bound water can all be achieved with optional add-on components. However, several doubts have been expressed regarding the precision of real-time monitors' ability to measure atmospheric particles because those that are run at high temperatures, like the TEOM, are known to lose semivolatile material, with the mass recorded by the TEOM being, on average, 18.3% lower than that of the reference method throughout Europe.

2. **Beta Gauge:** One popular continuous monitoring technique for measuring PM_{10} concentrations is the beta gauge. It is based on the measurement of the attenuation of ionizing radiation through particulate mass deposited on a filter, which is achieved by beta-ray absorption in a sample captured on filtering material. With a time resolution of roughly 0.5–2 hours, it enables long-term unattended operation. The particulate's beta absorption coefficient, which varies depending on its chemical composition, determines the instrument's response. The majority of monitoring applications do not significantly suffer from this limitation because the variation is

not very large. Information on the concentration of the suspended particle matter is provided by the difference in beta-ray absorption between the exposed and nonexposed filtering material, which is proportional to the mass of the suspended particle matter.

3. **Light-Scattering Instruments:** Although light-scattering instruments have been around for a while, their primary application has been in tracking dust exposure at work. Some of these instruments have been modified for ambient monitoring over the past few years, with varying degrees of success. The "workplace" units provide a direct readout of particle concentrations and are comparatively inexpensive and portable. Nevertheless, they typically have very low measurement sensitivity and precision. They are therefore only appropriate for simple survey work. However, there are currently some instruments that offer a significantly better level of performance, to the point where they can be used for monitoring against ambient guidelines and in regional networks. One such device is the GRIMM aerosol spectrometer, for which the makers anticipate receiving equivalency status from the USEPA soon.

   The primary drawback of light-scattering instruments is that, as opposed to the total mass of airborne PM, the instrument response is contingent upon the size distribution and particle count. This can be partially mitigated by performing routine manual filter sampling calibrations. However, due to variations in the makeup and characteristics of the airborne particles, these calibration "factors" are probably going to change depending on the monitoring site and the season. Particle size distributions can also be inferred from certain light-scattering devices. This could be useful in certain research projects.

10.2. AIR POLLUTANT MEASUREMENT

Three different types of measurement scales, such as big, medium, and small, are generally used to monitor air pollutants. When selecting an appropriate technique for detecting air pollutants, one of the characteristics that must be taken into account is the measurement scale's size. In general, large measurement scales work very well at detecting or monitoring air pollutants because of their high resolution; however, they are expensive and challenging to use. Some methods, like Ozone Monitoring Instrument (OMI), Infrared spectroscopy, Differential Optical Absorption Spectroscopy (DOAS), and Absorption Spectroscopy technique, use large measurement scales. The medium-type measurement scale is inexpensive, lightweight, and portable for detecting air pollutants. Some examples of medium measurement scale sensors are the capacitor gas sensor, semiconductor gas sensor, and solid electrolyte gas sensor. Because of its poor accuracy, the small type of measurement scale is uncommonly used to detect air pollutants.

10.2.1. Air Pollutant Gases Measurement Techniques

10.2.1.1. Absorption Spectroscopy

With a higher transition in the ultraviolet (UV) spectral region, sulfur dioxide is the primary gas that pollutes the environment. One useful technique for keeping an eye on airborne sulfur dioxide levels is absorption spectroscopy. Differential optical absorption spectroscopy (DOAS) and correlation spectroscopy (COSPEC) are two techniques for detecting sulfur dioxide in the air.

The improved method, also known as the Multi-Axis Differential Optical Absorption Spectroscopy (MAX-DOAS), is an improvement over the DOAS method. (MAXDOAS) is a technique that uses scattered sunlight observations from the ground to monitor trace gases, such as sulfur dioxide, in the lower troposphere. With a typical spectral resolution of 0.5 nm in both the horizontal and vertical dimensions, MAXDOAS instruments monitor scattered solar radiation in the UV and visible regions. One of the well-known techniques for identifying pollutants in the atmosphere is absorption spectroscopy. This method is primarily applied in the near-infrared region (0.6–2 μm), where weaker overtone and combination bands occur, and the mid-infrared spectral region (2–15 μm), where many species of interest have fundamental vibrational absorption bands. Strong electronic transitions are present in a large number of atoms and molecules in the ultraviolet (UV) spectral region (200–400 nm). The technique known as absorption spectroscopy is derived from Beer-Lambert's law. This approach is used for real-time measurement, specifically for monitoring sulfur dioxide pollution in the ultraviolet spectrum through broadband absorption spectroscopy. The devices used to set up the real-time sulfur dioxide monitoring system include a deuterium lamp, lens, sulfur dioxide cell, multimode optical fiber, monochromator charge coupled device (CCD) array, and computer (Table 10.1).

The efficacy of this method for measuring sulfuric acid emissions from coal-fired boilers is tested in Harbin, China. The concentration that was measured ranged from 100 to 200 ppm. The higher value concentration observed indicates that China's bituminous coal has a higher sulfur content. This demonstrates that, as predicted, the spectrometer can detect the sulfur emitted. This experiment suggests that the Doppler Effect alone can affect the absorption spectrum through temperature.

Table 10.1. *Order of World Importance for Air Pollutants, Variables Used and Air Quality Indexes Detected as Per a Review*

Phase	Keywords	Science Direct		Springer Link		Google Scholar		Taylor and Francis		Total Documents Detected	Average Q Index	Mean Quartile	Quartile Variation
		Detected Documents	Q Index	Detected Documents	Q Index	Detected Documents	Q Index	Detected Documents	Q Index				
Information universe	Index, Air Pollution and Asthma	4359	1.00	4497	1.00	20,800	1.00	2509	1.00	32,665	1.00	Q4	Q4, Q4, Q4, Q4
Air pollutants	Index, air pollution, asthma, and nitrogen dioxide	1502	0.33	1469	0.33	17,900	0.86	748	0.30	21,719	0.66	Q3	Q2, Q2, Q4, Q2
	Index, air pollution, asthma, and ozone	1651	0.34	1503	0.33	16,500	0.79	809	0.32	20,463	0.63	Q3	Q2, Q2, Q4, Q2

	Index, air pollution, asthma, and nitric oxide	892	0.18	785	0.17	17,600	0.85	479	0.19	19,756	0.60	Q3	Q1, Q1, Q4, Q1
	Index, air pollution, asthma, and par-ticulate matter	2636	0.54	2001	0.44	17,200	0.83	1021	0.41	22,858	0.70	Q3	Q3, Q2, Q4, Q2
	Index, air pollution, asthma, and sulfur dioxide	1206	0.25	1159	0.26	15,700	0.75	651	0.26	18,716	0.57	Q3	Q2, Q2, Q4, Q2
	Index, air pollution, asthma, and carbon monoxide	1191	0.25	1070	0.24	17,100	0.82	627	0.25	19,988	0.61	Q3	Q2, Q1, Q4, Q2
Climate variables	Index, air pollution, asthma, and precipita-tion	782	0.16	1007	0.22	15,200	0.73	494	0.20	17,483	0.54	Q3	Q1, Q1, Q3, Q2

	Index, air pollution, asthma, and wind speed	898	0.13	1233	0.27	14,900	0.72	320	0.13	17,351	0.53	Q3	Q1, Q2, Q3, Q1
	Index, air pollution, asthma, and wind direction	681	0.14	1181	0.26	16,800	0.81	195	0.08	18,857	0.58	Q3	Q1, Q2, Q4, Q1
	Index, air pollution, asthma, and tem-perature	2806	0.53	2812	0.63	17,100	0.82	1275	0.51	23,993	0.73	Q4	Q3, Q3, Q4, Q3
	Index, air pollution, asthma, and atmo-spheric pressure	931	0.19	1218	0.27	17,900	0.86	651	0.26	20,700	0.63	Q3	Q1, Q2, Q4, Q2
	Index, air pollution, asthma, and solar radiation	498	0.10	710	0.16	17,600	0.85	172	0.07	18,980	0.58	Q3	Q1, Q1, Q4, Q1

Air quality indexes	Index, air pollution, asthma, and Common Air Quality Index	2672	0.55	3167	0.70	18,300	0.88	1286	0.51	25,425	0.78	Q4	Q3, Q3, Q4, Q3
	Index, air pollution, asthma, and Air Quality Index	3899	0.80	3926	0.87	19,000	0.91	1831	0.73	28,656	0.88	Q4	Q4, Q4, Q4, Q3
	Index, air pollution, asthma, and Air Quality Health Index	3805	0.78	3873	0.86	9010	0.43	1273	0.51	17,961	0.55	Q3	Q4, Q4, Q2, Q3
	Index, air pollution, asthma, and Air Pollution Tolerance Index	743	0.15	1067	0.24	15,400	0.74	266	0.11	17,476	0.54	Q1	Q1, Q1, Q4, Q1

	Index, air pollution, asthma, and Arithmetically Aggregated Pollutant Index	25	0.01	13	0.00	83	0.00	1	0.00	127	0.00	Q1	Q1, Q1, Q1, Q1

Source: Barrera-Heredia AD, Zafra-Mejía CA, Cañas Arboleda A, Fernández Sánchez MJ, López-Kleine L, Rojas Moreno A. Trends in the Use of Air Quality Indexes in Asthma Studies. Atmosphere. 2024; 15(7):847. https://doi.org/10.3390/atmos15070847.

This approach will not take into account the impact of dust and other gases' scattering effects or gas absorption conclusions. This approach is suggested due to its simplicity in evaluating data, affordability, and greater dynamic range in CDD array.

10.2.1.2. Ozone Monitoring Instrument (OMI)

In July 2004, NASA's Earth Observing System AURA satellite unveils the ozone monitoring instrument (OMI) to the global community. With a 13 km × 24 km spatial resolution, OMI is an ultraviolet/visible (UV/VIS) nadir solar backscatter spectrometer that can cover almost the entire planet in a single day. Gases including ozone, sulfur dioxide, and nitrogen oxide can be found using this technique. OMI can also measure UV irradiance at the surface, cloud coverage, cloud top heights, and aerosol characteristics. Because it makes use of a 2-D CCD detector, the OMI technique is contemporary. OMI is made up of three components: an optical assembly that includes the detector modules and optical bench, an electronics unit, and an interface adaptor module. A UV/VIS spectrometer measures the amount of solar radiation that is absorbed and scattered by the elements that make up the earth's atmosphere. OMI is composed of three channels: VIS, UV1, and UV2. With a spectral resolution of roughly 0.42 nm, OMI can measure the reflected solar radiation in the ultraviolet and visible portions of the spectrum in the UV1 channel, ranging from 270 nm to 314 nm. With a spectral resolution of roughly 0.45 nm, OMI can measure the reflected solar radiation in the ultraviolet and visible portions of the spectrum in the UV2 channel between 306 nm and 365 nm. With a spectral resolution of roughly 0.63 nm, OMI can measure the reflected solar radiation in the ultraviolet and visible portions of the spectrum in the 350–500 nm range for the final channel, VIS.

Because it can measure trace gases like ozone, nitrogen dioxide, sulfur dioxide, formaldehyde, chlorine dioxide, and hypobromite, the OMI technique is a general-purpose technique. This method's daily worldwide coverage will aid the authority in learning more about the chemistry of the troposphere and stratosphere, as well as climate change. OMI's high spectral resolution allows for the detection of air pollutants at an urban scale. Regretfully, the cost of using this method is high because it necessitates the newest technology and modern equipment.

10.2.1.3. Solid State Gas Sensor

Although there are many different kinds of solid state gas sensors, the most commonly used types are semiconductor, non-dispersive infrared absorption, and solid electrolyte. Sulfur dioxide and nitrogen dioxide can be detected or tracked using a solid electrolyte sensor. Weppner has classified the ionic species derived from the gas questionable into three categories based on whether they coincide with the mobile (Type I), immobile (Type II), or neither of them (Type III) of the solid electrolyte used. An auxiliary phase that was affixed to the solid electrolyte's surface was required by the Type III sensor. Ionic species extracted from the gas are encapsulated in compounds used in this auxiliary phase.

This auxiliary phase functions as a solid electrolyte with poor ion conductivity that combines from either a Type I or Type II half cell. The electrochemical cell is then created after that. Depending on the type of combined half cell, a few subgroups of type III sensors exist. Oxygenic gases like carbon dioxide, nitrogen dioxide, and sulfur dioxide are detected by a Type III sensor.

10.2.1.4. Infrared Absorption

A straightforward technique with a specific focus on gas detection is non-dispersive infrared absorption. The energy that gas molecules absorb is similar to their quantized energy, which is primarily found in the infrared spectrum. The degree of gas absorption correlates with the desired gas concentration for detection.

In the Nondispersive Infrared Sensor (NDIR) gas detection technique, a monochromatic beam with a broad range of 4 to 5 μm is retrieved from an infrared light source using a filter. In the low temperature range of 77 K to 196 K, indium antimonide (InSb), lead selenide (PbSe), and mercury cadmium telluride (HgCdTe) are typically utilized as sensing materials for infrared detectors. PbSe and Lead(II) sulfide (PbS) have a broad range of 3 to 5 μm, and they are typically used as sensing materials for infrared detectors that measure hydrocarbon (HC) gases. Along with the chromatic sensor used in the NDIR method, an infrared sensor is used to detect gases such as carbon dioxide, CO, and hydrogen chloride. Because of the Beer-Lambert law, which states that the absorptivity of the incident beam when the gas passes through the cell is proportional to the path length and the gas concentration, the output voltage of an infrared sensor is directly proportional to the concentration of gas.

10.2.1.5. Semiconductor Gas Sensors

Semiconductor gas sensors are well known for their ability to identify hazardous and combustible gases in the atmosphere. The density of the conducting electrons in the polycrystalline sensor element changes as a result of the gas being absorbed by the surface semiconductor material. Four steps make up a chemical reaction: pre-adsorption of oxygen on the surface of a semiconducting material, adsorption of a particular gas, reaction between the adsorbed gas and oxygen, and lastly, desorption of the reacted gas on the surface.

Air pollution can be detected using semiconductor sensors like the MQ7, MQ3, and MQ2. The MQ7 semiconductor gas sensor detects CO, and its sensitive material is SnO_2, which has a lower conductivity in clean, fresh air.

10.2.2. Mobile and Network Measurement Techniques

10.2.2.1. Mobile Air Quality Monitoring Network (MAQUMON)

Subsequently, the Mobile Air Quality Monitoring Network (MAQUMON) is presented, which consists of sensor nodes installed on automobiles or other vehicles. The device takes a minutely sample of the pollutants while the car is moving and stores the results along with a time and location stamp. Samples are only taken a few times per hour if the car is stopped. All data are uploaded to the server to be processed and published on the Sensor Map port if a car is in the Wi-Fi-accessible area. The system's benefits include being portable with a USB cable, having the ability to charge the integrated Li-ion battery and power the board for online operations, and having a 2-axis microelectromechanical systems (MEMS) accelerometer to detect motion. All of the system's power-hungry parts, including the gas sensor, GPS, and Bluetooth, will be off when it is idle. This system uses an on-board 20 channel SiRF-III-based GPS module with a 1 Hz sampling rate to provide location and time data. This technique can be used to measure the following pollutants: nitrogen dioxide, CO, ozone, and volatile organic compounds.

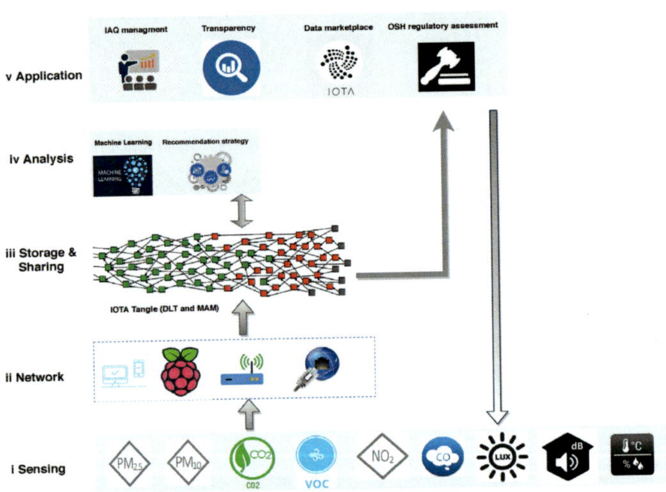

Figure 10.2. *IAQ data-monitoring application framework supported by Internet of Things (IoT) and Distributed Ledger Technology (DLT). IAQ: Indoor Air Quality; MAM: Masked Authenticated Messaging; OSH: Occupational Safety Health.*

Source: Sun S, Zheng X, Villalba-Díez J, Ordieres-Meré J. Indoor Air-Quality Data-Monitoring System: Long-Term Monitoring Benefits. Sensors. *2019; 19(19):4157. https://doi.org/10.3390/s19194157.*

The data, GPS data, and temperature and relative humidity data are stored in a 2MB serial flash device. Feedback regarding the state of the system, including connected interfaces, GPS lock, time, motion detection, and sensor readings, is shown on a 2 × 16 character LCD panel. The Intel 8051-based microcontroller is responsible for managing all system functions, such as USB protocol, analog to digital conversions, and battery charging. An example of another IOT-based application is illustrated in Figure 10.2.

10.2.2.2. GPRS-Sensors Array

An additional technique for identifying air pollutants is the GPRS-Sensors Array. This approach is a development of the sensor node method. This system made use of a fixed Internet-enabled Pollution Monitoring Server (Pollution-Server) and hardware known as the Mobile Data-Acquisition Unit (Mobile-DAQ). The components of the Mobile Data-Acquisition Unit (Mobile-DAQ) are a single-chip microcontroller, a Global Positioning System Module (GPS-Module), a General Packet Radio Service Modem (GPRSModem), and an array of air pollution sensors, including CO, nitrogen dioxide, and sulfur dioxide (Table 10.2).

A GPS physical, such as position, time, and date, is integrated with air pollution sensors. A fixed Internet-Enabled Pollution Monitoring Server (Pollution-Server) is composed of a laptop with internet access. This sensor is mounted on top of the car.

When it moves, a microcontroller attached to a GPS module creates a flame that indicates the amount of air pollution detected by the array of sensors. Next, the public mobile network uploads the pollutant frame to the (GPRS-Modem) and transmits it to the Pollution Server. Google Maps are linked to the Pollution Server, allowing it to show the levels of pollutants in real-time.

An IoT system for air pollution monitoring is shown in Figure 10.3.

Structure of air quality monitoring system main part.

Communication flow in the proposed system.

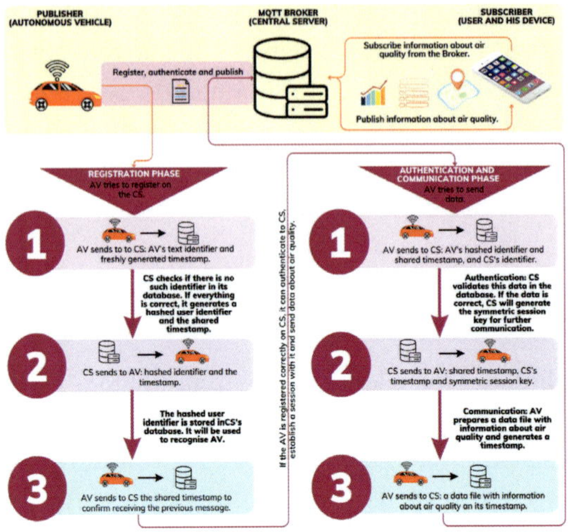

Vehicle used

Instrument used in the experiment.

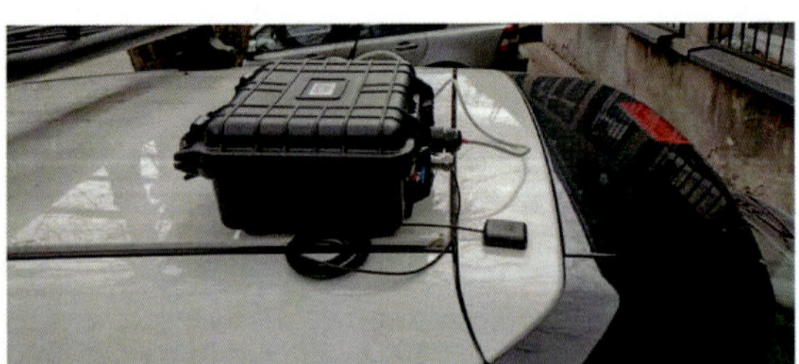

Figure 10.3. *An IoT system for air pollution monitoring.*

Source: Bobulski J, Szymoniak S, Pasternak K. An IoT System for Air Pollution Monitoring with Safe Data Transmission. Sensors. *2024; 24(2):445. https://doi.org/10.3390/s24020445.*

Table 10.2. *The Pollution Measurement System*

The data of air quality—day 1												
Time	**LAT**	**LON**	**ALT**	**PM10PM10**	**PM2.5PM2.5**	**NO_2**	**SO_2**	**NO**	**PID**	**Temp**	**Press**	**Hum**
	[°]	[°]	[m]	[µg/m³]	[µg/m³]	[ppm]	[ppm]	[ppm]	[ppm]	[°C]	[hPa]	[%]
14.11.2022 08:46	51.253555	22.563112	0	9.2	5.4	0.129	0	0.097	0.199	8.88	1007.26	68.89
14.11.2022 08:45	51.253555	22.563112	0	9.2	5.4	0.129	0	0.097	0.199	8.88	1007.26	68.89
14.11.2022 08:44	51.253555	22.563112	179.6	8.1	4.3	0.12	0	0.11	0.195	8.76	1007.06	68.9
14.11.2022 08:43	51.253135	22.560657	183.2	6.9	3.7	0.136	0	0.223	0.198	9.04	1007.17	68.27
14.11.2022 08:41	51.251875	22.551793	191.4	9.9	4.8	0.078	0	0.123	0.187	9.51	1006.38	66.89
14.11.2022 08:40	51.250048	22.552458	198.5	7.6	4.4	0.142	0	0.171	0.182	9.63	1004.89	65.52
14.11.2022 08:39	51.249043	22.554722	192.2	11.2	4.9	0.249	0	0.858	0.207	9.68	1005.28	65.99

14.11.2022 08:38	51.249638	22.55818	199.4	8.6	4.6	0.116	0	0.284	0.183	9.82	1004.94	65.06
14.11.2022 08:37	51.251273	22.558928	182.1	9	4.8	0.097	0	0.353	0.181	9.94	1006.06	64.2
14.11.2022 08:36	51.253255	22.561388	172.2	9.1	4.7	0.071	0	0.106	0.173	10.21	1007.28	62.32
14.11.2022 08:35	51.253255	22.561388	172.2	19.5	10.9	0.076	0	0.259	0.176	10.61	1007.33	61.47
14.11.2022 08:34	51.253093	22.563277	171.9	7.9	4.7	0.157	0	0.12	0.172	11.1	1007.52	58.85
14.11.2022 08:33	51.253477	22.56299	177.8	7.1	3.8	0.073	0	0.101	0.17	11.86	1007.43	55.57

The results of measurements—day 2

Time	LAT	LON	ALT	PM_{10}	PM2.5PM2.5	NO_2	SO_2	NO	PID	Temp	Press	Hum
	[°]	[°]	[m]	[μg/m³]	[μg/m³]	[ppm]	[ppm]	[ppm]	[ppm]	[°C]	[hPa]	[%]
2023–05–08 14:00	50.823513	19.11048	240.1	4.8	1.9	0	0.067	0.25	0.175	19.96	998.15	24.68
2023–05–08 13:59	50.823378	19.110498	243.6	5.9	1.7	0	0.063	0.263	0.177	19.3	998.13	25.94

2023–05–08 13:58	50.820803	19.111305	254.5	5.6	2.2	0	0.043	0.29	0.172	18.86	998.68	24.56
2023–05–08 13:56	50.823888	19.112418	315.1	4.5	1.8	0	0.058	0.319	0.179	20.78	998.37	22.78
2023–05–08 13:55	50.817292	19.114335	258.9	10.2	4.4	0	0.04	0.606	0.186	21.38	999.04	22.15
2023–05–08 13:54	50.817657	19.119252	263	11.1	2.8	0	0.027	0.399	0.185	22.84	999.26	20.24
2023–05–08 13:53	50.817695	19.12604	264.9	5.6	2.2	0	0.077	0.395	0.189	24.51	999.19	19.46
2023–05–08 13:52	50.817007	19.1292	260.0	7.8	2	0	0.076	0.332	0.188	24.52	999.57	18.93
2023–05–08 13:51	50.811723	19.130872	255.1	6.3	2.3	0	0.054	0.98	0.196	25.28	999.51	18.29
2023–05–08 13:50	50.811863	19.129823	254.8	5.4	2.1	0	0.069	0.324	0.191	25.42	999.44	17.82
2023–05–08 13:48	50.811735	19.123877	272.7	5.7	2.1	0	0.086	0.452	0.203	26.69	998.62	17.21
2023–05–08 13:47	50.811838	19.12062	264.3	8.5	2.8	0	0.07	0.409	0.205	26.85	998.57	17.18
2023–05–08 13:37	50.811457	19.113898	263.5	4.9	2.2	0	0.117	0.456	0.24	19.33	998.48	24.41

2023–05–08 13:31	50.806778	19.103398	273.5	5.5	2.3	0	0.233	0.69	0.305	16.54	997.53	29.27
2023–05–08 13:30	50.805333	19.103503	271.7	7.9	3	0	0.253	0.662	0.319	16.56	998.03	28.75
2023–05–08 13:29	50.806027	19.103187	270.9	7.1	2.3	0	0.3	0.882	0.337	17.09	997.9	28.29
2023–05–08 13:28	50.80842	19.103825	270.2	6.1	2.2	0	0.355	0.998	0.368	17.37	997.4	28.09
2023–05–08 13:27	50.810742	19.104067	268.0	6.3	2.1	0	0.404	0.729	0.37	17.69	997.61	26.86
2023–05–08 13:26	50.812808	19.104128	266.6	5.4	1.8	0	0.509	1.033	0.405	18.42	997.45	25.82
2023–05–08 13:25	50.817035	19.105248	267.4	5.3	1.7	0	0.687	0.881	0.44	19.13	998.57	25.4
2023–05–08 13:24	50.81727	19.112653	269.5	4.8	1.4	0	0.992	1.118	0.459	20.35	998.96	23.28
2023–05–08 13:23	50.820793	19.11131	282.5	4.8	1.7	0	2.172	1.68	0.509	22.02	998.56	22.18
2023–05–08 13:22	50.822162	19.110895	275.6	5.9	2.1	0	14.246	3.834	0.63	23.52	998.31	23.01

Source: Bobulski J, Szymoniak S, Pasternak K. An IoT System for Air Pollution Monitoring with Safe Data Transmission. Sensors. 2024; 24(2):445. https://doi.org/10.3390/s24020445.

Reducing emissions into the atmosphere is costly and technically challenging. Good air quality, however, is dependent on preventing pollution and limiting emissions, as rain is the only effective air-cleaning mechanism. There are several ways to achieve air emission control.

10.3. AIR POLLUTION CONTROL

10.3.1. Source Correction

It is frequently simpler to alter or stop a process that emits pollutants into the air than it is to try to capture the pollution. Although it might be required, a product or process could be altered to reduce emissions. For example, high lead levels in urban air were once caused by car exhaust. Lead was also decreased when gasoline was made free of lead, which was necessary for a catalytic converter to function properly. Similarly, less SO_2 is released when sulfur is removed from coal and oil before the fuel is burned. The source of the air pollution was fixed in these situations.

It is also possible to change processes to lessen air pollution. By running the incinerator at a temperature high enough to cause more thorough oxidation of odor-producing organic compounds, odors from municipal incinerators can be reduced. To reduce CO emissions from cars, the 1990 Clean Air Act requires the use of oxygenated fuels in urban areas. Controls are, strictly speaking, actions taken to meet emission standards, such as changing the equipment, raw material substitution, and process. On the other hand, abatement encompasses all instruments and techniques used to reduce the amount of pollution that enters the atmosphere after it has been produced by the source. We refer to all of the procedures as controls for the sake of simplicity.

10.3.2. Collection of Pollutants

The most significant issue with air pollution control may be the collection of pollutants for treatment. The main reason car exhaust is a known polluter is that it is very hard to contain and clean up. Treatment could be more effective in managing individual cars if vehicle exhaust could be directed to a central facility. One control mechanism is the recycling of exhaust gases. Even though recycling exhaust and blow-by gases alone won't allow cars to meet 1990 exhaust emission regulations, this approach proved to be a useful beginning for the automotive emission debate. Since even CO releases heat when burned to CO_2, many stationary industries recycle exhaust gases—typically CO and volatile organic compounds—as fuel for the process.

Process exhaust gases directed through one or more stacks are comparatively simple to gather; however, dust generated during the on-site transportation of partially processed materials, fugitive emissions from windows and doors, and wall fissures present a collection challenge. For proper control, some industries need to completely redesign their plant's air flow system.

10.3.3. Cooling

It is sometimes necessary to cool the exhaust gases before treating them because they are too hot for the control machinery. In order for certain pollutants to condense and accumulate as liquids, cooling may also lower the temperature below their condensation point. Dilution, quenching, and heat exchange are all appropriate cooling techniques. In addition to eliminating certain gases and particulates, quenching can produce a hot, dirty liquid that needs to be disposed of. Cooling coils are perhaps the most popular type of cooling technology and work particularly well in situations where heat conservation is possible.

10.3.4. Treatment

It is necessary to match the features of the treatment device with the properties of the pollutant in order to select the appropriate one. Pollutant particles range widely in size from microscopic particles with several mm in diameter to ideal gas molecules. Not every pollutant, or even every pollutant coming from the same stack, can be cleaned effectively and efficiently by a single device. The choice of a control mechanism may also be influenced by the chemical behavior of the contaminants. It is easy to distinguish between the different air pollution control devices that regulate gaseous pollutants and particulate matter.

10.3.4.1. Cyclones

The cyclone is a well-liked, cost-efficient, and successful particle-control method. Unclean air enters the cyclone off-center at the bottom creating a violent air swirl in the cone that accelerates particles centrifugally in the direction of the wall. Clean air exits at the center of the top of the cone as a result of the particles being slowed down by friction at the wall and sliding to the bottom, where they can be collected. Cyclones are frequently used as the initial step in the removal of dust because they are comparatively efficient at gathering large particles.

10.3.4.2. Fabric Filters

Particulate matter-controlling fabric filters work similarly to a vacuum cleaner. A fabric filter bag is used to catch dust particles as they are blown or suctioned through dirty gas. Periodically shaking the bag helps to remove the dust. In industrial settings, fabric filters are frequently employed as highly effective particle collectors, even for particles as small as sub microns.

It is believed that the fundamental mechanism of fabric filters' dust removal is comparable to that of sand filters used in the control of water quality. Surface force causes dust particles to stick to the fabric, trapping them. They come into contact with the fabric through Brownian diffusion or impingement. Simple sieving cannot be the removal method because fabric filters typically have an I to 1 air-space to fiber ratio.

Sometimes it's necessary to control very hazardous or toxic PM to better than 99.9% when its diameter is less than 1 gm. This level of control is achievable with a single stage of High Efficiency Particle Attenuation (HEPA) micropore or glass frit filters, through which the precleaned gas is forced or sucked by vacuum. 99.9999% control can be attained with four to six HEPA filter stages in series. For example, HEPA filters are frequently used to reduce the emission of radioactive particles.

10.3.4.3. Wet Collectors

For efficient removal of larger particles, use the scrubber, also known as the spray tower. By exerting force in a small throat area where water is introduced, more effective scrubbers increase the contact between air and water. In general, smaller gas bubbles or water droplets result from a more violent encounter, which makes cleaning more effective.

A venturi scrubber is a common high-energy wet collector. Water is introduced as high pressure streams perpendicular to the gas flow, and gas flow is restricted via a venturi throat section. The efficacy of the venturi scrubber in eliminating particles larger than 5 μm is nearly 100%.

Wet scrubbers have drawbacks despite their high efficiency and ability to capture both extremely fine PM and gaseous pollutants. Scrubbers use a lot of water, which after being used to scrub dirty gas either has limited use or needs additional treatment. Scrubbers might not be given the same priority as other uses of water in areas like the Colorado Basin, where water resources are scarce. Scrubbers also require energy consumption and are costly to build and maintain. Finally, a visible plume of water vapor is typically produced by scrubbers.

10.3.4.4. Electrostatic Precipitators

Fine PM is commonly trapped using electrostatic precipitators in situations where treating a large volume of gas is necessary but a wet scrubber is not the best option. Electrostatic precipitators are commonly used in coal-burning electric generating plants, primary, and secondary smelters, and incinerators. The process removes particles from the dirty gas stream as it travels over high-voltage wires, which are typically carrying a significant negative DC voltage.

After passing through these electrodes, the particles pick up an electrical charge and move through the electrostatic field to a grounded collection electrode. The high-voltage charging wire can be encircled by a cylindrical pipe or a flat plate, which can serve as the collection electrode. Either way, to remove the accumulated particles from its surface, it needs to be rapped on a regular basis with tiny hammer heads.

The efficiency of collection may be reduced when a layer of dust accumulates on the collecting electrode, especially if the electrode is inside a cylindrical pipe.

Furthermore, some dust adheres to the collection electrode rather than discharging due to its extremely resistive surface. Water-flushed or heated electrodes could be the solution to this problem. Very fine particles can be effectively collected by electrostatic precipitators. However, an electrostatic precipitator can use a significant amount of electrical energy, leading to a high operating cost, because the amount of dust collected is directly proportional to the current drawn.

10.3.4.5. Comparison of Particulate Control Devices

The approximate collection efficiencies, as functions of particle size, for the devices vary. Costs of collection also vary widely.

10.3.5. Control of Gaseous Pollutants

Gaseous pollutants can be captured, their chemical composition altered, or the process that generates them altered in order to remove them from the effluent stream. Pollutants can be eliminated using the wet scrubbers mentioned above by dissolving them in the scrubber solution.

This is a common method of controlling SO_2 and NO_2 in power plant off-gases. Packed scrubbers—spray towers filled with glass frit or platelets—perform such solution processes more effectively than standard wet scrubbers; the extraction of fluoride from aluminum smelter exhaust fumes is one such example. The process of eliminating organic compounds using an adsorbent such as activated charcoal is known as adsorption, or chemisorption.

When an organic pollutant can be oxidized to CO_2 and water, or when H_2S can be oxidized to SO_2, incineration or flaring is used. A type of incineration known as catalytic combustion occurs when surface catalysis facilitates the reaction energetically and at a lower temperature. It is covered in more detail in relation to mobile source control (Table 10.3).

Table 10.3. *Mitigation Strategies and Technology Adopted to Reduce Air-Polluting Emissions From Pyrolysis Plants*

Article ID	Feedstock	Study Scale	Mitigation Strategy	Removal Efficiency
4	Shredded tire rubber	Laboratory	FGD to reduce the HCl and SO_2 concentration	-
8	Scrap tires	Small [1]	WSB and a flare	PAHs by WSB = 76.2%, Flare = 66.8%
23	Various feedstocks	Small [1]	Flue gas recirculation, SNCR, or SCR to reduce the NO_x emissions	-
28	RDF	Small [1]	Caustic Scrubber and electrostatic precipitator (ESP) to reduce PM emissions	-
60	MSW	Industrial [2]	SNCR unit for NO_x control, a baghouse for PM control, and a scrubber unit for control of acid gases and volatile metals	-
61	Waste tires	Industrial [2]	Baghouse filter to reduce PM emissions (below 5 mg/Nm^3)	-
62	MSW	Industrial [2]	Ion exchange scrubber followed by a baghouse with carbon dosing before discharge of the flue gases to the atmosphere	Filtration system = 90%, carbon based gas cleaning system = 95%

63	Bark	Industrial [2]	SNCR/SCR for NO_x reduction, high-temperature ceramic process gas filter adsorbs >99% of the fine dust, for SO_x, additional gas cleaning module (which is standard for sewage sludge applications) to reduce acidic components of the exhaust gas with caustic soda	For fine dust is >99%

Source: Pivato A, Gohar H, Antille DL, Schievano A, Beggio G, Reichardt P, Maria FD, Peng W, Castegnaro S, Lavagnolo MC. Air-Polluting Emissions from Pyrolysis Plants: A Systematic Mapping. Environments. *2024; 11(7):149. https://doi.org/10.3390/environments11070149.*

[1] Feeding rate ≤ 70 kg/h; [2] Feeding rate > 70 kg/h.

10.3.5.1. Control of Sulfur

Sulfur dioxide (SO_2) is a common gas and a major source of pollution. The main source of it in the U.S and most likely the rest of the industrialized world is electricity generated from sulfur-containing fuels like coal and oil. Numerous options and methods for lowering SO_2 emissions have been developed in response to increasingly stringent control standards. Among these options are:

1. **Switch to fuel with less sulfur.** The sulfur content of natural gas is very low, whereas the sulfur content of oil burned for industrial heat and electric power generation ranges from 0.5 to 3%, and that of coal from 0.3 to 4%. However, low-sulfur fuel is costly and not always available.
2. **Elimination of sulfur.** Chemical methods akin to those employed to reduce the H_2S content of crude oil can be employed to extract sulfur from heavy industrial oil. Sulfur can be bound either organically or inorganically in coal as pyrite (FeS_2). By grinding the coal and using a detergent solution to wash it, pyrite can be eliminated. Sulfur that has been bound organically can be eliminated by washing with strong acid. The best techniques are solvent extraction for low-sulfur liquid fuel or coal gasification for pipeline-quality gas.

3. **Tall stacks.** Although a tall stack does reduce ground-level concentration of gaseous pollutants (*see* Equation 10.2), it is not an appropriate pollution control measure.

4. **Flue gas desulfurization**. The off-gases from combustion or other SO_2-producing processes are called flue gases. SO_2 may be cleaned from flue gas by chemical processes. The reaction of SO_2 to form sulfuric acid or other sulfates is the most frequently used flue-gas desulfurization method. It has two limitations: The flue gas must be cleaned of PM before entering an acid-producing plant, and acid formation is energetically favorable only for a fairly concentrated gas stream (about 3%--30,000 ppm--SO_2). The reactions for acid formation are

$$SO_2 + \tfrac{1}{2}O_2 \rightarrow SO_3 \tag{10.1}$$

$$SO_3 + H_2O \rightarrow H_2SO_4 \tag{10.2}$$

Industrial-grade sulfuric acid (98%) can be produced in a double-contact acid plant. The industry that uses this control method the most is the nonferrous smelting and refining sector. Similar processes can be used to create gypsum, $CaSO_4$, and the fertilizer $(NH_4)2SO_4$. The amount of SO_2 in flue gas produced during the burning of fossil fuels is too low to allow it to be trapped as gypsum, acid, or commercial fertilizer. The off-gases from burning coal are likewise too filthy.

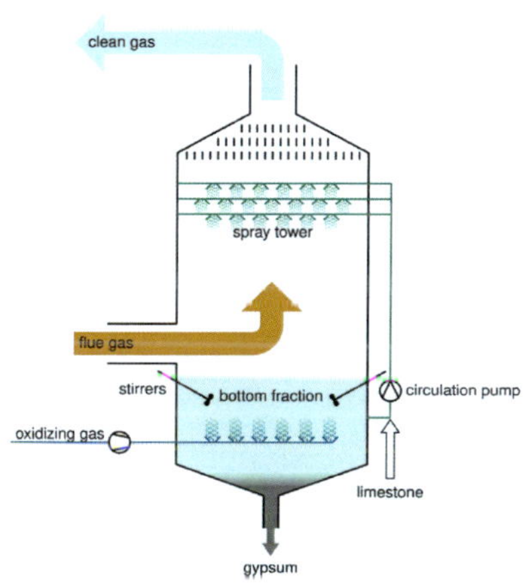

Figure 10.4. *Simplified diagram for scrubbing of flue gas with regeneration.*

Source: By translated by: Sponk – Own work based on: File:Flue gas desulfurization unit DE.svg by User: Sponk, CC SA 1.0, https://commons.wikimedia.org/w/index.php?curid=11430489.

A typical method is single-alkali scrubbing, for which the reactions are

$$SO_2 + Na_2SO_3 + H_2O \rightarrow 2NaHSO_3 \quad (10.3)$$

$$2NaHSO_3 \rightarrow Na_2SO_3 + H_2O + SO_2 \quad (10.4)$$

The concentrated SO_2 that is recovered from this process can be used industrially in pulp and paper manufacture and in sulfuric acid manufacture. Figure 10.4 is a diagram for scrubbing.

SO_2 may also be removed from flue gas by dissolution in an aqueous solution of sodium citrate by the reaction

$$SO_2(g) + H_2O(I) \rightarrow HSO_3^- + H^+ \quad (10.5)$$

Instead of directly entering the reaction, citrate buffers the solution at a pH of 4–5. Regenerating the citrate buffer is simple. The removal efficiency of the citrate method is between 80 and 99%, which is significantly higher than the 75% that can be attained through alkali scrubbing.

Control of Nitrogen Oxides

Although they are typically not installed primarily for NO_2 control, wet scrubbers can absorb both NO_2 and SO_2. Off-stoichiometric burning is a useful technique for reducing nitrogen oxides that is frequently employed in fossil fuel-burning power plants. By keeping the amount of air (or oxygen) in the combustion process to slightly above that which is required to burn the hydrocarbon fuel in question, this technique reduces the production of NO_2. Take the reaction that occurs when natural gas is burned.

$$CH_4 + O_2 \rightarrow 2O_2 + CO_2 \quad (10.6)$$

competes favorably with the high-temperature combination of nitrogen in the air with oxygen in the air to form NO (eventually oxidized to NO_2):

$$N_2 + O_2 \rightarrow 2NO \quad (10.7)$$

The stoichiometric ratio of oxygen needed in natural gas combustion is

$$32 \text{ g of } O_2\text{:}16 \text{ g of } CH_4 \quad (10.8)$$

Practically all of the oxygen in the combustion air will mix with fuel rather than nitrogen if there is a small excess of oxygen present. Off-stoichiometric combustion is accomplished by varying the air flow to the combustion chamber until any discernible plume vanishes.

10.3.5.2. Control of Volatile Organic Compounds and Odors

Since volatile organic compounds are only weakly soluble in aqueous scrubbing media, they are controlled by thorough oxidation through either incineration or catalytic combustion.

10.3.6. Control of Moving Sources

Particular attention has been paid to one mobile source in particular—the automobile—which presents unique pollution control challenges. Emissions controls for cars are similar to those needed for other mobile sources of pollution, like heavy and light trucks and diesel-powered cars. The important pollution control points in an automobile are:

1. Evaporation of hydrocarbons (HC) from the fuel tank.
2. Evaporation of HC from the carburetor.
3. Emission of unburned gasoline and partly oxidized HC from the crankcase.
4. CO, HC, and NO/NO_2 from the exhaust (Table 10.4).

Table 10.4. *Limit Values for Emissions of Toxic Exhaust Gases from Engines of Two-Wheeled Vehicles of Particular Categories in Europe; Euro 1–3*

Euro 1							
Category	**Category Name**	**Classifica-tion**	**Euro Stan-dard**	**Emission [g/km]**			
				CO	**HC**	**NO_x**	**HC + NO_x**
all vehicles	mopeds	<150 cm³	1	13	3	0.3	-
	motor-cycles	≥150 cm³	1				
Euro 2 + 3							
L1e	mopeds	<50 cm³	2 + 3	1	-	-	1.2
L3e	motor-cycles	<150 cm³	2	5.5	1.2	3	-
		≥150 cm³	2	5.5	1	3	-
		<150 cm³	3	2	0.8	1.5	-
		≥150 cm³	3	2	0.3	1.5	-
		v_{max} < 130 km/h	3	2.62	0.75	1.7	-
		v_{max} ≥ 130 km/h	3	2.62	0.33	2.2	-

Source: Szymlet N, Rymaniak Ł, Lijewski P. Two-Wheeled Urban Vehicles—A Review of Emissions Test Regulations and Literature. Energies. 2024; 17(3):586. https://doi.org/10.3390/en17030586.

When the engine is off and hot gasoline in the carburetor evaporates, evaporative losses from the gas tank and carburetor frequently happen. These vapors could be contained in an activated carbon canister, periodically purged with air, and then burned in the engine. It is possible to seal off the crankcase vent from the environment and recycle the blow-by gases back into the intake manifold. A tiny check valve called the positive crankcase ventilation (PCV) valve keeps pressure from building up in the crankcase.

Approximately 60% of the hydrocarbons and nearly all of the NO, CO, and lead released into the atmosphere are attributed to the exhaust. It therefore presents the most challenging control issue for mobile sources. Engine operation affects exhaust emissions. High compression generates a lot of NO/NO_2, efficient combustion, low CO and HC, and efficient combustion occur during acceleration. On the other hand, because of the unburned fuel still in the exhaust, deceleration causes low NO/NO_2 and high HC. Because of this variation in emissions, the EPA established a common acceleration-deceleration cycle for emissions measurement. After a cold start, the engine is tested by acceleration, constant speed cruising on a dynamometer to load it, deceleration, and a hot start.

Emission control techniques include engine tune-ups, engine modifications, exhaust gas recirculation, and catalytic reactors. A well-tuned engine is the first line of defense for emission control.

It is possible to modify an engine in a variety of acceptable ways. Water injection can lower NO emissions, and fuel injection can lower CO and HC emissions by omitting or bypassing the carburetor. Water injection, however, is incompatible with fuel injection because it could clog the injectors. Although the stratified charge engine runs on a very lean air/fuel mixture, it does not significantly raise NO. Instead, it reduces CO and HC. This is achieved by the engine's two compartments, or "stratification": the first takes in and ignites the fuel/air mixture, while the second produces a wide flame for an effective burn. This engine is capable of reducing CO by more than 90%.

Recirculating the exhaust gas through the engine can reduce hydrocarbons and CO by roughly 60%. A system for cooling the exhaust gas before recirculation is the only significant modification to a standard engine that exhaust gas recirculation (EGR) requires, aside from the required fittings. This is done to prevent heat deformation of the piston surfaces. Up until 1980, exhaust gas recirculation was a common and approved method of controlling emissions. However, modern emission standards demand a 90% reduction in carbon emissions, which can't be achieved with this method, despite increasing the rate of engine wear. Since 1983, new automobiles sold in the U.S. have been required to meet exhaust emission standards through the use of a catalytic reactor, also known as a "catalytic converter." As a result, the device is now standard equipment on new cars. The two processes carried out by the contemporary three-stage catalytic converter are the reduction of NO to N_2 and the oxidation of CO and hydrocarbons to CO_2 and water. Using a platinum-rhodium catalyst, the first step of NO reduction is achieved by burning a fuel-rich mixture, which lowers the oxygen content at the catalyst. In the second stage, hydrocarbons, and CO are oxidized at a lower temperature when air is introduced. Cars equipped with catalytic converters must run on unleaded gasoline because inorganic lead compounds render them inoperable.

The three primary pollutants produced by diesel engines are similar to those of gasoline engines, albeit in slightly different amounts. Furthermore, heavy-duty diesel-powered cars emit irksome black soot, which is essentially unburned carbon. Few studies on diesel exhaust emission control have advanced to the point of operational devices since control of diesel exhaust was not mandated in the United States until the 1990 Clean Air Act was passed, nor is it required anywhere else in the world.

It seems almost paradoxical to have an internal combustion engine that produces no emissions. A nearly pollution-free engine can be produced with an external combustion engine capable of controlling all three major exhaust pollutants to an extent better than 99%, thereby allowing for drastic reduction of emissions. Even though development on this engine started in 1968, a commercial model has not yet been constructed. Some cities use natural gas to power buses and fleets of cars, such as those owned by the local utility, but there are several competing uses for the scarce natural gas supply. A complete switch from gasoline to natural gas would necessitate the development of a new refueling infrastructure. Electric vehicles are environmentally friendly, but their range and power storage are constrained. Pollution is also produced during the electricity generation process used to power these vehicles, and a shift in the world's battery material supply would be difficult to implement. Cities that do not comply with the National Ambient Air Quality Standards are required by the 1990 Clean Air Act to sell oxygenated fuel in the winter. When gasoline with 10% ethanol (CH_3CH_2OH) is used, the conversion of CO to CO_2 is somewhat more efficient. This fuel is known as oxygenated fuel.

10.3.7. Control of Global Climate Change

Compounds that deplete the stratospheric ozone layer by producing free halogen atoms through photochemical reactions and those that absorb energy in the near-infrared spectral region, which may ultimately lead to global temperature changes, are the two types of compounds implicated in global climate change. Most of the substances in the first group are chlorofluorocarbons (CFCs). The

Montreal Protocol suggests banning the use of certain substances and controlling leaks, such as those from refrigeration systems, in order to reduce the emission of chlorofluorocarbons. Aerosol deodorant, cleaners, paint, hairspray, and other products that once used chlorofluorocarbon propellants are no longer made. It has been discovered that atomized liquids, hair mousse, brushed-on or rolled-on paint, roll-on deodorant, wipe-on cleaners, and brushed-on paint all work well without harming the ozone layer.

10.4. AIR POLLUTION LAW AND REGULATIONS

Similar to water pollution, the application of technologies for mitigating air pollution is subject to a complicated web of rules and regulations. This chapter describes how this system has changed over time, starting with common law and ending with the enactment of federal statutes and administrative initiatives. The issues faced by regulatory bodies and those who release pollutants are discussed, with a focus on the potential effects of the system on future economic growth. A path through this maze can be found in Figure 10.5.

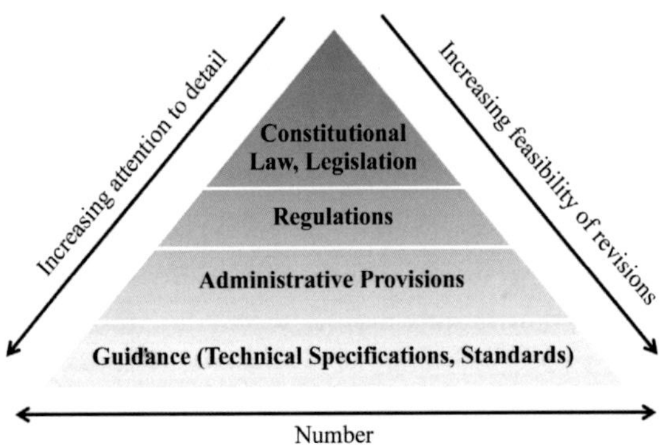

Figure 10.5. *Construction of the "legal pyramid" describing the body of normative documents comprising documents resulting from legislation processes.*

Source: Saffell J, Nehr S. Improving Indoor Air Quality through Standardization. Standards. *2023; 3(3):240-267. https://doi.org/10.3390/standards3030019.*

10.4.1. Air Quality and Common Law

An individual or group of individuals harmed by an air pollution source may invoke general principles from the tort and property branches of common law when relying on that legal framework. These branches may be applicable in specific situations and have evolved over time. The aggrieved party, known as the plaintiff, has the right to appear in court and demand remedies from the defendant for harm done to their person or property.

10.4.1.1. Tort Law

An injury sustained by one or more people is called a tort. This area of common law covers negligent mishaps and exposure to dangerous chemicals in the air. There are three main ways in which a polluter can be held accountable for harm to human health: strict liability, negligence, and tort liability. Proof of intentional wrongdoing against another party is necessary for intentional liability to exist. In the case of damages resulting from air pollution, this proof is particularly difficult. First, it is necessary to prove that a "wrong" actually happened. This can be done by strong inference or direct statistical evidence, like the outcomes of rat experiments conducted in a lab. Furthermore, it is necessary to prove intent to do the "wrong," which entails presenting written documentation or direct testimony from the accused person or group of people. Such proof is difficult to come by. The injured plaintiff may be granted both actual and punitive damages if the courts are satisfied that intentional liability was demonstrated.

The air polluter who permitted the injury to happen may have been negligent merely by being inattentive. In court, evidence primarily focuses on the defendant's lack of due care.

Examples of air pollution include failing to size and design an appropriate abatement technology or failing to inspect the maintenance and operation of electrostatic precipitators. Once more, the plaintiff may receive damages. Strict liability disregards the defendant's mental state or degree of fault. In extreme circumstances, a common law court has ruled that certain actions are abnormally dangerous and that those who perform them are strictly liable for any injuries sustained. The court does have a tendency to weigh the potential risks of a given action against its potential benefits to society. A radioactive or extremely toxic gas escaping from an industrial smoke stack could serve as an example.

Once more, in the event that a known source of air pollution causes personal harm to an individual, the aggrieved party may file a common law lawsuit seeking an injunction to prevent the polluter from continuing to pollute or monetary damages to be paid by the defendant. It is frequently difficult, if not impossible, to gather sufficient evidence and precedent, and tort law has frequently been found to be insufficient for limiting air pollution and determining damages.

10.4.1.2. Property Law

The theories of nuisance and property rights, on the other hand, are the main focus of property law; nuisance is defined as interference with the use or enjoyment of property, while property rights are defined as actual invasion of the property.

Property law takes into account things like damage and trespassing and is based on long-standing agreements between landowners. When a plaintiff bases their case on property law, they assume some risk, roll the dice, and hope that the court will find in their favor as it weighs individual property rights against the greater good of society.

The most common law action pertaining to the environment that is utilized is nuisance. Unlawful interference with a common public right, like the "right to clean air," is referred to as a public nuisance. The public that is harmed by the air pollution must be represented in court by a public official. Conversely, unreasonable interference with the use and enjoyment of private property is the foundation of private nuisance. The way the courts interpret "unreasonable" is crucial to a nuisance action. "The common law court weighs the equity, hardships, and injuries in the specific case and rules in favor of the plaintiff or the defendant based on precedents and the arguments of the parties involved."

Nuisance is closely associated with trespass. The main distinction is that any physical intrusion—no matter how small—is considered a trespass. Remember that the nuisance theory requires an unreasonable interference with land, and the way a court interprets "unreasonable" will determine the result of a given case. "Trespassing is not too complicated. Physical presence on the land, vibrations from adjacent surface or subsurface strata, and perhaps gases and microscopic particles emanating from a single smoke stack are a few examples. In conclusion, common law has typically shown to be insufficient to address air pollution issues. Due to the stringent burdens of proof that are necessary in court, decisions are frequently made in favor of the defendant, which contributes to smoke stacks that worsen air pollution. Furthermore, a court's ability to take action is frequently restricted by the particularity and intricacy of each case; difficult-to-find experts and convoluted tests frequently leave both the plaintiff and the court powerless. Additionally, for the plaintiff to have standing in court, they must have experienced tangible or physical harm as a result of the air pollution (i.e., the right to a judge's hearing).

The degree of variation among these common law principles is an important feature. Every state has its own body of common law, and those who depend on the legal system are usually limited to applying the relevant state's common laws. Congress passed the federal Clean Air Act to address the flaws in the common law approach to cases involving air pollution.

10.4.2. Statutory Law

The federal statutory legislation governing air pollution dates back to the Clean Air Acts of 1963 and 1967. These laws only applied air pollution controls in particularly dirty communities, not across the entire United States, even though they provided broad goals and funding for research. The EPA was established to issue clean air regulations and carry out the act's enforcement after the Clean Air Act was modified in 1970 to encompass the entire country. Subsequently, in 1977, the act was amended to include provisions for the protection of extremely clean areas (known as PSD, or protection against significant deterioration), the enforcement of noncompliant areas, and the extension of the automobile emission standards compliance dates. Amendments aimed at controlling all vehicle emissions and harmful air pollutants were made in 1990. But the current federal clean air laws are based on the amendments of 1970.

10.4.2.1. National Ambient Air Quality Standards

The EPA has the authority to establish the permissible ambient concentrations of pollutants in the air across the entire country. The national strategy to protect air quality has been centered around these National Ambient Air Quality Standards (NAAQS). The primary NAAQS aim to safeguard human health, while the secondary NAAQS is designed to "protect welfare.." In actuality, the later levels are those that are required to preserve the vegetation.

The setting of NAAQS is based on the analysis of large-scale data sets that show how these air pollutants affect vegetation, ecosystems, human health, and materials. The EPA refers to these compilations as criteria documents, and the pollutants that are part of the NAAQS are occasionally referred to as criteria pollutants. Data show that there is a threshold for all criteria pollutants below which no harm occurs. The majority of enforcement authority under the Clean Air Act is given to the states by the EPA; however, the states have to demonstrate that they are able to clean up the air to the NAAQS standards. This display is made in a State Implementation Plan (SIP), which is a document that includes all local, municipal, and state regulations controlling air pollution control. The SIP has the same legal force as federal law, but it needs to be approved by the EPA.

10.4.2.2. Regulation of Emissions

The EPA is only permitted to establish emission standards (also known as performance standards) for new or significantly altered sources of the criteria pollutants under Section 111 of the Clean Air Act. The states have the power to enforce the EPA's new source performance standards (NSPS) and may set performance standards for sources that are already in operation. Additionally, the EPA has given some states permission to create their own NSPS. Since 1971, a priority list of industries has been in place for which NSPS are to be set; revisions to the NSPS may also be motivated by new technology. The EPA can establish national emission standards for hazardous air pollutants (NESHAPS) for all sources of those pollutants under Section 112 of the Clean Air Act. The EPA had to create a list of chemicals that would need to be regulated as part of the 1990 Clean Air Act Amendments. Research is ongoing, so substances may be added to or removed from the list. There are hundreds of pollutants on the list as of right now.

Categories of industrial facilities that emit these listed pollutants must use removal technologies if the facility emits 10 tons per year of any single hazardous substance or 25 tons per year of any combination. Categories of industrial sources, taken from the list of hundreds, include:

1. Industrial external combustion boilers.
2. Printing and publishing.
3. Waste oil combustion.
4. Gasoline marketing.
5. Glass manufacturing.

As per the 10 ton/yr or 25 ton/yr limit, every "major source" needs to attain a maximum achievable control technology (MACT) for emissions. A State Implementation Plan may, on an individual basis, mandate that a facility install control equipment, modify an industrial or commercial process, substitute materials during production, alter work procedures, and certify and train workers and operators. As the new Clean Air Act is put into effect throughout the 1990s, the MACT requirements will have an impact on sources of air pollution.

10.4.2.3. Prevention of Significant Deterioration

When the EPA failed to maintain air quality in areas of the country where the quality of the air exceeded the NAAQS, the Sierra Club filed a successful lawsuit against the agency in 1973. Congress responded by adding prevention of significant deterioration (PSD) to the Clean Air Act amendments of 1977 and 1990. The U.S. is split into class I and class II areas for PSD purposes, with some areas potentially being designated as class III. Class I comprises the so-called "mandatory class I" areas, all national parks and monuments larger than 6000 acres, all national wilderness areas larger than 5000 acres, and any area designated as class I by a state or Native American tribe. With the exception of a state or tribe that may petition the EPA to be reclassified to class III, the remainder of the United States is class II.

To date, the only pollutants covered by PSD are sulfur dioxide and PM. PSD limits the increases in these. In addition, visibility is protected in class I areas.

If an industry wants to construct a new facility, it must demonstrate that it won't go over the permitted increment using dispersion modeling and a year's worth of weather data. The EPA grants the industry a PSD permit after such a demonstration. The PSD permitting process has significantly impacted the location of new facilities. Furthermore, any new project in a nonattainment area needs to ensure that there are "offsets" for any emissions from the facility in addition to using the strictest control technology. That is to say, the new facility might still release 15 tons of particles annually even if it uses MACT. Before the new facility can start up, some other facility has to cut its annual emissions of particulates by 15 tons. The PSD limits are obviously meaningless in situations where ambient concentrations are already approaching the NAAQS and they would be able to surpass it.

10.4.2.4. Nonattainment

A nonattainment area for a given pollutant is defined as an area where the NAAQS for one or more criteria pollutants are exceeded more frequently than once annually. A traffic reduction plan and an inspection and maintenance program for exhaust emission control are necessary if the lead, CO, or ozone standards are not met. The state loses federal funding for highway construction if they don't comply.

An offset program needs to be started if emissions from stationary sources are the cause of nonattainment. For a program like this to be implemented, current stationary sources must reduce their emissions so that the total emissions following the operation of the new source will be lower than before. In nonattainment areas, new sources are required to achieve the lowest achievable emission rate (LAER) without taking the cost of that emission control into account. Although the offset program permits industrial expansion in nonattainment areas, it presents a unique set of difficulties for air pollution control engineers. Emission reductions that would not otherwise be necessary are represented by the offset. Tighter controls on current operations at the same site, a legally binding agreement with another facility to reduce emissions, and the purchase of another facility to reduce emissions by installing control equipment or shutting down the facility are all actions that have the potential to produce offsets.

10.4.3. Moving Sources

The emissions from cars, trucks, and buses are regulated under the 1990 Clean Air Act. Phase I of the car standards beginning in the year 1994 for CO, nonmethane hydrocarbons, and NO_x are

3.4 g/mile for CO.

0.4 g/mile for NOx.

0.25 g/mile for nonmethane hydrocarbons.

Phase II standards, being considered for implementation in the year 2003, are

1.7 g/mile for CO.

0.2 g/mile for NO_x.

0.125 g/mile for nonmethane hydrocarbons.

Standards allowing slightly higher emissions for these pollutants, as well as a standard of 0.08 grams of PM per mile, will be set for light-duty trucks. Hazardous air pollutants emitted from vehicles, particularly benzene and formaldehyde, will also be regulated.

10.4.4. Tropospheric Ozone

Each state must create ozone nonattainment area categories in accordance with the 1990 Clean Air Act. Depending on the level of nonattainment, all air polluters in such areas must use reasonably available control technologies (RACT). Figure 10.6 provides a summary of management. Every state needs to categorize regions based on their ozone concentration nonattainment.

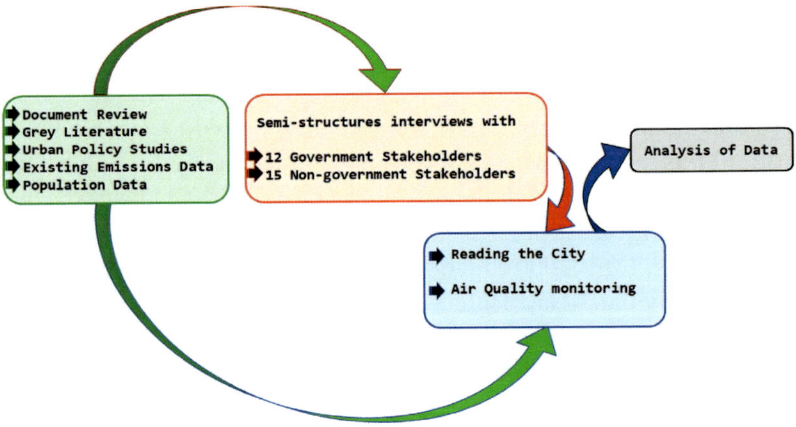

Figure 10.6. *An example of urban diagnostics.*

Source: Singh A, Bakare H, Mazzeo A, Avis WR, Ng'ang'a D, Gatari M, Bartington SE, Thomas GN, Bryson JR, Andres L, Quinn A, Burrow M, Ndegwa EN, Mwaniki G, Randa T and Pope FD (2022) Urban diagnostics and a systems approach to air quality management: Pathways towards sustainable economic development and a healthy nairobi, Kenya. Front. Environ. Sci. 10:978002. doi: 10.3389/fenvs.2022.978002.

10.4.5. ACID RAIN

Acid rain is primarily caused by fossil fuel electric generating facilities. The 1990 Clean Air Act recognizes the significance of these sources of nitrogen oxides and sulfur dioxides, and its regulations mandate a reduction of 2 million tons in NO_2 emissions and a reduction of 9 million tons in SO_2 emissions from the 1980 levels of national emissions.

Reductions are planned at 111 major fossil fuel facilities spread across 21 states, including the Appalachian states, the Ohio River Valley, New England and the northeastern seaboard, and the North Central states. According to Section 404 of the act, regulated facilities are granted "allowances" for sulfur dioxide emissions, which they are free to use, sell to other polluters, or store for later use. Based on historical operation data for the calendar years 1985 through 1987, allowances have been made for 111 facilities. These transferable discharge permits will be put up for auction each year, and anybody can place a bid, including local and national environmental organizations (Boubel et al., 1994; Ayers et al., 1999).

REFERENCES

1. Ahmet, S., & Van Dijk, M. (1994). *Monitoring of Air Quality in the Port Phillip Control Region, 1979–1991*. Environment Protection Authority, Government of Victoria (EPAV), Publication 421.
2. Allen, G., Sioutas, C., Koutrakis, P., Reiss, R., Lurman, F. W., & Roberts, P. T. (1997). Evaluation of the TEOM method for measurement of ambient particulate mass in urban areas. *Journal of the Air and Waste Management Association, 47*, 682–689.
3. Artaxo, P., Oyola, P., & Martinez, R. (1999). Aerosol composition and source apportionment in Santiago de Chile. *Nuclear Instruments and Methods in Physics Research, Section B, 150*, 409–416.
4. Ayers, G. P., Keywood, M. D., & Gras, J. L. (1999). TEOM vs. manual gravimetric methods for determination of PM2.5 aerosol mass concentrations. *Atmospheric Environment, 33*, 3717–3721.
5. Bayen, S., Gong, Y. H., Chin, H. S., Lee, H. K., Leong, E. Y., & Obbard, J. P. (2004). Determination of polybrominated diphenyl ethers (PBDEs) in marine biological tissues using microwave-assisted extraction (MAE). *Journal of Chromatography A, 1035*, 291–294.
6. Boubel, R. W., Fox, D. L., Turner, D. B., & Stern, A. C. (1994). *Fundamentals of Air Pollution* (3rd ed.). Academic Press.
7. Chan, T. W., & Mozurkewich, M. (2007a). Simplified representation of atmospheric aerosol size distributions using absolute principal component analysis. *Atmospheric Chemistry and Physics, 7*, 875–886.
8. Chan, T. W., & Mozurkewich, M. (2007b). Application of absolute principal component analysis to size distribution data: Identification of particle origins. *Atmospheric Chemistry and Physics, 7*, 887–897.
9. Chan, Y. C., Simpson, R. W., Mctainsh, G. H., Vowles, P. D., Cohen, D. D., & Bailey, G. M. (1999). Source apportionment of PM2.5 and PM10 aerosols in Brisbane (Australia) by receptor modeling. *Atmospheric Environment, 33*, 3251–3268.
10. Chang, S. N., Hopke, P. K., Gordon, G. E., & Rheingrover, S. W. (1988). Target transformation factor analysis of airborne particulate samples selected by wind-trajectory analysis. *Aerosol Science and Technology, 8*, 63–80.
11. Cheng, F. L., Hsiao, J. H., Cheng, S. J., & Hsieh, H. H. (2007). Identification of regional air pollution characteristic and the correlation with public health in Taiwan. *International Journal of Environmental Research and Public Health, 4*, 106–110.

12. Chow, J. C. (1995). Measurement methods to determine compliance with ambient air quality standards for suspended particles. *Journal of the Air and Waste Management Association, 45*, 320 382.
13. Chow, J. C., Watson, J. G., Lowenthal, D. H., & Countess, R. J. (1996). Sources and chemistry of PM10 aerosol in Santa Barbara County, CA. *Atmospheric Environment, 30*, 1489–1499.
14. Chueinta, W., & Hopke, P. K. (2001). Beta gauge for aerosol mass measurement. *Aerosol Science and Technology, 35*(4), 840–843.
15. Chuen-Jinn, T., & Hsin-Ying, H. (1995). Atmospheric aerosol sampling by an annular denuder system and a high volume PM10 sampler. *Environment International, 21*, 283–291.
16. Cooper, J. A., & Watson, Jr., J. G. (1980). Receptor oriented methods of air particulate source apportionment. *Journal of Air Pollution Control Association, 30*, 1116–1125.
17. Cooper, J. A., Watson, J. G., & Huntzicker, J. J. (1984). The effective variance weighting for least squares calculations applied to the mass balance receptor model. *Atmospheric Environment, 18*, 1347–1355.
18. Eatough, D. J., Long, R. W., Modey, W. K., & Eatough, N. L. (2003). Semi-volatile secondary organic aerosol in urban atmospheres: Meeting a measurement challenge. *Atmospheric Environment, 37*, 1277–1292.
19. Facchinelli, A., Sacchi, E., & Mallen, L. (2001). Multivariate statistical and GIS-based approach to identify heavy metals sources in soils. *Environmental Pollution, 114*, 313–324.
20. Federal Register. (1997). Revised requirements for designation of reference and equivalent methods for PM2.5 and ambient air quality surveillance for particulate matter; Final Rule: 40 CFR Parts 53 and 58. *Federal Register, 62*, 38764.
21. Fisher, G. W., Graham, B. W., & Bell, M. J. (1995). Design of a national ambient air quality monitoring network for New Zealand. *NIWA Report AK95001 to the Ministry for the Environment*, 107 pp.
22. Friedlander, S. K. (1973). Chemical element balances and identification of air pollution sources. *Environmental Science and Technology, 7*, 235–240.
23. Gao, Y., Nelson, E. D., Field, M. C., et al. (2002). Characterization of atmospheric trace elements on PM2.5 particulate matter over the New York–New Jersey harbor estuary. *Atmospheric Environment, 36*, 1077.
24. Guo, H., Wang, T., & Louie, P. K. K. (2004a). Source apportionment of ambient non-methane hydrocarbons in Hong Kong: Application of a principal component analysis/absolute principal component scores (PCA/APCS) receptor model. *Environmental Pollution, 129*, 489–498.

25. Guo, H., Wang, T., Simpson, I. J., et al. (2004b). Source contributions to ambient VOCs and CO at a rural site in eastern China. *Atmospheric Environment, 38*, 4551–4560.
26. Henry, R. C., Lewis, C. W., Hopke, P. K., & Williamson, H. J. (1984). Review of receptor model fundamentals. *Atmospheric Environment, 18*(8), 1507–1515.
27. Hering, S., Fine, P. M., Constantinos, S., et al. (2004). Field assessment of the dynamics of particulate nitrate vaporization using differential TEOM and automated nitrate monitors. *Atmospheric Environment, 38*, 637–651.
28. Hien, P. D., Binh, N. T., Truong, Y., Ngo, N. T., & Sieu, L. N. (2001). Comparative receptor modeling study of TSP, PM2, and PM2–10 in Ho Chi Minh City. *Atmospheric Environment, 35*, 2669–2678.
29. Hopke, P. K. (1980). Source identification and resolution through application of factor and cluster analysis. *Annals of New York Academy of Sciences, 338*, 103–115.
30. Johnson, D. L., Davis, B. L., Dzubay, T. J., et al. (1984). Chemical and physical analyses of Houston aerosol from interlaboratory comparison of source apportionment procedures. *Atmospheric Environment, 18*, 1539–1553.
31. Kaya, G., & Tuncel, G. (1997). Trace element and major ion composition of wet and dry deposition in Ankara, Turkey. *Atmospheric Environment, 31*, 3985–3998.
32. Kowalczyk, G. S., Choquette, C. E., & Gordon, G. E. (1978). Chemical element balances and identification of air pollution sources in Washington, DC. *Atmospheric Environment, 12*, 1143–1153.
33. Larsen, G., & Schaug, J. (1995). *EMEP Workshop on Quality Assurance of Measurements*, Berlin, November 20–23, 1995.
34. Liu, C. K., Roscoe, B. A., Severin, K. G., & Hopke, P. K. (1982). Source identifications and spatial distributions of aerosol measured at multiple sites across St. Louis. *American Industrial Hygiene Association Journal, 43*, 314–318.
35. Lodge, J. P. (1989a). Method 406: Determination of the NO2 content of the atmosphere (Griess–Saltzman reaction). In *Methods of Air Sampling and Analysis* (3rd ed.), InterSociety Committee (APCA, ACS, AIChE, APCSA, ASME, AOAC, HPS, ISA), Michigan, USA, Lewis Publishers Inc., pp. 389–393.
36. Lodge, J. P. (1989b). Method 878: Determination of the NO2 content of the atmosphere (Griess–Saltzman reaction). In *Methods of Air Sampling and Analysis* (3rd ed.), InterSociety Committee (APCA, ACS, AIChE, APCSA, ASME, AOAC, HPS, ISA), Michigan, USA, Lewis Publishers Inc., pp. 389–393.

37. Malinowski, E. R., Howery, D. J., Weiner, P. H., et al. (1980). FACTANL-target Transformation Factor Analysis. Quantum Chemistry Program Exchange, Indiana University Bloomington, Inc., 1979, Program 320.
38. Manoli, E., Voutsa, D., & Samara, C. (2002). Chemical characterization and source identification/apportionment of fine and coarse air particles in Thessaloniki, Greece. *Atmospheric Environment, 36*, 949–961.
39. Massey, D., Masih, J., Kulshrestha, A., Habil, M., & Taneja, A. (2009). Indoor/outdoor relationship of fine particles less than 2.5 μm (PM2.5) in residential homes locations in central Indian region. *Building and Environment, 44*(10), 2037–2045.
40. Meyer, M. B., Patashnick, H., Ambs, J. L., & Rupprecht, E. (2000). Development of a sample equilibration system for the TEOM continuous PM monitor. *Journal of the Air and Waste Management Association, 50*, 1345–1349.
41. Miller, M. S., Friedlander, S. K., & Hidy, G. M. (1972). A chemical element balance for the Pasadena aerosol. *Journal of Colloid and Interface Science, 39*, 65–176.

INDEX

C

D

E

M

N

O

P

FSC
www.fsc.org
MIX
Papier | Fördert
gute Waldnutzung
FSC® C083411

Kolibri 360 GmbH
Ferdinand-Jühlke-Straße 7
99095 Erfurt, Deutschland
produktsicherheit@kolibri360.de